SEEING THE UNSEEN

Seeing the Unseen

Geophysics and Landscape Archaeology

Editors

Stefano Campana
*Department of Archaeology and History of Arts,
University of Siena, Italy*

Salvatore Piro
*Institute of Technologies Applied to Cultural Heritage -
National Research Council, Italy
University "La Sapienza", Rome, Italy*

CRC Press
Taylor & Francis Group
Boca Raton London New York

CRC Press is an imprint of the
Taylor & Francis Group, an **informa** business

A BALKEMA BOOK

CRC Press
Taylor & Francis Group
6000 Broken Sound Parkway NW, Suite 300
Boca Raton, FL 33487-2742

First issued in paperback 2019

Typeset by Vikatan Publishing Solutions (P) Ltd., Chennai, India

No claim to original U.S. Government works

ISBN-13: 978-0-415-44721-8 (hbk)
ISBN-13: 978-0-367-38680-1 (pbk)

Visit the Taylor & Francis Web site at
http://www.taylorandfrancis.com

and the CRC Press Web site at
http://www.crcpress.com

This book is dedicated to the memory of the late Professor **Riccardo Francovich**, of the University of Siena. His inspiration, encouragement and ever-helpful criticism were an inestimable gift to all who took part as teachers or students in the International Summer Schools project

Seeing the Unseen – Campana & Piro (eds)
© 2009 Taylor & Francis Group, London, ISBN 978-0-415-44721-8

Table of Contents

Seeing the Unseen – Campana & Piro (eds)
© 2009 Taylor & Francis Group, London, ISBN 978-0-415-44721-8

List of lecturers

Helmut Becker – Bavarian State Conservation Office-Munich-GERMANY

Stefano Campana – University of Siena-ITALY

Lawrence Conyers – University of Denver-USA

Michel Dabas – University of Paris-FRANCE

Chris Gaffney – Remote Vision Research-UK

Dean Goodman – Geophysical Archaeometry Laboratory-USA

Albert Hesse – University Pierre et Marie Curie-FRANCE

Gianfranco Morelli – Geostudi Astier-ITALY

Iacopo Nicolosi – National Institute of Geophysics and Vulcanology-ITALY

Yasushi Nishimura – National Research Institute for Cultural Properties-JAPAN

Salvatore Piro – ITABC - National Research Council-ITALY

Dominc Powlesland – Landscape Research Centre-UK

Armin Schmidt – University of Bradford-UK

Alain Tabbagh – University Pierre et Marie Curie-FRANCE

Gregory Tsokas – Aristotle University of Thessaloniki-GREECE

Meg Watters – University of Birmingham-UK

List of students

Andrea Anibaldi – University of Urbino, Italy

Federica Boschi – University of Ravenna, Italy

Miguel A. Bru – University of Autónoma de Madrid PhD, Spain

Anna Caprasecca – University of Siena, Italy

Mario Carnevale – Hager GeoScience Inc., USA

Mickaël Chemin – University of Rennes 1, France

Daniela Degan – University of Ca' Foscari di Venezia, Italy

Marco Didò – University of Milan, Italy

Ernesto Enriquez – University of Lecce, Italy

Fabrizio Fantini – Techgea Service Sas, Italy

Lorenzo Facco – University of Padua, Italy

Daniele Farina – University of Urbino, Italy

Cristina Felici – University of Siena, Italy

Dennys Frenez – University of Bologna, Italy

Barbara Frezza – University of Siena, Italy

Elisa Girmonte – University of Calabria, Italy

Romas Jarockis – Klaipeda University, Lithuania

Katia Luzio – University of Lecce, Italy

Kate McAnally – University of Calgary, USA

Jaroslaw Majewski – GeoRADAR society, Poland

Emanuele Magnani – University of Venice, Italy

Domenico Marchetti – Geo 3D Sas, Italy

Adela Mates – "1 December 1918" University, Romania

Alexandre Novo – University of Vigo, Spain

Gaia Lombardo Pijola – University of Milan, Italy

Paola Pacchiarotti – Sapienza University of Rome, Italy

Mariarita Paffetti – University of Siena, Italy

Giovanna Pappalardo – Geoscan, Italy

Francesco Pericci – University of Siena, Italy

Daniela Peloso – ITABC CNR, Italy

Pietro Pepe – Apogeo Soc. Coop, Italy

Fabrizio Pieri – University of Pisa, Italy

Valeria Poscetti – Sapienza University of Rome, Italy

Palmina Pratillo – University of Napoli, Italy

Alessandro Rizzo – Università di Lecce, Italy

Isabel Rodriguez – Polytechnique University of Valencia, Spain

Alessandro Russel – University of York, UK

Tanel Saimre – Tartu University, Estonia

Stefano Sfarra – University of L'Aquila, Italy

David Simpson – Ghent University, Belgium

Till Sonnemann – Politecnico di Zurigo, Switzerland

Matteo Sordini – University of Siena, Italy

Guglielmo Strapazzon – University of Padua, Italy

Immo Trinks – University of Cambridge, Sweden

Seeing the Unseen – Campana & Piro (eds)
© *2009 Taylor & Francis Group, London, ISBN 978-0-415-44721-8*

Program of the school

DAY 1 (10.07.06)

15:00–15:15 **Official Opening**
PRO-RECTOR OF THE UNIVERSITY OF SIENA AND MAYOR OF GROSSETO

15:15–15.30 **Introduction to the School**
R. FRANCOVICH - S. PIRO - S. CAMPANA

15:30–17:00 **The process of archaeological mapping: some thoughts on the relationship between Remote Sensing and Geophysical Prospection applied to Archaeology**
S. CAMPANA (LANDSCAPE ARCHAEOLOGY - UNIVERSITY OF SIENA)

17.00–17.30 **Break**

17.30–19.00 **Introduction to Geophysical Prospection applied to Archaeology**
S. PIRO (ITABC-NATIONAL RESEARCH COUNCIL OF ITALY)

DAY 2 (11.07.06)

9:00–11:00 **Earth Resistance and Magnetic Susceptibility in Archaeological Prospection**
A. SCHMIDT (UNIVERSITY OF Bradford)

11:00–11:30 **Break**

11:30–13:30 **Short history, strategies and practical aspects of electromagnetic detection based on the description of field experiments**
A. Hesse (Unité Mixte de Recherche-Sisyphe)

13:30–15:00 **Lunch**

15:00–17:00 **Theory and practice of the new fast electrical imaging system: ARP, application to different archaeological sites and their Hinterlands**
M. DABAS (Université Paris VI)

17:00–17:30 **Break**

17:30–19:30 **The complementary nature of magnetic and resistivity methods**
M. WATTERS (University of Birmingham)

20:30 **Opening dinner**

DAY 3 (12.07.06)

9:00–11:00	**Electrical Resistivity Tomography ERT: large scale exploration using ERT, simple systems to perform ERT along with resistance mapping, use of flat base electrodes to investigate the walls and floors of standing monuments** G. TSOKAS (University of Thessaloniki)
11:00–11:30	**Break**
11:30–13:30	**Caesium-magnetometry for large scale archaeological prospection** H. BECKER (Bavarian State Conservation Office-Munich)
13:30–15:00	**Lunch**
15:00–17:00	**Ground Penetrating Radar: theory, acquisition, elaboration and representation techniques. Examples from international case histories.** D. GOODMAN (GAL-Los Angeles USA)
17:00–17:30	**Break**
17:30–19:30	**Ground penetrating radar case histories: successes and potential pitfalls** L. CONYERS (University of Denver)

DAY 4 (13.07.06)

9:00–11:00	**Comparative Geophysical Survey results in Japan** Y. NISHIMURA (National Research Institute for Cultural Properties-Japan)
11:00–11:30	**Break**
11:30–13:30	**Why bother? Large scale geomagnetic survey and the quest for 'Real Archaeology'** D. POWLESLAND (LANDSCAPE RESEARCH CENTRE, WEST HESLERTON)
13:30–15:00	**Lunch**
15:00–17:00	**Electromagnetic methods** A. TABBAGH (Unité Mixte de Recherche-Sisyphe)
17:00–17:30	**Break**
17:30–19:30	**Large scale archaeological investigations using geophysical methods** C. GAFFNEY (GSB Prospection-Bradford)

DAY 5-6-7-8 (14-17.07.06)

FIELD DATA ACQUISITION AND DATA PROCESSING

The 35 students attending the course will be divided into four groups (G1-G2-G3-G4). In the morning two groups have operated in the field, while the other two have worked in the laboratory. In the afternoon the groups have been switched.

DAY	GROUPS	AM 9:00–13:00	PM 15:00–19:00
5 (14.07.06)	G1 & G2	ERT & MAG	Data processing in the Lab
	G3 & G4	Data processing in the Lab	ERT & MAG
6 (15.07.06)	G1 & G2	MAG & ERT	Data processing in the Lab
	G3 & G4	Data processing in the Lab	MAG & ERT
7 (16.07.06)	G1 & G2	GPR & ERT	Data processing in the Lab
	G3 & G4	Data processing in the Lab	GPR & EM
8 (17.07.06)	G1 & G2	EM + GPR	Data processing in the Lab
	G3 & G4	Data processing in the Lab	ERT + GPR

DAY 8 (17.07.06)

20:30 Social dinner

DAY 9 (18.07.06)
10:30   Presentation of certificates and closing session of the XVth International Summer School in Archaeology

Preface

In history and prehistory artefacts and structures are often concealed beneath the surface of the earth as a result of geological processes and perhaps more frequently as a result of human activity. A key aspect of landscape archaeology lies in the identification and interpretation of this hidden evidence or in the broader sense of the buried landscape.

Recent decades have seen landscape archaeologists concentrating mainly on the collection of vast numbers of sites, for the most part in isolation from one another. We might call this a *site*-based approach. But neither present-day nor past landscapes consist only of houses, settlements, cemeteries, industrial areas and the like. More recently archaeologists have become aware that there is a great range of evidence (on-site as well off-site or non-site), from scatters of artefacts to road systems, plough-marks and field boundaries, that can provide important information, not only about human exploitation of the environment but also about cultural, social and economic developments. This has created a 'new' challenge. We are called to face the inherent complexity of landscapes and their internal relationships—often hidden beneath or between 'sites' and for the most part represented by relatively 'weak' evidence. We might call this a *landscape*-based approach.

The site-based approach has produced limited results in landscape terms, not least because the main investigative method has been restricted to surface collection. Reconnaissance survey of this kind is of course essential, and can be very productive, but like every other method it has its limitations. We would point particularly to its inadequacies in the identification of structures, features, chronological phases and social groupings. It goes without saying that some archaeological phases are more readily 'visible' than others and that some cultures, such as those of mobile hunter-gatherers or pastoralists, leave a very different imprint on the landscape than those of agricultural or urban societies. It is equally obvious that surface artefact collection allows us to recognise only a small range of the potentially available evidence. In Mediterranean landscapes, for instance, the definable site typologies are limited to such things as 'grave', 'farm', 'roman villa', 'industrial area', 'off-site finds' etc. On the basis of reconnaissance survey it is virtually impossible to collect a wider range of evidence or to achieve a more precise and detailed classification of site typology. We are in no sense averse to field-walking survey, which we see as probably the best available method in regional studies for detecting sites that produce surface evidence in the form of artefact scatters, building material and the like. We are, on the other hand, convinced that the limitations of each and every research method should be openly acknowledged. It is our contention that without the *integration* of a variety of information-recovery techniques we cannot begin to confront and comprehend the inherent complexity of past landscapes.

Remote sensing, and in particular aerial reconnaissance, can play a crucial role in the discovery, recording, interpretation and monitoring of sites. Satellite imagery, airborne scanning (multi-spectral, LiDAR), vertical air photography and exploratory aerial survey have developed into some of the archaeologist's most valued tools. But, just like field-walking survey, these sources and techniques have their own particular limitations—LiDAR and multi-spectral sensors, for instance, each pose different kinds of problems. Post-depositional processes can also affect the results because of thickness, weakness and size of evidence. And with all techniques there is also the imponderable affect of 'serendipity'—good fortune (or otherwise) in the local situation or in the luck of the moment.

Any non-destructive method capable of reducing or offsetting the uncertainness of field-walking survey, trial excavation or the varying capacities of different types of remote sensing should be greeted with open arms by archaeologists if it leads to an increased probability of information-recovery. One such method, now the subject of new or revived attention, is near-surface geophysical survey.

In recent years a wide range of scientific disciplines have provided useful tools for an integrated approach to data-collection, interpretation and conservation in the field of the cultural heritage. In this context archaeological prospection presents a whole range of non-invasive techniques, including various kinds of geophysical survey, satellite imagery and aerial photography, as well as a variety of digital site-recording systems and numerical techniques for processing, analysing and representing the different data-sets that can be collected through ground-based survey.

The last twenty years have seen great technological advances in these non-invasive methods. New instruments, data-acquisition techniques, geophysical methods and processing strategies have made the fieldwork much faster, more sophisticated and more effective. From a site-based outlook, geophysical prospection increases the visibility of features but also greatly enhances the complexity and sophistication of site interpretation, providing high-resolution maps of the subsoil—not merely the surface—and, depending on the technique applied, even maps at different depths. The capacity to see the sub-surface pattern of archaeological features makes it possible to refine the generalised types of site classification achievable through traditional field survey, and gives the opportunity to draw a wider variety of conclusions about questions of cultural, social and economic background.

Equally important is the increase in the **speed** of measurement achieved by a small number of pioneer scholars—many of them represented in this volume—making it possible to move geo-physical survey from a site-based to a landscape-based approach. This sort of perspective is not entirely new. We should recall, for instance, the extraordinary revelations of John Bradford on the Foggia plain in Apulia, Sothern Italy (Bradford 1957), where he discovered probably one of the most outstanding stratified landscapes visible from the air. Nearly sixty years ago he wrote in the British journal Antiquity: '*We can now go towards peopling this landscape, not in a sand-table world of theory, but in a stetting of actual fields and farms which provide unrivalled opportunities for the direct study of roman agriculture*" (Bradford 1949). But it must be recognised that it is the geological and land-use character of the Foggia plain, rather than its cultural pattern, that gives it an extremely high archaeological visibility when viewed from the air. The main innovation and advance offered by large-scale continuous geophysical survey within a landscape-based approach lies in the opportunity to overcome or circumvent the limitations of archaeological visibility caused by such things as clay soils, unfavourable land-use or unhelpful cultivation patterns—along, of course, with the 'vagaries of the moment' that affect the success or otherwise of aerial survey. Bradford said about the roman landscape on the Foggia plain: "*Never before has the actual landscape of the Roman farmer, described in the Georgics, been better preserved for direct study. One can walk along the grassy roman roads from one farmstead to the next and enter up its ditch-flanked drive bordered by vines or trees*" (Bradford 1949). In the present day this could be said, too—despite inevitable cultural differences—about the Vale of Pickering in north-east England, studied over the past thirty years by Dominic Powlesland, another contributor to this volume. Of course we are not maintaining that geophysical prospection provides the ultimate panacea for the archaeologist concerned with the cultural heritage. In some ways it is, but like all other techniques it can fail in some circumstances to spot significant archaeology—features that are too small, too deeply buried or too weak in the signals or reflections that they produce.

It is in our view fair to claim that the results achieved in the last decade through the devel-opment and application of geophysics should be acknowledged by the archaeological commu-nity as amongst the most important methodological changes of recent times. It is our hope that, through this book, these developments can be more fully understood by archaeologists, who often approach 'new technology' with a degree of suspicion. There is a striking paradox here.

Archaeologists define society and the major phases of prehistory from an explicitly technological point of view: Stone Age, Bronze Age, Iron Age etc. Moreover, the most important cultural revolutions in the history of humanity have been defined on the basis of technological developments: the introduction of agriculture, the Industrial Revolution and now the advent of computer science. How can we deny the extraordinary contribution of geophysical science, and more generally of technology, in the search for archaeological understanding?

Stefano Campana
Salvatore Piro

Introduction

Geophysical techniques are based on well known physical principles. These principles involve indirect measurements of earth related physical attributes and in the case of archaeological structures, contrasts resulting from subsurface disturbances associated with past cultural activities. Where contrasts between the sought after archaeological features and surrounding soils are small or non-existent the limitations of geophysical methods and instrumentation become apparent.

The actual enhancement and success achieved in a geophysical investigation is subject to numerous site specified related conditions. Important factors should be recognized in:

1. The survey methods, with methods stressed, since it is believed multiple survey methods should be considered a requisite for all site investigations.
2. Site compatibility.
3. The persons conducting the survey must be knowledgeable of selection of proper methods, survey techniques and the nature of results to be expected (i.e., experience factor).

Therefore, considered the increasing interest in the topic and the related need of knowledge, we decided first to organize an International Summer School and to put together this volume for use by both students and researchers.

The volume is divided in two sections: theory of geophysical prospection and practice into the field. The theoretical section groups chapters dealing with similar topics related to: introduction to landscape archaeology, mapping and geophysical prospection, electrical methods, magnetic methods, electromagnetic methods, ground penetrating radar. Many of the papers in one group, however, also touch on research applications considered in one or more of the other sections, and several deal also with issues of data integration.

Theoretical Section

Introduction to landscape archaeology, mapping and geophysical prospection

The first paper by Stefano Campana (University of Siena-Italy)—*Archaeological Site Detection and Mapping: some thoughts on differing scales of detail and archaeological 'non-visibility'*—deal with problems, strategies and main methods actually available for the study of landscape archaeology and some potential contribution in the process of archaeological mapping of geophysics.

The contribution of Salvatore Piro (National Research Council-Italy)—*Introduction to Geophysical Prospection applied to Archaeology*—present an overview of all geophysical methods employed for archaeological prospection. The many field techniques deployed in archaeological research today include geophysical and related methods of non-invasively detecting usually shallow-buried structures and remains. The probability of a successful application rapidly increases if a significant multi-methodological approach is adopted, according to a logic of objective complementarity of information and of global convergence toward a high quality multi-parametric imaging of the buried structures.

Electrical methods

Armin Schmidt (University of Bradford-UK)—*Electrical and Magnetic Methods in Archaeological Prospection*—discuss the use of the two main geophysical techniques used in landscape archaeology, namely earth resistance and magnetic surveying.

Gregory Tsokas (Aristotle University of Thessaloniki-Greece)—*Electrical Resistivity Tomography: A Flexible Technique In Solving Problems Of Archaeological Research*—highlight the valuable contribution of information about the depth extend of the buried targets. This issue was one of the main reasons for the development of the electrical resistivity tomography (ERT). The term implies that automated multiplexers are used for data acquisition which are subsequently inverted by some mathematical scheme. The present lesson presents the merits of resistance mapping and also its drawbacks which created the need to apply ERT in archaeological prospection.

Michel Dabas (University of Paris-France)—*Theory and practice of the new fast electrical imaging system ARP©* —shows the principle of a new towed system devoted to electrical mapping of soils: the ARP© system (*Automatic Resistivity Profiling*) and to give some examples obtained with this system in Archaeology. The author demonstrate clearly that nowadays it is also possible to apply electrical prospection (traditionally the slower method) with a landscape-based approach.

Magnetic method

Helmut Becker (Bavarian State Conservation Office-Munich-Germany)—*Caesium-magnetometry for large scale archaeological prospection*—in the first part of his paper provides a survey the history of studies of magnetic methods. In the second part he present the development of his high speed, sensitivity and spatial resolution magnetic probe system and the related field work procedures. Becker presents also several case histories collected in some of the most important world heritage site.

Dominc Powlesland (Landscape Research Centre-UK)—*Why bother? Large scale geomagnetic survey and the quest for 'Real Archaeology'*—discusses the results of the project on landscape archaeology centred on the Vale of Pickering (UK). He applies various comolementary remote sensing methodologies and archaeological excavation in an attempt to discover the 'real' archaeology of the Heslerton research area. He issues regarding archaeological 'quality' with reference to the sustainability of the resource within an aggregate bearing landscape. The employment of magnetic survey to scan 1000 hectares continuous landscape should be highlighted nowadays as the most important application of geophysical technique to landscape archaeology.

Meg Watters (University of Birmingham-UK)—*The complementary nature of Geophysical survey methods*—presents a brief introduction to geophysical survey methods, discusses details of geophysical survey project planning. She also shows a case study from the Catholme Ceremonial Complex located in the UK. The geophysical survey methods presented include magnetometry, resistivity and GPR. As an important part of geophysical surveys, project planning is discussed in detail and aimed to outline a comprehensive framework from which any type of geophysical survey can be designed.

Chris Gaffney (Remote Vision Research-UK)—*The Use of Geophysical Techniques in Landscape Studies: Experience from the Commercial Sector*—faces the commercial geophysical work for archaeological purposes has in the United Kingdom background. He underline how in the last years has been possible to change a belief that was commonly held in British archaeology; it was thought that geophysical survey was too slow and the interpretation too imprecise to be of any great value in the commercial arena. While a cursory look through any text book on geophysics will suggest that about 10 ground based geophysical techniques are on offer for near surface work, only a few are regularly used in the commercial field and even fewer are used to investigate landscapes. Most commonly the techniques of earth resistance, magnetometry, magnetic susceptibility sampling and GPR are used, although low frequency EM systems are also valuable. Magnetometry, in its various forms, is the most widely used geophysical technique for detecting buried archaeology.

Electromagnetic methods

Albert Hesse (University Pierre et Marie Curie-France)—*Short history, strategies and practical aspects of electromagnetic detection based on the description of field experiments*—starts his paper from a critical viewpoint arguing as follow: electrical and magnetic methods for archaeological surveying have reached nowadays such a high level of efficiency in terms of speed and legibility of maps that other methods often appear to be of less interest. This is the case for electromagnetic methods despite their unique ability for detection of specific targets or under special circumstances. These methods can be disregarded for several reasons among which the complexity of theoretical aspects, the large variety of available instruments with not well-known specific abilities, a rather slow rate of data collection, etc. must be recalled. This paper emphasize some historical aspects and case histories of significant interest.

Alain Tabbagh (University Pierre et Marie Curie-France)—*Electromagnetic methods*—presents how E.M. low frequency slingram instruments can measure both the electrical conductivity and the magnetic susceptibility of the ground. They also detect metallic objects. Moreover the coupling between magnetic and E.M. data opens interesting paths for a better characterisation of feature geometries and magnetisations.

Ground penetrating radar

Dean Goodman (Geophysical Archaeometry Laboratory-USA)—*GPR Methods for Archaeology*—introduces the method of Ground Penetrating Radar (GPR) for archaeology using examples from simulation software and subsurface imaging software. Examples of what are referred to as reflection multiples, velocity pull-ups, radar shadow zones and several other geometric effects from recording with broad beam single channel GPR equipment is used as guide in avoiding interpretation pitfalls. Basic signal and image processing for GPR are also introduced along with examples of successful GPR imaging at Roman, Japanese and Native American Indian sites are presented.

Lawrence Conyers (University of Denver-USA)—*Ground-penetrating Radar for Landscape Archaeology: Method and Applications*—based the subject of his lesson on general technical aspects of the method and its employment to survey archaeological sites.

Yasushi Nishimura (National Research Institute for Cultural Properties-JAPAN)—*Comparative Geophysical Survey Results in Japan: Focusing on Kiln and Building Remains*—give an overview of the application of geophysical methods to survey different archaeological sites in Japan.

Field Work Section

During the International Summer School students and lectures—after three days theoretical lessons—spend four days into the field collecting geophysical measurement. The archaeological test-site is placed on lowland quite close to Grosseto, the city where the University of Siena recently open a satellite of the Department of Archaeology and History of Arts and where there was the headquarter of the Summer School. The site is a further big roman villa starting form the first century AD that we estimated occupied until the sixth century AD. Grid collection and pottery analysis allow us to recognize the re-occupation of the site during the late ninth and tenth century AD. The site measure about 4 hectares in extent so we decide to focus our attention on four square sample of 50 by 50 meters where we rotated four different geophysical methods: magnetometry (fluxgate, Overhouser, Cesium), ground penetration radar (GPR), electro-magnetometry (EM), electrical resistivity tomography (ERT).

In this section we repot the work achieved during the Summer School but also we publish the data collected starting from the 2001 when the site has been discovered during the Aerial Archaeology Research School (Culture 2000 project) to the last survey we did in autumn 2007. One intention working on the Aiali test-site is to apply the highest available level and intensity of archaeological prospection methods on a large, complex and stratified site, producing material from the from Etruscan, Roman and Medieval periods. As our good friend Chris Musson has argued—during one of our long discussions on this topic—the archaeological objective and outcome of the Aiali project has to take account of the critical impact of the kinds of information that are available for recording: to use his own words "in assessing the potential or interpretation of a landscape it is at least as important to know what may not be visible as to appreciate what is visible".

The first and in the second chapters of section 2 Stefano Campana introduces the site background and the results obtained through remote sensing tools from Quickbird-2 satellite imagery, to vertical air photographs and aerial survey. In chapter 3 Emanuele Vaccaro and Mariaelena Ghisleni provide the results of field walking survey, grid collection and the detailed study of pottery. Chapter 4 is addressed by Stefano Campana and Salvatore Piro to introduce the geophysical surveys of the site. In the next chapter Helmut Becker, Stefano Campana, Thomas Himmler and Iacopo Nicolosi to discuss the results in the sample areas of different magnetic sensors (Fluxgate, Overhouser And Caesium-Magnetometry). Chapter 6 handles with Ground Penetrating Radar (GPR) surveys by Dean Goodman and Salvatore Piro. Chapter 7 discuss the results obtained through the application of electromagnetic survey by Alain Tabbagh while the next one deals with the last methods applied during field work, electrical survey and is summarized by Michel Dabas and Gianfranco Morelli.

As conclusion Stefano Campana and Salvatore Piro, an archaeologist and a geophysicist, try to combine all the information together. Through a GIS-based analysis the editors integrate different sources geophysical measurement as well satellite imagery, aerial photograph, archaeological information collected during field walking survey and archaeological knowledge. The critical impact of the work is addressed to show the improvement of available archaeological information as a consequence of the improvement of the survey methods.

Stefano Campana
Salvatore Piro

Acknowledgments

People

The authors owe an enormous debt of gratitude to the late Professor Riccardo Francovich, of the University of Siena, for his inspiration, encouragement and ever-helpful criticism throughout all stages of the International Summer School project.

We are very grateful to Chris Musson for his great passion, wide experience and critical comments on the practice of archaeological research, and also for help with the English version of the text.

Special thanks are due to Dr. Barbara Frezza who helps a lot the production of this book in the format of the texts and Daniele Verrecchia (ITABC-CNR) for the editing and format of the figures.

Helmut Becker (formerly of the Bavarian State Department of Historical Monuments, Germany) and Iacopo Nicolosi (National Institute of Geophysics and Vulcanology, Italy) contributed greatly in the field and also helped to establish the best configuration for the employment of the magnetometer equipment. Dean Goodman (Geophysical Archaeometry Laboratory, Los Angeles, USA) did great work during GPR data acquisition and processing. Our thanks also go to all of the teachers who contributed so effectively to the success of the school: M. Dabas, A. Tabbagh, A. Schmidt, C. Gaffney, L. Conyers, M. Watters, D. Powlesland, Y. Nishimura and D. Goodman.

Special thanks are also due to the research team of the Laboratory of Landscape Archaeology and Remote Sensing at the University of Siena: Dr. Cristina Felici, Dr. Francesco Pericci, Dr. Matteo Sordini, Dr. Emanuele Vaccaro, Dr. Barbara Frezza and Dr. Mariaelena Ghisleni.

Institutions

The organization and support for this International School drew on the experience and help of many people, projects and institutions: the University of Siena, the Monte dei Paschi Foundation, the National Research Council (ITABC-Geophysical team), the Culture 2000 project *European Landscapes: past present and future*, English Heritage and Bob Bewley, the Italian National Research Project *Integrated technologies for the mapping and management of the Tuscan Archaeological Heritage* (PRIN 2004—Project Leader prof. M. Guaitoli, Unit Coordinator prof. R. Francovich), the International Society for Archaeological Prospection (ISAP) and the international Aerial Archaeology Research Group (AARG).

Companies

We gratefully acknowledge the partnership, support, expertise and genuine friendship of the geophysical company Geostudi Astier (Livorno) in the organization of the School.

The bulk of the funding was provided by Geostudi Astier (Livorno), Terranova (Paris), Foerster, GEM Systems, Geonics Limited, Geosoft Software, IRIS Instruments, Leica Geosystems and ESRI Italia.

Funding and logistical support were also provided by University of Siena, the National Research Council (ITABC), the Culture 2000 project, English Heritage, the Italian National Research Project PRIN 2004 and the Aerial Archaeology Research Group (AARG).

Theoretical Section

*Introduction to landscape archaeology,
mapping and geophysical prospection*

Seeing the Unseen – Campana & Piro (eds)
© 2009 Taylor & Francis Group, London, ISBN 978-0-415-44721-8

Archaeological site detection and mapping: Some thoughts on differing scales of detail and archaeological 'non-visibility'

S. Campana
Department of Anthropology and History of Arts, University of Siena, Italy

ABSTRACT: Archaeological mapping shares affinities with topographical mapping but there are also significant differences. This contribution will concentrate on two basic aspects: differences in the scale of representation, and the question of archaeological visibility. The traditional subdivision between micro, semi-micro, mid- and macro scale tends to omit the 'local' level which more closely matches the characteristics and needs of field archaeology and landscape studies. As regards archaeological visibility and non-visibility the key point is that, in contrast to topographical cartography, the majority of the features depicted in archaeological mapping are not directly recognisable in their own right but reveal themselves as micro variations in the top-soil or as surface reflections of things buried beneath the ground. This contribution aims to present a summary of the main archaeological survey methods, along with their key characteristics and limitations, while also outlining the potentialities that can arise from their integration with one another in the creation of cartography at the macro, local and micro-territorial scale.

1 INTRODUCTION: ARCHAEOLOGICAL CARTOGRAPHY AND THE QUESTION OF SCALE

Cartography constitutes an indispensible instrument for the representation, management and communication of geographical data. Archaeological information is no exception. On the contrary, the complexity inherent in differentiating or bringing together the intricate and deeply stratified palimpsest of information about any particular area leaves no alternative to the use of maps within the creative framework of geographical information systems (GIS).

The kind of information that can be depicted in archaeological maps, just as in topographical maps, is in essence determined by the scale of representation. Various basic levels of scale have been recognised in archaeology. Clarke, for instance, identied three: 'macro', 'semi-micro' and 'micro' (Clarke 1977). Starting from the scale definition of Clarke it is possible to deduce that the 'micro' scale, aimed at intra-site analysis, works at the level of points in space, geographically circumscribed and definable by single identifiable cultural characteristics. The elements represented might include building structures, artifact scatters, remotely sensed features, etc. The mapping scales might vary from lifesize to about 1:200. The 'semi-micro' level, still focusing on intra-site problems, represents the bringing together of the various elements that constitute a multiple-activity area. The scales here might range from 1:200 to roughly 1:1000. The 'macro' scale deals with larger territories, from sub-regions and regions to national states, and aims at analysing relationships between sites, or perhaps one should say between 'clusters of evidence'. The scales here might vary from approximately 1:1.000 to 1:1.000.000 or more. Butzer (1982) proposed a more detailed graduation of scales, including 'mid-scale'. The 'mid-scale' is used for within-structure aggregation areas (site) and macroscale for intersite patterning related to environmental features in or around a node defined also on a cultural basis. It should be clear that the scale size graduation is completely throw off balance, with three different within-site level and only one addressed to inter-site mapping and analysis. However, no one scale that is better than the others; the key point is that it is the purpose of the mapping that should determine its scale (Raffestin 1987;

Sydoriak Allen 2000; Lock, Molyneaux 2006). For this reason, too, the transition from one scale to another during synthesis can have a very significant impact on the understanding of landscape patterns (Marquard, Crumley 1987).

The transition from the 'micro' to 'macro' level, for instance, does not consist of a simple mathematical and graphical process of reduction, simple enough to achieve in the age of GIS. Rather, it involves complex procedures of simplification, generalisation and blurring of distinctions which have significant effects on the quality and quantity of the information transmitted. The transition in the opposite direction, from the 'macro' to the 'micro' scale for instance, entails even more complex problems. In topographical mapping, for example, an increase in detail of this kind involves the revision or supplementing of the contour lines and spot heights. In this case the difference of scale does not affect the strategy of work in any substantial way, nor the technical means or basic methodology for achieving it (generally related with photogrammetry). The archaeologist who has to cope with the transition to a more detailed scale must, however, give thought to the availability or introduction of instruments that are barely applicable today at the smaller scale. Archaeological mapping at 'macro' scale depends for its support most of all on literary, bibliographical and documentary sources, on toponomy, iconography, epigraphy, historical cartography, aerial photography, satellite imagery and occasionally field survey (Cambi 2003). The 'micro' scale, on the other hand, is traditionally concerned with strategies for the recovery of material within a site and its subsequent examination and analysis (Haselgrove et al. 1985; Schofield 1991); in the past decade, however, it has also drawn heavily on the contribution of geophysics (Gaffney, Gater 2003). The 'local' scale, a term used here to indicate the shadowy zone between mid- and macro scale, represents in our view the cognitive level which is most problematical. Up until now there has been a prevalent tendency for the simplistic superposition of the macro and micro scales. The results are almost invariably disappointing or illusory. Satisfacotory results can only be achieved when there are contexts that are particularly favourable to the conservation and visibility of indications from the past. If we take, for example, a region such as Tuscany, in thirty years of active research about 18.000 archaeological sites have been identified. The representation of the evidence, through symbols, at the regional or provincial scale, from about 1:2,000,000 and 1:200,000, presents seemingly high or very high densities (Fig. 1, right). The translation to the 'local' scale, for instance that of a moderately small river catchment area, at 1:50,000 or 1:25,000 scale, produces on the other hand an expanse of 'near-emptiness', in which it is easy to see the apparent scarcity of the available data (Fig. 1, left).

Integration with the micro/semi-micro scale, often involving understanding of the intricate inter-relationship between individual sites or contexts, makes even more obvious the profound lacunae in our archaeological mapping at the inter-site level. The result at the 'local' scale is to present multi-period archaeological landscapes as a series of points (the sites), usually lacking any kind of linking physical relationship (this topic is taken up by Powlesland elsewhere in the volume). The result is totally inadequate either for the writing of history or for heritage conservation. This way of working does not in the great majority of cases allow us to perceive and understand the transformations through time of the missing 'connective tissue'. This forms an indispensible element in the comprehension of landscapes made up not only of settlements and cemeteries but also of agricultural activity, communication systems and infrastructure element, ecofacts, morphology, hydrology, natural resources, economics and so on. The omission of this level of scale would mean in effect the abandonment of landscape archaeology, at least in terms of its original aim of integrating the cultural tradition related to field archaeology and local history (Fleming 2006; Aston, Rowley 1974).

Another problem—directly related to the last and relatively common in archaeological mapping—concerns the relationship between the micro and the macro scales. In the absence of the missing 'local' scale, contexts which can be studied comprehensively at the micro or semi-micro scale have been generally discovered in mapping at the macro scale level. As already mentioned, the instrumentation and methodology used in the process of archaeological mapping change with the variation in scale. The jump from macro to micro scale, without the benefit of the intermediate variations, risks the loss of many significant pieces of information because these—depending on

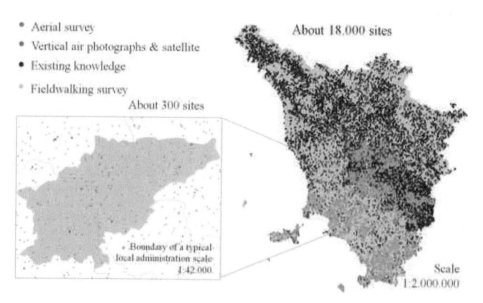

- Aerial survey
- Vertical air photographs & satellite
- Existing knowledge
- Fieldwalking survey

About 18,000 sites

About 300 sites

Boundary of a typical local administration scale 1:42,000

Scale 1:2,000,000

Figure 1. Right: representation at a scale of 1:2,000,000 of the known archaeological sites in Tuscany. Left: the same data at a scale of 1:42,000 for a single local administration studied in the archaeological mapping project of the University of Siena. Please note that the map represents the diachronic distribution of sites, from prehistory to the late middle age. That means in every period map site density is definitively more and more lower.

the nature of the individual context—become effectively invisible at the macro scale or due to the peculiarity of the involved methods. The prevalence of this kind of thing is particularly damaging because it generates a sort of short circuit, giving prominence only to those sorts of archaeological evidence that properly belong to the macro scale. Amongst the consequences of this approach to landscape archaeology there is the tendency (often not specifically declared) in the stages of synthesis to treat the recovered information and its distribution as if it represents the whole of the original reality—a totally unacceptable position undermined by recent studies which suggest that mapping at the macro scale (without taking into account field walking survey) allows the recovery, even on the most optimistic estimate, of no more than 5% (national Italian average) of the potentially surviving archaeological evidence (Guaitoli 1997).

2 ARCHAEOLOGICAL VISIBILITY AND NON-VISIBILITY GETTING STARTED

Before turning to the various methods used in the search for the basic evidence it is necessary to talk briefly about the concept of archaeological 'visibility'. In practice, and in contrast to the situation in topographical mapping, the great majority of the items recorded in archaeological mapping are not visible in their own right but appear as one kind of reflection or another of buried deposits. In the study of landscape archaeology the concept of visibility has usually been associated with questions of land-use and sedimentation (Cambi, Terrenato 1994). Discussion has also been focused, for the most part, on the impact of visibility and non-visibility on the results of surface collection survey (Terrenato 2000). Moreover, the question of visibility has been addressed more often as a means of pointing out the limitations of archaeological documentation based on surface observation, rather than in an attempt to correct its shortcomings, perhaps by integrating surface observation with other methods of data collection (Cambi 2000). This kind of discussion took place mainly in the 1980s (Schiffer 1987) and the early 1990s (Schofield 1991; Allen 1991), only to be virtually forgotten later. The thoughts presented in the following sections represent the fruit of the last decade in the study of medieval archaeology in Tuscany by the University of Siena.

First and foremost the term visibility—especially if used in the widest sense of the term—represents in archaeology a particularly ambiguous concept which can have a wide variety of different meanings or (perhaps more properly) which reflects a wide variety of different factors. Apart from problems connected with land-use and the local geology and soils, experience suggests a number of other factors that can sometimes have a significant impact on archaeological visibility.

The transformation of the rural and suburban landscape in the recent past can play a determining role in the observer's perception of a context, or of its complexity, as well as in the identification and conservation of archaeological deposits. A striking example is provided by the Val d'Orcia, a district covering about 500 Km^2 south-east of Siena, today characterized by an expanse of gently rolling hill-country dominated by cereal cultivation and (in the southern parts) by vineyards. This kind of monoculture, however, is the result of drastic and often radical transformations that began to take effect in the Fascist era, continuing in the post-war years and on to the present day (Fig. 2). Val d'Orcia, in the past, was an area of summer drought and winter floods, its bleakness dominated by biancane (eruptions of clay from the underlying subsoil) and calanchi (deep erosion gullies in the clay substrata) (Mangiavacchi 2004). Its morphology and heavy clay soils (too intractable for widespread cultivation by traditional pre-mechanized methods) restricted agricultural activity and productivity. Similar patterns can be seen in the areas of Maremma and Chianti, now radically transformed by land improvement and widespread monoculture.

A phenomenon not to be under-estimated is the contribution of agricultural activity, not so much in terms of land use as in the differing impacts of intensive, biological or traditional farming on subsurface deposits and on archaeological material brought to the surface by the plough. As a generality it is worth noting the ever-increasing problem of the progressive degradation of surface deposits above archaeological features after up to half a century of mechanized cultivation. Also the present tendency, under EU influence, towards less deep ploughing, resulting in

Figure 2. Panoramas from the same points in the landscape of Pienza (Val d'Orcia-Siena) in 1930 and in 2000 (from Felici 2004). The white 'eruptions' of the biancane (left) and the erosion scars of the calanchi (right) have now been totally erased by decades of mechanized arable cultivation.

less archaeological material being brought to the surface. There is an additional problem in the present-day difficulty of gaining access to some agricultural holdings.

Another key question concerns the extent to which the things that we are seeking are likely to be visible. Material culture changes over time, of course. The physical manifestation of settlements, communication systems and agricultural patterns can sometimes be monumental in character, at other times modest and ephemeral. The trappings of everyday life in one period may be highly durable, with well-made and hard-fired pottery, for instance, but such things may have been radically different in the preceeding or following phases, with the use of poor-quality clay, inadequate firing or even the substitution of durable materials such as pottery in favour of wood or other perishable materials. Different material cultures may therefore present differing levels of archaeological visibility (Fig. 3). The less intensive incidence of one culture in a particular context, and the less durable character of the materials used, can also give rise to fundamental difficulties in the archaeologist's recognition of crucial pieces of evidence (Hamerow 2004; Liebeschuetz 2007).

Figure 3.   Difference in the surface scatter of artefacts between a Roman villa (above) and an Early Medieval village (below).

Continuing along the subject of visibility and the peculiarities of cultural choice, a question that is very familiar to medieval archaeologists in particular is the relationship between visibility and continuity of settlement. In contexts where there is continuous occupation of the same location for a long period of time there are many difficulties in identifying the evidence from the earlier phases because of the degradation and interruption of deposits and features by centuries of continuous settlement. This is so, for example, in the case of Tuscany where the researches of Riccardo Francovich have shown that after abandonment of the Late Roman settlement pattern traces of early medieval settlement can be traced beneath the many castles that dot the hills of Tuscany (Francovich, Hodges 2003).

These kinds of questions, revolving around archaeological visibility, have often been discussed in general terms but any proper examination of the concept would have to stress the arbitrary risks of research that does not take account of the problem. In the face of these problems and their affect on research it is hard to maintain the idea that samples recovered from surface collection survey are nevertheless representative. On the other hand it is important to recognise that problems of visibility are not necessarily incapable of solution, in that they vary according to the survey method used in any particular piece of research. It is therefore legitimate to speak, for example, of 'ground visibility' (Schiffer 1987; Schofield 1991; Allen 1991), or 'aerial visibility' (Mills, Palmer 2007) or even 'cultural visibility' (Francovich 2004). Every method of research is selective in one way or another, in the sense that it aims to track down particular forms of evidence rather than others. Surface collection survey, for example, cannot hope to reveal traces of timber structures that aerial photography, by contrast, can delineate with great precision, as for instance at Woodhenge, in southern England (Cunnington 1927). Aerial photography, however, runs into significant problems when it tries to uncover archaeological evidence on heavy clay soils, while magnetometry and geo-eletrical methods might produce excellent results in the same situation. And so on. In drawing this discussion of archaeological visibility (or non-visibility) to a close it should be emphasized that the intention is not in any sense to deny the credibility of territorial research, nor of any one or more of its methods in particular. The objective is to encourage a robustly critical approach to the inherent problems of this kind of research and to promote efforts to offest or moderate their distorting effects so as to attain a higher level of reliability and credibility in the study of archaeological landscapes.

3   MACRO-ENVIRONMENTAL SCALE: THE REGION

There are many methods for the identification of archaeological evidence at the macro-environmental scale, ranging from the analysis of written sources, through iconography, archival and place-name studies to cultural anthropology, geomorphology and remote sensing and various forms of fieldwork. Discussion in the following paragraphs will focus on the principal methods in field archaeology and remote sensing, as experienced in the work of the Department of Archaeology at the University of Siena. For other methods readers should refer to publications such as Gamble 2007 and Renfrew & Bahn 2008.

Tuscany covers an area of about 22,900 km^2, a huge area characterized by wide variability in its landscapes and hence a lack of homogeneity in the acquisition of archaeological data. Within the regional context the parameters which define the various grades of archaeological visibility, on the ground and in the air, interweave with one another, producing areas with extremely high levels of visibility and others with extremely low levels, interspersed with innumerable variations in between. For example, as concerns land use, we have already seen that 50% of Tuscany is given over to woodland. Woodland vegetation, clearly, constitutes a serious obstacle to field survey and is equally inimical to the recovery of archaeological evidence from the air. So, as we look at the region today, at least half of its area presents extremely low levels of visibility, both on the ground and from the air. Turning to the region's soils, the Provinces of Siena and Pisa are particulary disadvantaged by the widespread presence of clay soils, widely recognised as producing relatively limited aerial visibility, with a poor response to the formation of cropmark and soilmark evidence.

Over the past ten years the University of Siena has been pursuing a programme of research which has been structured so far as possible to reduce the influence of this kind of contextual factor on the archaeological return, in pursuit of more homogeneous results. The strategy, inevitably, has to remain flexible and open, founded on the conviction that only the integrated use of a wide range of survey methods and technologies, applied at varying scales of detail, will make it possible to confront these problems and respond adequately to the innate complexity of studying settlement dynamics in the landscapes of the past (Campana, Francovich 2007).

3.1 *Vertical air photography: historical and recent*

For the better part of a century aerial photography has been one of the archaeologist's most important instruments for the detection of archaeological features and the representation of the resulting evidence through cartography (Bewley 2005). In particular, 'historical' photographs from the 1930s, 1940s and 1950s show a landscape profoundly different from that of today. Building development, infrastructure projects, land-use change and mechanised agriculture have radically changed the landscape, completely destroying or partially concealing significant parts of the archaeological resource. Only through the detailed analysis of this historical air photo coverage will it be possible to recover, in part at least, the presence, siting and inter-relationship of settlements and other man-made features and natural phenomena (Bradford 1957). Further reasons for interest in this source lie in the breadth of coverage represented in vertical aerial photography (extending in most cases to at least the provincial or regional scale) and in the huge amount of photographic material available, often without charge, in provincial, regional and military archives and in private collections (Going 2002; Guaitoli 2003). More recent vertical images offer an up-to-date record of the landscape for comparison with the earlier coverage, thereby defining in detail the transformations that have taken place in any chosen study area in recent years. In addition, 'new' vertical photography, if carefully timed to coincide with the best visibility of cropmarks or other traces of the past, can prove a highly effective research tool. In this context one might point to the research carried out in Italy at the University of Lecce (Guaitoli 2003), in Austria at the University of Vienna (Doneus 2001) and in Great Britain by a variety of bodies. It has been demonstrated that vertical coverage which coincides with a window of high aerial visibility can documment many hundreds of archaeological sites and features in a few tens of minutes (see, for example, Coleman in Mills & Palmer 2007). Any kind of cost-benefit comparison with other investigative techniques would see this as an extraordinarily favourable return.

3.2 *Exploratory (oblique) aerial survey*

Amongst the methods available for territorial research aerial survey through oblique photography occupies a place of its own. It permits the collection of data at the regional scale and in the process makes it possible to carry out more detailed analysis of particualr locations or areas through repeated observation, with the capacity (within limits) to vary the degree of detail. The long history of this technique in the study of the landscape over very wide areas links it inevitably with the macro-territorial scale. In particular one might point to the ongoing National Mapping Programme for England, started in the late 1980s with the aim of systematically mapping all no-longer-used features that can be seen on vertical and oblique aerial photography across the whole of England (Bewley 2005).

The basic principles of oblique aerial survey are substantially the same as those which apply in vertical air photography (Wilson 1982, 2000; Piccarreta, Ceraudo 2000). Particular features of the method have been described and discussed by Musson et al. 2005 and Brophy & Cowley 2005. A key point feature of the technique is its flexibility in the choice of time of year or day to carry out the photography (subject, of course, to regional or national variations in response to the local climate). This allows the photography to take place, at relatively economical cost, when the cropmarks or other traces of the past are at their most visible. Also important is the aerial archaeologist's capacity to vary the point of view in flight, and thereby to capture the photographs in such a way as to obtain the best return from the local conditions of lighting and crop developments etc. In addition, oblique images, in presenting a perspectival view of the landscape (albeit sometimes relatively distorted), present a picture that is closer to everyday reality and therefore for most observers easier to appreciate in comparison with vertical images. This renders oblique air photography particularly useful both for the initial documentation and for its subsequent communication to others. Moreover, the technique can provide effective documentation not only of cropmarks and similar anomalies but also of classical and medieval structures and settlements in their rural or urban landscape settings. Real advantages accrue when the photographer is also the archaeologist who is himself studying the landscape, whether in the initial recognition of the evidence or in the stimulus that the aerial viewpoint gives to new ideas about its character or development. Among the limitations of active aerial survey there should be mentioned the inherent subjectivity and selectivity of the method, depending as it does on the personal abilities of the archaeologist-photographer. Oblique air survey also suffers in comparison with vertical photography in lacking the latter's inherent steroscopic properties and total coverage of the survey area. As a result there is difficulty in deriving precise and large-scale numerical comparisons for statistical or other analysis. These limitations, however, can be largely offset by combining oblique aerial survey with other techniques of remote sensing and/or with direct observation on the ground.

Within the strategic study of the landscape the capacity of exploratory aerial survey to reveal previously unrecorded sites or features, or to increase knowledge about known sites, varies according to the conditions of archaeological visibility from the air. The Province of Siena, for instance, without doubt represents the least favourable zone in any part of Tuscany for the occurrence traces of the past in the form of cropmarks, soilmarks or micro-refief—the University of Siena has documented around 450 archaeological sites during more or less systematic coverage of the province's medieval castles but in the process has recorded only two cropmarks. The role of exploratory aerial survey in this province is therefore likely to remain fairly marginal in terms of new discoveries, even allowing for the fact that the work was started less than ten years ago (the need for perseverance in aerial survey is discussed in Palmer 2007). In Tuscany the opposite end of the scale of aerial visibility applies in some parts of the Province of Grosseto. In Maremma, for example, aerial survey has a more even balance between the documentation of already known stuctures and the discovery of previously unrecorded features. Since 2005, therefore, the University of Siena has concentrated its attention on this area, with encouraging results, in particular for the Etrusco-Roman and medieval periods. Etrusco-Roman features are especially evident in the Ardegna Valley, in Val d'Orcia and in the valley of the Ombrone, particularly in relation to communication systems and domestic or semi-industrial structures such as villas and farms. More specifically, near Roselle there lies an area which so far represents the most favourable stretch of Tuscany for exploratory aerial survey. Immediately to the west of the town there are clearly visible traces of buildings of various dimensions, along with communication systems and agricultural land-divisions. It is becoming clear that this area offers conditions of visibility, conservation and archaeological richness sufficient for aerial survey in the next few years to play a leading role in reconstructing the area's intricate multi-period landscapes (Fig. 4).

3.3 *Satellite imagery*

Amogst the major limitations of exploratory aerial survey we have noted subjectivity on the part of the surveyor. Recognition (and therefore recording) of the evidence during flight is entirely dependent on the ability and experience of the archaeologist, since he documents only those things which he thinks archaeologically or historically significant, omitting all the rest. To overcome this problem it is necessary to supplement oblique aerial survey with some form of 'total' recording at those times when archaeological visibility is at its best. A little earlier we implied that it is possible to programme vertical air photography to achieve this objective, though it is complex, costly and sometimes difficult to this quickly enough to match the speed of change in the visibility of the aerial evidence.

Figure 4.   Landscapes in the valley west of Roselle, with close-ups of some of the more significant feature. From left, features related with road system, buildings, a small farm, field systems.

The same considerations apply to the latest generation of satellite imagery. In appropriate circumstances images captured by high-resolution satellites are beginning to rival the results of medium-scale vertical photography. The level of detail in Ikonos-2, Quickbird-2 and Orbview-3 satellite images makes it possible to distinguish features with a minimum width of between 50 cm and 1 m or of polygonal features with a surface area of around 500–1000 m². In addition to being the most important instrument for remote sensing in parts of the world where it is difficult to access traditional images such as vertical or oblique air photography, satellite imagery today constitutes a potentially valuable source for archaeological exploration in the western world (Wiseman, El-Baz 2007). Experience in Italy suggests that reasons for the increasing interest in high-resolution satellite imagery for exploratory survey and archaeological mapping lie in the GIS-ready and multi-spectral characteristics of the resulting data, the presence of the infra-red channel, the capacity for stereo-viewing and the possibility of planning (within certain limits) the moment of acquisition. Progress in appreciation of the multispectral and diagnostic features of the near-infrared channel will probably depend on the success of particular research initiatives but the possibility of timing data acquisition in response to specific archaeological needs will derive entirely from the way in which the aerospace industry works. In 2007, after a hiatus of around five years, we seem to be on the cusp of a revival in this sector, probably stimulated at least partly by new means of public access such as geographical browsers such as Google Earth. In the very near future, too, the launch of a new satellite (Geoeye-1) will bring the frontier of resolution down to 0.41 m in the panchromatic spectrum and 1.64 m in the multispectral spectrum (http://www.geoeye.com/).

3.4   *Archaeological field survey: surface collection*

In the Mediterranean area surface collection survey is considered one of the most fruitful methods for the discovery and characterisation of archaeological sites and deposits (Francovich, Patterson 2000). Archaeologists have always used the collection of surface material as a means of identifying the chronological and topographical characteristics of a site prior to excavation. For many decades now, however, surface collection has gone beyond this simple pre-excavation function, serving also for the survey of wider areas defined by geographical or cultural boundaries or related to random or mathematically-based sampling strategies. Following definition of the search area and the choice between total or sample coverage, the fieldwork is placed in train (Orton 2000). Systematic survey requires the field-workers to walk at predetermined distances from one another across the ploughed area. The archaeologist's task is to examine and document the area for the presence of smaller or larger concentrations of archaeological material brought to the surface by the plough (Thomas 1975; Foley 1981). Surface collection has assumed a significant role following the demonstration in regional studies of its capacity to identify, primarily, settlement areas (Aston 1985; Brown 1987). Surface collection, however, suffers from significant limitations, including its inefficacy in identifying ancient field patterns, communication systems

13

and other aspects of the landscape's infrastructure. Moreover, there are often difficulties in presenting or discussing the method's results in quantitative terms. Controversy also surrounds the representativeness of the results (Banning 2002). The relationship between the evidence present on the surface and that buried beneath the soil is undoubtedly complex, with a host of variations from site to site. The results are perhaps more trustworthy in the case of long-term projects which are structured to provide repeated survey of the same areas. In addition to variations in visibility from one year to another repetition allows the survey to be repeated with a different group of field-workers, thereby reducing or offsetting the possible of bias or differing ability in one group compared with another. In essence, however, the method still has its limitations, and it is always advisable to combine it with aerial survey and/or geophysical prospection of the whole or sample parts of the study area as a basis for test excavation to establish chronological or functional relationships more precisely.

4   POINT-ENVIRONMENTAL SCALE: INDIVIDUAL EVIDENCE

In the past the individual study of a single site has more often than not been focused on diagnostic work preparatory to excavation. Today, intra-site analysis is increasingly aimed at the recovery of information as a substitute for excavation when the latter is precluded for bureaucratic or (more often) financial problems. Modern excavation is time-consuming and expensive, and detailed intra-site analysis by other means can be faster and more economical. The results, of course, are very different. The kind of detail recovered through excavation cannot be reproduced by the alternative methods but in spite of this the information recoverable through autoptical analysis and examination of various chemical and/or physical characteristics of the soil make it possible to locate aspects of the evidence more precisely and on some occasions to formulate quite complex interpretations. In general archaeologists choose this kind of approach as a source for understanding broad-scale transformations across space, whether at the micro, local or macro scale.

4.1   *Surface collection survey*

In the previous section it has been suggested that the quickest, most economical and effective means of gaining a preliminary understanding of a buried archaeological site is by direct survey on the ground. After the identification of a finds scatter the quantity and quality of the information that the archaeologist subsequently acquires depends on the objectives of the research and the method of collection and documentation adopted in response to this. A detailed analysis will first require the accurate topographical survey of the area, nowadays easily achieved directly through DGPS survey (Campana, Francovich 2006) or indirectly through air photography and photogrammetry (Ceraudo, Piccarreta 2000). The topographical information is essential in allowing the interpreted archaeological evidence to be placed within its local and broader landscape setting. The next necessity is to decide how to carry out the survey and what material to collect. This is not in any sense a casual choice but one which will have a significant influence on the kind of analysis that can be undertaken subsequently. It is widely acknowledged by archaeologists that the distribution of surface material across a site does not always reflect the underlying stratification. Nevertheless, study of the surface distribution is a widely used research technique. It effective conduct, however, requires the positional recording of every single find (Ebert 1992) or group of finds through collection within some kind of grid (Campana 2005). Knowledge of the location of the material, albeit of objects in a state of continuous movement through ploughing and other agricultural activities, allows the later stages of analysis to generate diachronic, synchronic and thematic distribution maps (of building materials, amphorae, table-ware, industrial by-products etc). This topic will be addressed again in the second part of the volume, dealing in more detail with work in the field, the treatment and analysis of the collected material and the results achieved through grid-collection in a study area at Aiali in southern Tuscany.

4.2 Geophysics

The contribution of geophysics to the archaeological study of sub-surface strata has increased notably over the past ten years, both qualitatively and quantitatively. Today, geophysical techniques represent an indispensable and complementary instrument alongside surface collection for the study of the archaeological relationships within (and in some cases outside) a site. To demonstrate, through a practical example, the special contribution of geophysical prospection in intra-site survey (other than referring readers to the Aiali discussion later in the volume) we can focus here on another area study, undertaken at Romitorio in central Tuscany by the University of Siena's Laboratory for Landscape Archaeology and Remote Sensing (LAP&T).

The site at Romitorio, in the District of Siena, lies within one of the sample areas in the project to create an Archaeological Map of the Province of Siena. In this area Late Medieval documents record 'uico nomini oracolo Santi Ampsani' (714–715); the church referred to in the documents has generally been ascribed to the place-name Sant'Ansano, today attached to a farm close to Romitorio (Schiaparelli 1929). Surface collection undertaken from 2001 onwards brought to light a wide range of archaeological material, interpreted as belonging to a village of the Late Republic—Early Imperial period (1st century BC—1st century AD). This had taken over an area used during at least the Archaic, Etruscan and Hellenistic periods (6th—2nd centuries BC), for domestic settlement in the latest period and perhaps as a sacred site in the earlier phases. This latter conjecture is based on the presence of a few fragments of painted black and red tiles, one of them bearing a horizontal red band parallel to the edge of the tile. Beneath this there can be made out two areas of probable geometric decoration, with alternating red and black chequers. A second fragment presents a less regular decoration, consisting solely of curvilinear red stripes. This type of decorated tile finds comparison with decoration on the roofs of temples in the Etruscan period. A fairly close parallel for the fabric as well as the red and black colouring and the repeating geometric motif can be found in the temple of Vigna Grande, in the Province of Orvieto, ascribed to the 5th century BC (Torelli 2000, p. 331, p. 620, no. 267). The same area at Romitorio has also produced material from later phases of the settlement, during the Imperial period in the 4th and 5th centuries AD. There is no archaeological evidence for later phases of occupation. The site was also surveyed from the air at intervals between 2001 and 2007. Both recent and historical vertical photographs were also examined but the analyses produced no firm evidence apart from traces of earlier agricultural field divisions. In all probability the lack of evidence from the air can be related to the clay subsoil, which is generally unfavorable to the development of cropmarks or soilmarks (Musson et al. 2005; Mills, Palmer 2007). Even a winter flight in 2004 after a light snowfall (which usually produces ideal conditions for the detection of eroded earthworks) failed to produce any positive observations. However, geophysical prospection over an area of about 10 hectares yielded more encouraging results—far better, indeed, than had been expected (Fig. 5). In the field immediately east of the farmhouse regularly-arranged anomalies can be seen, some of them suggesting the outlines of buildings, with varying alignments which show them to be of more than one chronological phase (Fig. 5, nos. 1–4). There were no clear anomalies that could be directly attributed to the religious building attested in the documents. One possible hint, however, lay in the presence towards the north of a fairly well-defined rectangular anomaly, lacking any curvilinear element that might have belonged to an apse but with dimension in width of about 10 m (by about 20 in extent; Fig. 5, no. 1) and an east-northeast/west-southwest orientation similar to that of the church at Pava, excavated only a few kilometres away to the north in recent years (Campana et al. 2005).

In the fields to the south and east, however, there were further magnetic anomalies that posed new and unexpected problems of interpretation. In particular, two anomalies characterised by linear dipoles took the form of regular circles, each measuring 50 m in diameter (Fig. 5, nos. 5–6). Their morphology and topographical position (at the top of a hill dominating the surrounding countryside) can be paralleled in the Siena area, and more generally in Etruria, in funerary monuments. Tentatively, the evidence could be interpreted as belonging to two Etruscan tombs, completely flattened by long-term ploughing and now only showing as variations in the local magnetic

Figure 5. Magnetic map showing well-defined anomalies belonging to a number of different archaeological features.

field. The picture becomes clearer, and the conjecture more secure, when account is taken of the painted tiles found during surface collection in the area. A key point, however, is that the geophysical prospection revealed archaeological features that had previously remained totally invisible to traditional archaeological research. Moreover, the integration—in this case—of magnetic and electrical survey (using the ARP©, Automatic Resistivity Profiler) produced even further evidence, as illustrated in figure 6. Other less distinct and incomplete circular anomalies can be seen in the graphical representation of the site (Fig. 5, nos. 7–8), though the dimensions are variable and the interpretation uncertain. Further elements emerge from analysis of the magnetic data, including linear dipoles which extend for a total of 320 m (Fig. 5, nos. 9–13). The dimensions and overall pattern of these anomalies suggest their possible attribution to agricultural boundaries, a curvilinear enclosure or to one or more trackways. In summary, the gradiometer survey played a decisive role in the investigation, adding information about many phases of the settlement evidence originally revealed by field survey and surface collection. The use of geo-electrical survey (ARP©) also served to increase the range of evidence and to facilitate a more detailed interpretation. Overall, the new evidence supplemented and reinforced the suggested use of the site as a focus of ritual activity. The development of these ideas owes much to the integration of the different survey methods—without the evidence from documentary sources and field-walking survey the interpretation of the magnetometer/ARP© evidence might have seemed weak or illusory, and vice versa.

4.3 Test-excavation

Sondages or test excavations have generally been brought into play in contexts lacking good conditions of visibility. Various strategies have been adopted in the field, including random, mathematical or targeted sampling (McManamon 1984). The technique has also been used to enhance understanding of particularly significant contexts already surveyed through surface collection or

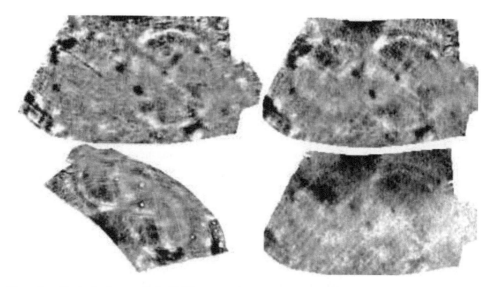

Figure 6. Clockwise from top left: ARP© maps at 50 cm, 1 m and 1.7 m depth, clearly showing two adjacent circular features, one more clearly marked than the other; 3D visualization and draping of the magnetic map on the DGPS digital terrain model, with overlaid archaeological interpretation.

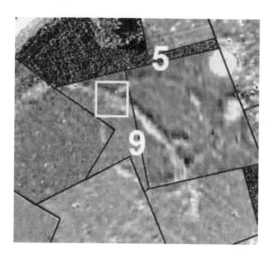

Figure 7. Detail from the magnetic map at Romitorio, super-imposed with the suggested areas of test excavation.

geophysical prospection. Returning for a moment to the Romitorio case study it is clear that the combined use of several different survey methods has not resolved the full complexity of the site, indeed the number of outstanding questions has been increased. For instance, the large linear anomaly, perhaps representing an agricultural boundary, a curvilinear enclosure or one or more trackways, appears to cut the southernmost of the two large circles (Fig. 7). If the interpretation of the circles as Etruscan tombs is accepted, the linear anomaly should presumably be attributed to a later phase—in the Late Etruscan, Roman, Late Antiquity or Medieval periods. Clearly there is only one way to resolve these uncertainties—the application, in the zones most in doubt, of archaeological excavation. A trial section would probably suffice, one at the intersection between the southern circle and the 'field boundary' (Fig. 7, nos. 5 and 9).

5 LOCAL-ENVIRONMENTAL SCALE: LANDSCAPES

In the introduction to this chapter we suggested that the 'local' level should be seem as the critical scale in 'landscape' research. If we try to translate data from the macro and micro scales and reproduce it at the 'local' scale the landscape appears to consist of widely separated sites (points) interspersed with large areas of 'empty' space (Fig. 1, left). The problem has a direct impact upon the type and objectives of historical reconstruction that it is intended to pursue. To explain more fully, mankind in the past, much like people of the present day, did not perceive the space around his settlements as being composed of 'emptiness'. In between any pair or group of settlements one finds the fields with their various crops, woodland, pasture, hunting areas, water sources, lagoons and ponds, quarries, mines, civil or religious administrative boundaries, streets and simple pathways etc. If the intention is to bring these elements into the historical interpretation it is essential to adopt a research strategy aimed at the recovery of the necessary information. The first problem revolves around the difficulty of identifying the indicators of such features through surface collection. This returns the discussion, to some extent at least, to the question of archaeological visibility. It is no coincidence that the few archaeologists who have attempted detailed investigation of areas outside the confines of traditional settlements or other sites with clearly defined functions or boundaries have included pioneers of aerial archaeology such as O.G.S. Crawford and John Bradford (Gaffney, Gaffney 2006; Bowden 1999). Their key research technique (aerial survey) and the favourable contexts in which they were working (the intricate and closely articulated landscape of Wessex in southern England and the cropmark-rich plain of the Tavoliere delle Puglie in southern Italy) played a determining role in the development of this approach. The majority of the features in the spaces between the settlement do not manifest themselves in the form of surface finds, and even when they do so the material is so difficult to interpret that it tends to be described as 'off-site' (Banning 2002). So, while at the micro and semi-micro scale the problem relates principally to the completeness of the sample and the complexity of articulation within a site, at the 'local' scale the debate is more concerned with the representativeness of the sample rather than its completeness. Amongst those who have tackled this problem in recent years the most significant results have been achieved by researchers who have made intensive use of remote sensing methods, in combination with geomorphological and palaeo-environmental analysis (cf: Powlesland in this volume).

5.1 *Airborne scanning: multispectral and LiDAR data*

In the earlier section on exploration methods appropriate to the macro scale mention was made of the limitations which affect various forms of remote sensing. That discussion, however, did not address the various techniques of airborne laser scanning, though these are in every way compatible with that level of scale. The reason for omitting them at that stage was the rarity with which archaeologists have made use of laser-scanned data on a large enough scale for this kind of work. In general when archaeologists manage to obtain laser-scanned data this is related to relatively limited areas. Therefore, the sample is generally more appropriate for the 'local' scale of analysis. The most widely used airborne scanning systems in present-day archaeology are hyperspectral imaging and LiDAR. The former allows the acquisition of data across a sunstantial part of the enormous electomagnetic spectrum, from blue to thermal infrared, by registering the informationn in a wide range of individual bands (Shell 2001; Donoghue 2001; Cavalli, Pignatti 2001). It has been shown that these systems, if properly used, offer significant advantages compared with traditional aerial photography. Briefly, the advantages are as follows:

- Hyperspectral sensors, in combination with image processing (IP), are less dependent on the brief periods of time when the aerial evidence is clearly visible (to the naked eye and to traditional photography).
- Cropmarks become more easily detectable in the near-infrared spectral band (760–900 nm).
- Soilmarks are more readily detectable in the red part of the spectrum (630–690 nm).

- In the thermal infrared band (8000–12000 nm) it is possible to record information that is not recoverable using other passive optical instruments (traditional cameras and films or digital camera).

In effect these systems, by registering chemical and physical properties that are different from those recorded through traditional air photography, can make a special and significant contribution to the study of archaeological landscapes, sometimes emphasising or revealing elements that appear only fleetingly (if at all) in traditional photography (Powlesland 2006). Their major limitations perhaps lie in their relatively poor geometric resolution (generally not less than 3 m/pixel) and in the relatively high cost of this kind of data.

Earlier in the chapter it was noted that the project to create an Archaeological Map of Tuscany suffers from a serious but relatively common weakness which compromises its capacity to document archaeological evidence evenly across the landscape. The problem lies in the near-impossibility of investigating areas of woodland, which cover about half the total area of Tuscany. A new development of great potential in confronting this problem is provided by recent work in Great Britain, Germany, France and Austria, using airborne laser scanning or LiDAR (Devereux et al. 2005; Shell 2005; Crutchley 2006; Doneus, Briese 2006; Sittler, Schellberg 2006). The LiDAR system consists of an airborne laser scanner capable of recording the morphology of the underlying ground surface with great precision (about the principle of the system and the nature of first and last pulse see Doneus, Briese 2006). After computer-processing of the data, using appropriate algorithms, it is possible to 'remove' the woodland vegetation and reveal in great detail the surface relief not of the tree canopy but of the underlying ground surface, along with any archaeological topography that might be present. A degree of caution is needed, of course, but the technique could prove absolutely revolutionary in its impact on the process of archaeological mapping by making it possible to record, without physical intervention, the previously hidden archaeological resource within woodland areas. In favourable circumstances it may even be possible to uncover whole 'fossil' landscapes (Bewley 2005). This could have a dramatic impact on opportunities for archaeological and landscape conservation, as well as on scientific investigation of settlement dynamics in various phases of our history.

It is worth emphasizing that interest in this technique is not limited to its potential for penetrating woodland areas but also for its contribution to the study of open contexts such as pastureland and arable areas. In these zones, as under woodland cover, the availability of extremely precise digital models of the ground surface will make it possible to highlight every tiny variation in level, by using computer simulations to change the direction or angle of the light and/or to exaggerate the value of the z coordinate (Doneus, Briese 2006). Moreover the method can fill a particularly obvious gap for work at the local scale by making available a numerical representation of the landscape's morphology, which through GIS can be readily integrated with data acquired though the other remote-sensing methods appropriate to this scale. We should seize on this as a very real opportunity to free the physical structure of the context from its former position as 'background information' and allow it to play a full part in the process of narrative interpretation (Gaffney, Gaffney 2006).

In 2005, through a Culture 2000 project of the European Union, entitled European Landscapes: past, present and future, the University of Siena took its first steps in LiDAR data acquisition, processing and interpretation for four sample areas in the provinces of Siena and Grosseto. This was made possible through the good services of colleagues in England at the Natural Environment Research Council and the Unit for Landscape Modelling at the University of Cambridge. The results were processed in the Department of Geography at Durham University (UK) under the supervision of Prof. Daniel Donoghue and Dr. Nikolaos Galiatsatos. Success was achieved in one of the case studies, aimed at penetrating the tree canopy so as to record underlying archaeological features long protected from plough-erosion or other human activity by the woodland cover (Fig. 8; Campana et al. forthcoming). We see this as only the tip of the iceberg, however, and are absolutely convinced that advances in the use of this technique will in the coming years have a decisive impact on our understanding of ancient landscapes.

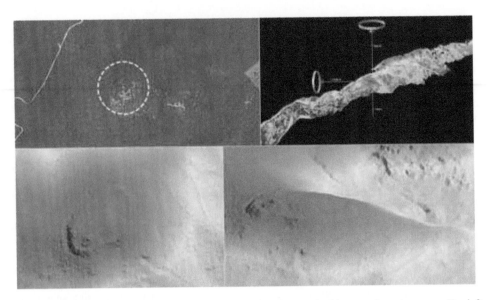

Figure 8.   Sample area in Maremma where the landscape is characterized by very dense tree cover. Top left: in the centre, under dense vegetation, are the well-known ruins of a medieval castle. Top right: point cloud collected by the UK Natural Environment Research Council and pre-processed by the Unit for Landscape Modelling at the University of Cambridge. Bottom left and right: data processing and filtering in the Department of Geography at the University of Durham, UK, has here allowed 'removal' of the dense vegetation to achieve a digital terrain model showing the previously hidden archaeological features.

5.2   *Large-scale geophysical prospection*

Recent years have seen an increasing availability of geophysical instruments technologically similar to those produced in the past but characterised by multi-sensor instrumentation (for an overview see Gaffney, Gaffney 2006; Kvamme 2003). This change has reflected the needs of a large number of professionals, including archaeologists, for the rapid acqisition of geophysical data over ever-larger areas of ground. In archaeology this has led, for the most part, to the application of these new instruments on progressively larger individual sites and on the more or less systematic integration of the resulting data with information derived from field survey and surface collection (Campana, Francovich 2005; Ciminale et al. 2007). Few archaeologists, however, have posed the following question: if geophysical methods enable us to gather such a significant mass of information for contexts initially identified through surface find scatters or aerial survey in times of good visibility, what would they be able to uncover in areas where other methods at present show only 'emptiness'? Even fewer archaeologists have been falling over themselves to seek an answer to this question! But see Powlesland elsewhere in this volume.

The University of Siena has been experimenting with several new systems in its chosen study areas and has recently launched a research initiative aimed at large-scale exploration without gaps. In particular, use has been made of the ground-penetrating radar system GSSI TerraVision, the Foerster gradiometer (in MULTICAT configuration and with a trolley pushed by an operator) and the ARP© system developed and managed by Terranova, a spin-off from the University of Paris (Fig. 9). The Terravison system consists of 14 radar antennae set 12 cm apart at varying inclinations, mounted on a trolley (Finzi et al. 2005). Limited experience on only a single context has revealed practical limitations in the instrument's need for extremely homogeneous soil conditions (rarely encountered in agricultural situations) along with its lack of a fast and reliable georeferencing system for the collected data and of software dedicated to the processing and future management of the recorded measurements. This is undoubtedly an instrument of considerable potential but one which still needs further development.

Figure 9.   From top to bottom: GSSI Terravision, Foerster Multicat, ARP© Terranova.

The FEREX® fluxgate gradiometer system has 4 (or up to 8) sensors with a resolution of 0.1 nT mounted in parallel on a robust fiberglass trolley or on a hand-cart (Campana 2006). Depending on the configuration used the instrument is either pulled by a quad bike or pushed by an operator. These instruments, in addition to reducing the acquisition time through the use of a large number of sensors, are able to work without physical reference systems placed on the ground to control the positioning of the measurements. In practice, the need in some geophysical applications to lay out physical reference grids during the data-acquisition stage constitutes one of the most time-consuming and wasteful parts of the process. The new generation of instruments for the most part (even if for instance this facility was not available in the case of the Terravision system) are provided with in-built navigation systems (such as the DATAMONITOR software of the Foerster Group) based on DGPS technology and real-time visualisation on a computer or data logger of each completed traverse. These latest instruments permit data acquisition in the order of 3–4 hectares per working day, or about 60 to 80 hectares of high resolution data for each month of work.

There are also innovative solutions in the field of geoelectrics. A case in point is the ARP© (Automatic Resistivity Profiler) system developed by the group co-ordinated by Michel Dabas at the University of Paris (Dabas in this volume) and experimented with by surveyors from the University of Siena in a variety of contexts, including the site at Aiali described later in the volume. Further discussion will be left to that description.

6   ARCHAEOLOGICAL MAPPING

A pre-requisite for the handling of this kind of territorial data is knowledge about each measurement's position in relation to a known system of geographical coordinates. Failure to satisfy this condition results in an inability to localise the acquired information. The entry of the data into an archaeological GIS is the basis for any attempt at integration of the information so as to facilitate a critical narration of the local history or conservation of the archaeological resource (Plate 1).

At scales such as 1:50,000 the mapped archaeological information is depicted by means of symbols so as to overcome the limitations of graphical representation. Also included as part of the background information are the site contours. In the case of surface scatters one records the concentrations of material, preferably through a GPS unit working either in simple or in DGPS mode. Entry into the GIS of oblique aerial photographs, and hence of the information that they contain, is also essential. Active aerial reconnaissance is a relatively recent development in Italy (Musson et al. 2005) but research based on the photogrammetric analysis of vertical imagery has reached high levels of sophistication in recent decades, as illustrated in numerous published examples of photo-interpretation and cartographic representation (see, for instance, Piccarreta, Ceraudo 2000; Guaitoli 2003). Building on the experience and basic principles which allow the restitution of archaeological and topographical features through optical photogrammetry, there have been developed by American and British scholars software packages which permit the geometric correction and georeferencing of oblique aerial photographs (see Scollar 2001; Haigh J.G.B. 2000). Geophysical measurements are georeferenced in relation to the topographical relief through DGPS or total station survey of the individual measurement points or of the corners of the grids within which the the measurements were acquired; as noted above, in the latest generation of instruments geolocation is achieved through advanced systems of interfacing between the instrument itself and a topographical recording unit.

Georeferencing of the remotely-sensed data does not represent the end of the archaeological mapping process but only an intermediate stage. On their own, aerial photographs or magnetic and geoelectrical maps signify little. It is the responsibility of the archaeologist (often in collaboration with specialists such as geophysicists) to give sense to the photographs or to the measurements of chemical and physical parameters in the soil. In summary, the interpretation of the data is made real and communicable through cartographic restitution of the elements perceived as anomalies. This is therefore the critical phase in landscape and archaeological research. In practice the

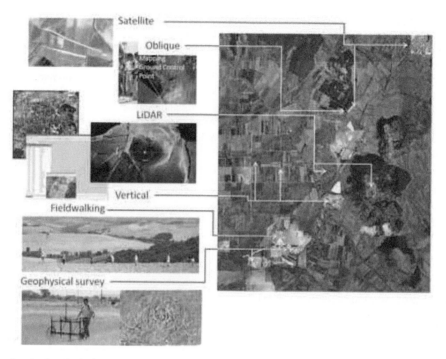

Plate 1. Archaeological mapping. Integration of different detection methods to draw the jigsaw puzzle in which we can measure and position each piece of information while at the same time perceiving the overall picture, whether synchronically or diachronically. (See colour plate section)

process advances through the drawing, in a digital way as well as in a hand way, of the anomalies and other elements deemed to be of archaeological interest. Georeferenced graphical restitution of the information contained in vertical or oblique aerial photographs, in high-resolution satellite imagery, in LiDAR data and in maps derived from geophysical measurements allow us to overlay on topographical maps the results of the various investigative methods, along with a mass of other data stratified layer upon layer over the years. The result is a jigsaw puzzle, a complex representation in which we can measure and position each piece of information while at the same time perceiving the overall picture, whether single-phase or spread across time, along with the overlapping and stratified fragments of whole systems of ancient and medieval landscapes. Through archaeological mapping and the use of GIS these become capable of study against other layers of archaeological and non-archaeological information in the writing of history or in heritage protection through the planning process, through conservation or through monitoring of the shared cultural inheritance.

ACKNOWLEDGMENTS

The author owes an enormous debt of gratitude to his mentor, the late Professor Riccardo Francovich (University of Siena), for his passion, help and criticism throughout all stages of the author's research experience. Special thanks are also due to Chris Musson (for help with the English version of the text) and Dominic Powlesland (Landscape Research Centre, UK) for their valuable comments on the practice of archaeological research. Bob Bewley (formerly of English Heritage), Daniel Donoghue (University of Durham), Dean Goodman (Geophysical Archaeometry

Laboratory, Los Angeles, USA), Darja Grosman (University of Ljubljana) and Salvatore Piro (ITABC - CNR) have helped in a variety of ways. Helmut Becker (formerly of the Bavarian State Department of Historical Monuments, Germany) and Iacopo Nicolosi (National Institute of Geophysics and Vulcanology, Italy) contributed greatly in the field but also helped me to understand the best configuration for our magnetometer work. Salvatore Piro and Dean Goodman did great work during acquisition and processing of GPR data. Many researchers and students have collaborated, and are continuing to collaborate, in the Siena and Grosseto archaeological mapping projects. Special thanks are also due to the research team of the Laboratory of Landscape Archaeology and Remote Sensing at the University of Siena: Anna Caprasecca, Cristina Felici, Barbara Frezza, Mariaelena Ghisleni, Francesco Pericci and Emanuele Vaccaro.

REFERENCES

Allen M.J. 1991, Analyzing the Landscape: a Geographical Approach to Archaeological Problems, in Schofield J. 1991, pp. 39–58.

Aston M., Rowley T. 1974, Landscape Archaeology: An Introduction to Fieldwork Techniques on post-roman Landscapes, Newtown Abbot.

Aston M. 1985, Interpreting the Landscape. Landscape Archaeology in Local Studies, London.

Banning E.B. 2002, Archaeological Survey, New York.

Bewley R. 2005, Aerial Archaeology. The first century, in Bourgeois J., Meganck M. (eds.), Aerial Photography and Archaeology 2003. A century of information, Ghent.

Bowden M. 1999, Unravelling the landscape. An inquisitive Approach to Archaeology, Stroud.

Bradford J. 1957, Ancient Landscapes, London.

Brophy K., Cowley D. 2005, From the Air. Understanding Aerial Archaeology, Tempus, Stroud.

Brown A. 1987, Fieldwork for Archaeologists and Local Historians, London.

Butzer 1982, Archaeology as Human Ecology, New York.

Cambi F. 2000, Quando i campi hanno pochi significati da estrarre: visibilità archeologica, storia istituzionale, multi-stage work, in Francovich R., Patterson H. (eds.), Extracting Meaning from Ploughsoil Assemblages, Oxford.

Cambi F. 2003, Archeologia dei paesaggi antichi: fonti e diagnostica, Roma.

Cambi F., Terrenato N. 1994, Introduzione all'archeologia dei paesaggi, Pisa.

Campana S. 2005, Tecnologie GPS e personal data assistant applicati all'archeologia dei paesaggi, Archeologia e Calcolatori, 16.

Campana S., Francovich R. 2005, Seeing the Unseen. Buried Archaeological Landscapes in Tuscany, Taylor & Francis, The Netherlands, pp. 67–76.

Campana S., Francovich R. 2006, Paesaggi Archeologici e Tecnologie Digitali 1: Laser scanner e GPS, Florence.

Campana S., Francovich R. 2007, Understanding archaeological landscapes: steps towards an improved integration of survey methods in the reconstruction of subsurface sites in South Tuscany, in Wiseman J., El-Baz F. (eds.), Remote Sensing in Archaeology, Springer, Boston Ma., pp. 239–261.

Campana, S., Donoghue, D., Galiatsatos, N. (forthcoming). Undercanopy archaeology using airborne laser scanner to overcome the Mediterranean vegetation, 37th International Symposium on Archaeometry.

Cavalli R.M., Pignatti S. 2001, Il telerilevamento iperspettrale da aereo per lo studio dei beni archeologici: applicazione dei dati iperspettrali MIVIS, in Campana S., Forte M., Remote Sensing in Archaeology, Florence, pp. 221–232.

Ciminale M., Becker H., GALLO D., Integrated technologies for archaeological investigation; the Celone Valley project, Archaeological Prospection 14, 3, pp. 167–181.

Clarke D.L. 1977, Spatial Information in Archaeology, in Clarke D.L. (ed.), Spatial Archeology, New York.

Coleman S. 2007, Taking advantage: vertical aerial photographs commissioned for local authorities, in Mills J., Palmer R., Stroud.

Crutchley S. 2006, Seeing through the trees: LIDAR and aerial archaeology in England, in Campana S., Francovich R. (eds.), GPS e laserscanner per l'archeologia dei paesaggi, Firenze.

Cunnington M. E. 1927, Prehistoric timber circle, Antiquity 1, plates I–II, pp. 92–95.

Devereux B.J., Amable G.S., Crow P., Cliff A.D. 2005, The potential of airborne lidar for detection of archaeological features under woodland canopies, Antiquity 79–305, pp. 648–660.

Doneus M. 2001, The impact of vertical photographs on analysis of archaeological landscapes, in Doneus M., Eder-Hinterleitner A., Neubauer W. (eds.), Archaeological Prospection, 4th International Conference on Archaeological Prospection, pp. 94–96.

Doneus M., Briese C. 2006, Full-waveform airborne laser scanning as a tool for archaeological reconnaissance, in Campana S., Forte M. (eds.), From Space to Place, Proceeding of the 2nd International Conference on Remote Sensing in Archaeology, pp. 99–105.

Donoghue 2001, Multispectral Remote Sensing for Archaeology, in Campana S., Forte M., Remote Sensing in Archaeology, Florence, pp. 181–192.

Ebert J. 1992, Distributional Archaeology, New York.

Felici C. 2004, Carta archeologica della provincia di Siena. Pienza, Volume VI, Siena.

Finzi, E., Francese, R.G., Morelli, G. 2005, High-resolution geophysical investigation of the archaeological site "Le Pozze" in the surroundings of the town of Lonato (Brescia, Northern Italy), in S. Piro (ed.), Proceedings of the 6th International Conference on Archaeological Prospection, Rome, pp. 215–219.

Fleming A. 2006, Post-processual Landscape Archaeology: a Critique, Cambridge Archeological Journal, pp. 267–280.

Foley R.A. 1981, Off-site archaeology: an alternative approach for the short-sited, in Hodder I., Isaac G., Hammond N., (eds.), Patterns of the Past: Studies in Honour of David Clarke, Cambridge, pp. 157–183.

Francovich R., Villaggi dell'altomedioevo: invisibilità sociale e labilità archeologica, in Valenti M., Insediamento altomedievale nelle campagne toscane. Paesaggi, popolamento e villaggi tra VI e X secolo, Florence.

Francovich R., Patterson H. (eds.) 2000, Extracting Meaning from Ploughsoil Assemblages, Oxford.

Francovich R., Hodges R. 2003, Villa to village: the Transformation of the Roman Countryside in Italy c. 400–1000, London.

Gaffney C., Gater J. 2003, Revealing the buried past. Geophysics for Archaeologists, Stroud.

Gaffney C., Gaffney V. 2006, No further territorial demands: on the importance of scale and visualisation within archaeological remote sensing, in From Artefacts to Anomalies: Papers inspired by the contribution of Arnold Aspinall, University of Bradford 1–2 December 2006, http://www.brad.ac.uk/archsci/conferences/aspinall/presentations/

Gamble 2007, Archaeology: The Basics, Second edition, London.

Going C. J. 2002, A Neglected Asset. German Aerial Photography of the Second World War Period, in Bewley R., Raczkowski W., Aerial Archaeology. Developing Future Practice, Amsterdam.

Guaitoli M. 1997, Attività dell'Unità operativa Topografia antica, in Metodologie di catalogazione dei beni archeologici, Beni archeologici-Conoscenza e tecnologie, Quaderno 1,2, Bari, pp. 9–50.

Guaitoli M. (ed.) 2003, Lo sguardo di Icaro. Le collezioni dell'Aerofototeca Nazionale per la conoscenza del territorio, Rome.

Haigh J.G.B. 2000, Developing rectification programs for small computers, Archaeological Prospection 7, 1–16.

Hamerow H., Anglo-Saxon settlements in a post-roman landscape, in Dopo la fine delle ville: le champagne dal VI al IX secolo, Proceedings of the workshop (Gavi, 8–11 May 2004), Mantova.

Haselgrove C., Millet M., Smith I. 1985, Archaeology from the Ploughsoil. Studies in the Collection and Interpretation of Field Survey data, Sheffield.

Kvamme K. 2003, Geophysical Survey as Landscape Archaeology, American Antiquity, vol. 68, pp. 435–457.

Liebeschuetz W., L'aristocrazia in occidente tra il 400 e il 700, in I Longobardi. Dalla caduta dell'impero all'alba dell'Italia (catalogue of the exibition), Milan.

Lock G., Molyneaux B.L. (eds.), Confronting Scale in Archaeology: Issues of Theory and Practice, Springer, New York.

Mangiavacchi M. 2004, Archivio del Consorzio di Bonifica della Val d'Orcia. Immagini fotografiche per la lettura del territorio, Firenze.

Marquard W.H., Crumley C.L. 1987, Regional Dynamics: Burgundian Landscapes in Historical Perspective, San Diego.

Mcmanamon F.P. 1984, Discovering Sites Unseen, in Schiffer (ed.), Advances in Archaeological Methods and Theory, pp. 223–292.

Mills J., Palmer R. 2007, Populating Clay Landscapes, Stroud.

Musson C, Palmer R., Campana S. 2005, In volo nel passato. Aerofotografia e cartografia archeologica, Florence.

Orton C. 2000, Sampling in Archaeology, Cambridge.

Palmer R. 2007, Seventy-five years v. ninety minutes: implications of the 1996 Bedfordshire vertical aerial survey on our perceptions of clayland archaeology, in Mills, Palmer, pp. 88–103.

Piccarreta F., Ceraudo G. 2000, Manuale di aereofotografia archeologica. Metodologia, tecniche, applicazioni, Bari.

Powlesland D., Redefining past landscapes: 30 years of remote sensing in the Vale of Pickering, in Campana S., Forte M. (eds.), From Space to Place, Proceeding of the 2nd International Conference on Remote Sensing in Archaeology, pp. 197–201.

Powlesland D. (in this volume), Why bother? Large scale geomagnetic survey and the quest for 'Real Archaeology'.

Powlesland D., Lyall J., Hopkinson G., Donoghue D., Beck M., Harte A., Stott D., Beneath the sand—remote sensing, archaeology, aggregates and sustainability: a case study from Heslerton, the Vale of Pickering, North Yorkshire, UK, Archaeological Prospection, Volume 13 Issue 4, pp. 291–299.

Raffestin C. 1987, Carta e potere o dalla duplicazione alla sostituzione, in Cartografia e Istituzioni in età moderna, atti del convegno, Genova, pp. 23–31.

Renfrew C., Bahn P. 2008, Archaeology: Theories, Methods and Practice, London.

Schiaparelli, L. 1929–1933, Codice Diplomatico Longobardo, Roma.

Schiffer 1987, Formation Processes of the Archaeological Record, Albuquerque.

Schofield J. 1991, Interpreting Artefact Scatters. Contribution to Ploughzone Archaeology, Oxford.

Scollar I. 2001, Making Things Look Vertical, in Bewley R, Rączkowski W. (eds.), Aerial Archaeology. Developing Future Practice, Amsterdam, pp. 166–172.

Shell 2001, Airborne High-Resolution Digital, Visible, Infra-Red and Thermal Sensing for Archaeology, in Bewley R, Rączkowski W. (eds.), Aerial Archaeology. Developing Future Practice, Amsterdam, pp. 173–180.

Shell C. 2005, High Resolution Digital Airborne Survey for Archaeological Research and Cultural Landscape Management, in Musson C., Palmer R., Campana S., In volo nel Passato. Aerofotografia e cartografia archeologica, Firenze, pp. 271–283.

Sittler B., Schellberg S. 2006, The potential of LIDAR in assessing elements of cultural heritage hidden under forest or overgrown by vegetation: Possibilities and limits in detecting microrelief structures for archaeological surveys, in Campana S., Forte M. (eds.), From Space to Place, Proceeding of the 2nd International Conference on Remote Sensing in Archaeology, pp. 117–122.

Sydoriak Allen K.M. 2000, Consideration of Scale in Modelling Settlement Pattern Using GIS: an Iroquois Example, in Wescott K.I., Brandon R.J. (eds.), Practical Application of GIS for Archaeologists. A predictive modelling kit, London, pp. 101–112.

Terrenato N. 2000, The visibility of sites and the interpretation of field survey results: towards an analysis of incomplete distributions, in Francovich R., Patterson H. (eds.), Extracting Meaning from Ploughsoil Assemblages, Oxford.

Thomas D.H. 1975, Non-site sampling in Archaeology: up the Creek without a Site?, in Mueller J.W. (ed.), Sampling in Archaeology, Tucson, pp. 61–81.

Torelli M. 2000 (ed.), Gli Etruschi, Catalogue of the exibition, Venice.

Wiseman J., EL-BAZ F. 2007, Remote Sensing in Archaeology, New York.

Seeing the Unseen – Campana & Piro (eds)
© *2009 Taylor & Francis Group, London, ISBN 978-0-415-44721-8*

Introduction to geophysics for archaeology

S. Piro

Institute of Technologies Applied to Cultural Heritage - National Research Council,
Monterotondo Sc.(Roma), Italy

ABSTRACT: It is now widely acknowledged that archaeological sites are almost everywhere increasingly threatened by many natural and anthropic agents. Dealing with essentially non-renewable cultural resources, archaeologists are as a consequence becoming more and more concerned about their husbanding. Many significant data can be obtained by applying remote sensing, visual and infrared aerial sensing, surface surveying and the more limited methods of pedestrian surveying, surface collecting and shovel testing. With specific symbols this kind of information can be reported on topographic maps. Their combination can provide some close correspondence between the remote sensing indications and the presence of sites frequented by men. The preliminary analysis of materials could also provide a dating, but it is generally not possible to establish extension and depth of any archaeological settlement. Only applying geophysical prospecting methods this further information can be properly achieved. Nowadays, non-destructive ground surface geophysical prospecting methods are increasingly used for the investigation of archaeological sites, which implies detailed physical and geometrical reconstructions of hidden ambients. The probability of a successful application rapidly increases if a consistent multi-methodological approach is adopted, according to a logic of objective complementarity of information and of global convergence toward a high quality multi-parametric imaging of the buried structures. Non-invasive geophysical prospecting methods are to-date the only means available for local reconnaissance and discrimination, prior to any excavation work. Geophysical methods can measure various physical properties of the subsurface soils and rocks. Such properties are not only shaped by geophysical processes but they can also reflect alterations caused by humans. Many methods were originally designed to measure geophysical features at the scale of several metres or kilometres, while archaeological features are of interest at the scale of centimetres or a few metres, at most. Thus, some methods can be readily adapted to archaeological sites, while others are of marginal or negligible value. Traditionally, the geophysical methods are classified into the two main groups of passive and active methods. Within the first group, the amplitude of nearly steady gravitational, magnetic and electrical perturbation fields, generated by buried features, are measured at the sensing device. In the second group, artificial electrical, electromagnetic and acoustic signals are emitted by the device, which then senses the return signals, more or less altered by the typical responses of the subsurface features.

1 INTRODUCTION

It is now widely acknowledged that archaeological sites are almost everywhere increasingly threatened by many natural and anthropic agents. Dealing with essentially non-renewable cultural resources, archaeologists are as a consequence becoming more and more concerned about their husbanding.

The wise management of the cultural heritages calls for the non-invasive assessment of sites not threatened, evaluation of sites in areas to be altered, and identification and recovery of information where sites are to be destroyed. In all cases, any rapid, non-destructive technique for site evaluation, providing archaeologically meaningful knowledges, may become of paramount importance.

Many significant data can be obtained by applying remote sensing, visual and infrared aerial sensing, surface surveying and the more limited methods of pedestrian surveying, surface collecting and shovel testing. With specific symbols this kind of information can be reported on topographic maps. Their combination can provide some close correspondence between the remote sensing indications and the presence of sites frequented by men. The preliminary analysis of materials could also provide a dating, but it is generally not possible to establish extension and depth of any archaeological settlement. Only applying geophysical prospecting methods this further information can be properly achieved.

Nowadays, non-destructive ground surface geophysical prospecting methods are increasingly used for the investigation of archaeological sites, which implies detailed physical and geometrical reconstructions of hidden ambients. The many field techniques deployed in archaeological research today include geophysical and related methods of non-invasively detecting usually shallow-buried structures and remains. Such methods, long used in civil engineering and geotechnical studies, form a well-known class of field techniques which are generally combined with other types of reconnaissance, such as aerial photography or satellite imagery. They are capable of providing high-quality information on the nature and depth of remains, they are flexible in their ability to be used in many different archaeological circumstances, and they may be fast as well as economical.

The probability of a successful application rapidly increases if a consistent multi-methodological approach is adopted, according to a logic of objective complementarity of information and of global convergence toward a high quality multi-parametric imaging of the buried structures. The fine representation of the static configurations of the explored areas and of the space-time evolutions of the interaction processes between targets and hosting materials, are fundamental elements of primary knowledge in the case of archaeological research. Non-invasive geophysical prospecting methods are to-date the only means available for local reconnaissance and discrimination, prior to any excavation work.

Geophysical methods can measure various physical properties of the subsurface soils and rocks. Such properties are not only shaped by geophysical processes but they can also reflect alterations caused by humans. Prospecting methods were originally designed to measure geophysical features at the scale of several metres or kilometres, while archaeological features are of interest at the scale of centimetres or a few metres, at most. Thus, some methods are readily adapted to archaeological sites, while others are of marginal or negligible value.

Geophysical techniques provide raw data which must then be evaluated. The first step is to present the data in a form which can be understand by archaeologists, either by constructing a model of the physical phenomena thought to be responsible and changing this until the measured data are accounted for within a minimum error, or by using information which provide some degree of separation of the components of the measurements which are due to archaeological sources from those of natural or modern origin.

Usually, the geophysical methods are classified into the two main groups of passive and active methods. Within the first group, the amplitude of nearly steady magnetic, gravitational and electrical perturbation fields, generated by buried features, are measured at the sensing device. In the second group, artificial seismic, electrical and electromagnetic (inductive and impulsive) signals are emitted by the device, which then senses the return signals, more or less altered by the typical responses of the subsurface features (Aitken, 1974; Weymouth, 1986a; Wynn, 1986; Scollar et al. 1990; Cammarano et al. 1997).

1.1 *Passive methods*

In this category, traditionally, the method of major consequence to archaeology is the magnetic prospecting (Aitken, 1974; Weymouth 1986b; Gibson, 1986). By this method, high sensitivity instruments are utilized for the measurement of the total magnetic field of the earth (Scollar et al. 1990; Becker, 1995; Becker et al. 2005; Neubauer et al. 1997a, 2005; Hesse et al. 1997; Tsokas et al. 1997a; Herbich, 2005; Fassbinder et al. 2005; Kuzma et al. 2005; Aminpour, 2005; Benech

28

2007). Ferrous metallic objects or inhomogeneities in the magnetic properties of soils will cause variations in the measured field.

It can be mentioned in passing that another passive method, namely the gravitational surveying, is often utilized by geophysicists in archaeology. The method depends on the mass difference between feature and its surrounding matrix. It is not sufficiently sensitive, however, for typical archaeological purposes (Linington, 1966; Di Filippo et al. 2005a,b). It is however very useful for large-scale site reconnaissance.

Finally, the Self-Potential (SP) method has proven to be the least expensive geophysical method available for archaeological purposes. The equipment consists of only a digital voltmeter, some wire, and several low noise unpolarizable electrodes (Black and Corwin, 1984; Corwin and Hoover, 1979). Wynn & Sherwood (1984) reported that the SP method often shows anomalies over archaeological targets in areas where one or more other geophysical methods fail to indicate anything unusual.

1.2 *Active methods*

This category includes several methods of great value in archaeology.

a. Seismic methods—These involve introducing a pulse of sound into the earth and measuring the time of return of the pulse reflected by discontinuities in mass density and elastic properties of the soil. Geophysicists have used the refraction seismic method in archaeology with relatively little success (Aitken, 1974; Carabelli, 1966; Carson, 1962). The refraction method works best in mapping undisturbed layers having velocities increasing with depth. The method, however, becomes much less useful and interpretation becomes very qualitative and difficult when there are velocity inversions, which are representative of human cultural disturbance, or highly three-dimensional target bodies, such as burial sites or stone walls and foundations. The reflection seismic method is also often used mainly to detect cavities in an otherwise homogeneous rock mass or in ancient stone structures (Stright, 1986; Tsokas et al. 1995a).

b. Electromagnetic methods—The next group, which we consider, can be lumped under the very general term of non-contacting electromagnetic methods (often called also EM or induction methods). In all of these methods a transmitter coil sends a primary signal into the ground causing a secondary signal to be emitted. The secondary signal is then picked up by a receiver coil. Objects and soils that conduct electrical currents or become magnetized alter the nature of the secondary signal (Scollar, 1962; Foster, 1968; Tite & Mullins, 1970; Tabbagh, 1974, 1986; Bevan, 1983; Frohlich and Lancaster, 1986; Dalan, 1991; Hesse 1991, 1992; Hesse and Doger 1993; Tsokas et al. 1994; Bozzo et al. 1991; 1994; Sarris et al. 1998).

c. Galvanic methods—They are the ones that have been and still are adopted fairly widely in archaeology. Galvanic or soil-conduction electrical methods have been used since 1950s (Aitken et al. 1958). The best known technique is the resistivity profiling for which typically a Wenner or pole-pole or dipole-dipole array is utilized. Resistivity methods are mainly helpful in detecting gross porosity changes caused by buried stone structures. Many authors, such as Aitken (1974), Hesse (1966), Bernabini et al. (1985, 1988), Weymouth (1986b), Hesse et al. (1986), Clark (1986,1990), Brizzolari et al. (1992c), Carabelli (1967), Noel and Xu (1991), Lapenna et al. (1992), Bozzo et al. (1995), Orlando et al. (1987), Patella (1978), Tsokas et al. (1994), Cammarano et al. 1998, Dabas et al. 2000, Piro et al. 2001b, Cardarelli et al. 2006b, 2007, have used these methods for archaeological prospecting. All the galvanic techniques measure the electrical resistivity of soils and rocks, and hence they respond to any local variation of this parameter produced by human activities. It is a rather slow but inexpensive method. Recent developments with conductivity-meters permit a more expensive but faster approach to similar information.

The development of the technology associated with automatically multiplexed electrode arrangements and automatic measuring systems facilitate the acquisition of a large number of measurements in a limited time. Further, the advent of fast computers allowed the development of automated resistivity inversion schemes, which aim to construct an estimate of a subsurface resistivity distribution that is consistent with the experimental data (Papadopoulos et al. 2006). Several two dimensional smoothness constrained inversion algorithms for ERT (Electrical Resistivity Tomography) data have been presented in literature (Sasaki, 1992; Xu and Noel, 1993; Loke and Barker, 1995; Tsourlous 1995; Cardarelli and Fishanger 2006a).

The Induced-Polarization (IP) method has been adopted with moderate success since the 1960s (Aspinall & Lynam, 1968, 1970). IP is useful because it can provide information on the presence of disturbed clay- or pyrite-rich horizons in areas where there has been human occupation. Limited field experiences suggest that, in some cases, the IP method may provide information of greater clarity than the resistivity methods (Aitken, 1974). The requirement of detecting voltage time decays or resistivity frequency spectra using unpolarizable electrodes slows down the field work considerably, however. The method is rarely used now, because of this time constraint and mainly of the cost of the sophisticated electronic equipments. Its use can be justified only for some very specialized research, including localization of some diffuse presence of processed clayey and/or metallic features of anthropic origin.

d. Ground Penetrating Radar—The Ground Penetrating Radar (GPR) offers a high resolution sounding capability with detection of features of the order of a few tens of millimetres thickness at ranges of several metres. The range decreases to a few metres in conductive materials such as clays, silts and soils with saline or contaminated pore water (Davis and Annan, 1989).

Radar signals are transmitted into the ground and then reflected by discontinuities in the electric properties of soils. The reflection times of the signals provide depth information (Moffatt, 1974; Morey, 1974; Ulriksen, 1982; Bevan, 1983; Vaughan, 1986; Finzi & Piro, 1991, 2000; Brizzolari et al. 1992d; Malagodi et al. 1996; Conyers and Goodman, 1997; Goodman & Nishimura, 1993; Nishimura and Goodman, 2000; Goodman et al. 2004a,b, 2007; Neubauer et al. 2002; Piro et al. 2001c, 2003). The GPR method can be of fairly rapid field use but neither the equipment nor the interpretation of radargrams are simple tools.

The resolution of the system will improve at higher frequencies if the ratio of the bandwidth to the centre frequency remains the same. This is one reason ground-penetrating radar systems are made to operate at a number of different frequencies. There must be a compromise between range and resolution for GPR systems. The attenuation decreases as the frequency decreases in wet geological materials. The resolution is increased as the bandwidth is increased and this usually requires that the centre frequency of the radar be increased (Davis and Annan, 1989).

The principal weakness of GPR is that it cannot normally penetrate below a clayey horizon. Often the use of incorrect antennas means that important features are obscured or missed entirely. This happens because of poor resolution of or excessive signal attenuation by the feature of interest. Efforts are now being made to convert move out correction and signal processing seismic techniques to radar data.

In the last ten years advances in GPR data visualization and processing have significantly increased the utility of this geophysical remote sensing tool, particularly for the field of archaeological prospection (Goodman et al. 1993, 2004b, 2007; Sigurdsson et al. 1998; Malagodi et al. 1996).

New integrated geophysical approach

Generally, it is difficult to apply geophysical prospecting methods to small subsurface bodies, because of the low values of the signal compared to the local noise. The main sources of noise are soil inhomogeneities and environmental disturbances. It is possible to overcome these problems by i) improving data acquisition methods, ii) improving data processing methods and iii) integrating results obtained with different geophysical methods (Brizzolari et al. 1992a).

The geophysical fundamental goal in near-surface investigations and environmental applications is to construct as complete as possible three-dimensional maps of subsurface targets. The main effort in archaeology is also the integration of different, absolutely non-invasive techniques, especially if they are used in the ultra-high resolution 3-D tomography mode. The integrated approaches must of course operate according to the principle of potential correlation among all those methods that have demonstrated the highest efficacy in investigating inhomogeneous media (Cammarano et al. 1998; Piro et al. 2000; Gaffney et al. 2004; Kvamme 2006).

2 MAGNETIC SURVEYING

The magnetic methods were first applied in the 1950s (Aitken et al. 1958) and have since then become the backbone in archaeological prospecting. Now, they are used even more frequently than electric prospecting methods. Typically, soils that have had campfires over them show increased magnetic susceptibility, resulting from the consequent reducing environment. This condition causes the formation of magnetite even when moderate amounts of iron are present. Magnetometers can easily detect variations of less than 0.1 percent magnetite content in soil.

The magnetic surveying is a passive technique, relying, for the source of its signal, on the presence of the earth's magnetic field. The appropriate unit by which to measure the field strength is the nanoTesla (nT) and picoTesla (pT), (Parasnis, 1986; Breiner, 1973; Becker, 1994). In magnetic surveying, the field strength is measured with a magnetometer placed a few tens of centimetres above the ground level. Measurements can be taken either along profiles or on a mesh of points. Above a flat, homogeneous medium the magnetic field would result quite uniform. However, the presence of local concentrations of soils, rocks and iron objects, magnetized by the earth's main field, slightly alter the magnetic field all nearby. These anomalies may be observed in the measured values. The aim is to interpret such kind of anomalies in terms of possible archaeological resources (Aitken, 1974; Breiner, 1973; Huggins, 1984; Weymouth, 1986a-b; Brizzolari et al. 1992e-1993; Scollar et al. 1990; Eder-Hinterleitner et al. 1996; Neubauer et al. 1997b; Piro et al. 1998; Becker and Fassbinder, 2001; Ciminale and Loddo, 2001; Godio and Piro 2005; Tsivouraki and Tsokas, 2007).

The degree of soil magnetization is a function of the external field strength and of a property of soils known as magnetic susceptibility. The sources of susceptibility are iron compounds in soils, principally hematite, magnetite and maghoemite.

If some demagnetized object is placed in a magnetic field, it can become magnetized. In absence of external field, if the object presents magnetism, this magnetization is called remanent. If this magnetism is a consequence of heating, we have thermoremanent magnetization. All soils have magnetic susceptibility to some degree, but what is fundamental in generating anomalies is the susceptibility contrast of a structure with respect to the surrounding matrix. Natural and anthropogenic causes can generate this contrast. Anthropic features with significant contrasts include pits with organic content, ditches filled with top soil, intrusive structures such as walls and foundations, fire hearths, burned houses and bricks.

In archaeological prospectings, three types of instruments are currently used, namely the fluxgate gradiometer, the alkali metal (cesium or potassium) optical pumping magnetometer and the proton free precession magnetometer. Under ideal conditions, if the magnetic field were measured on a row of points above a flat, uniform earth, all the values would be the same. If the uniform earth contained even one small, isolated feature having a magnetic susceptibility greater than that of the surrounding rock, the feature would be more magnetized than the surrounding soil and would set up its own local magnetic field. This extra induced field, which is similar to that of the earth, but at a notably smaller scale, is known as a dipole field. This local dipole field then adds vectorially to the field of the earth to produce an anomaly. Fig. 1 shows a profile produced by the earth's main field combined with an anomalous magnetic field due to a local feature.

Anomalies exist in archaeological sites as a consequence of the contrast in magnetic properties between the archaeological features of interest and the surrounding medium, both of which are usually composed of material of natural origin such as rocks or soil or even empty space. This magnetic contrast is a function of both the concentration and the thermal and mechanical history of magnetite in either the archaeological target or its hosting material.

The remanent magnetization of archaeological objects is particularly significant not only because of its high relative intensity, but also because it is intimately associated with many objects of ancient habitations, as bricks, tiles, pottery, kilns, hearths and similar features (see Fig. 2).

This remanent magnetization, which as above said is a thermoremanent one, is created when the magnetite-bearing clay is heated to a relatively high temperature and then cooled in presence of the earth's magnetic field.

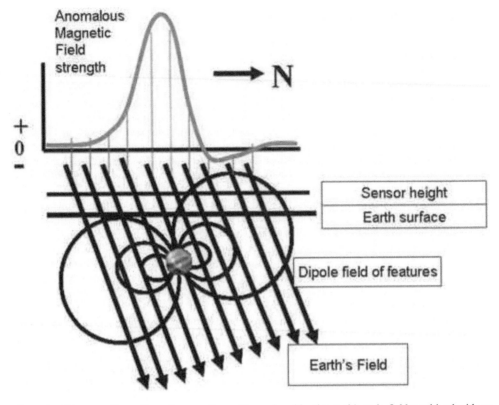

Figure 1.   Diagram of a south-north magnetic profile produced by the earth's main field combined with an anomalous magnetic field due to a local feature (After Weymouth, redrawing 1986b).

To estimate the expected maximum effect due to an archaeological object, we can consider the anomaly due to a single cube of rock representing a buried body, which is given by

$$T = (KFD3)/r3, \tag{2.1}$$

where T is the anomaly amplitude in nT, K is the susceptibility contrast per unit volume, F is the earth's field intensity also in nT, D is the dimension of one side of the cube in whatever distance unit and r is the distance between the magnetometer and the center of the cube in the same units as D.

The anomalies observed in archaeological sites are in most situations very complex as a consequence of several factors. The sources of the anomalies are relatively shallow and therefore very close to the magnetometer, which of course automatically tends to emphasize the highly complex nature of the near field of any magnetic object.

The various sources of magnetic anomalies from soils, near surface rocks and the clutter of ancient and/or modern human habitations, including the many objects of real interest, are usually very strong. However, the measures collected in archaeological contexts are often masked by noise. For such main reason, it is advisable during the survey to employ portable gradiometers. This instrument is a differential magnetometer in which the spacing between the two sensors is fixed and small with respect to the distance to the sources, whose magnetic gradient effects are to be measured. The difference in intensity divided by the distance between the two sensors is then the gradient attributed to the midpoint of the two sensors spacing (Breiner, 1973). The most

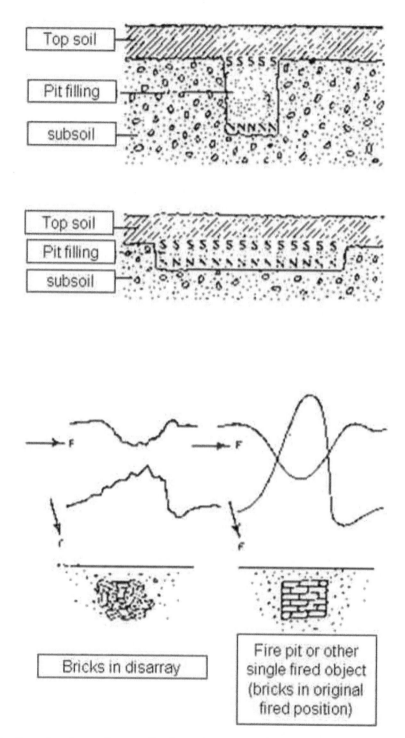

Figure 2. Examples of magnetic anomalies due to the presence of a pit (a) and bricks (b) (After Breiner, 1973).

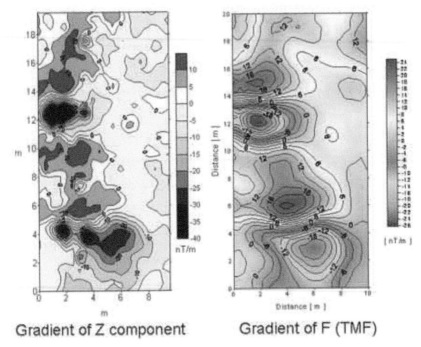

Gradient of Z component   Gradient of F (TMF)

Figure 3. Sabine Necropolis at Colle del Forno. Contour map of the gradient of Z component (of TMF—total magnetic field) for the Test_area1 and 2.

significant constraint included in the above definition is hence the requirement that the spacing r between the sensors be small with respect to the distance r to the source of the anomaly. For example, if one considers a dipole, one sensor at r would measure an anomaly T. A second sensor at $2r$ would only measure 1/8 of T. The second sensor is essentially not sensing the anomaly at all and may as well be at infinity.

Therefore, the magnetic field gradient can be expressed as:

$$\frac{dT}{dr} = \lim_{\Delta r \to 0} \frac{T_{r+\Delta r} - T_r}{\Delta r} \approx \frac{\Delta T}{\Delta r} \qquad (2.2)$$

where ΔT is the differential total field between the two sensors positions, spaced Δr apart, and dT/dr is the derivative or gradient of T in the direction of r.

The gradiometer automatically removes the regional field and thus it notably increases the resolution of the local anomalies. A composite source can be therefore portrayed as a more resolved anomaly, since different edges of the source can this way be separated into two or more discrete anomalies.

In Figs. 3 and 4 examples of contour maps of the gradient of the vertical component Z of the total magnetic field are shown.

In Figs. 5a and 5b, the results after the application of numerical technique to enhance the S/N ratio and to obtain the inversion are shown.

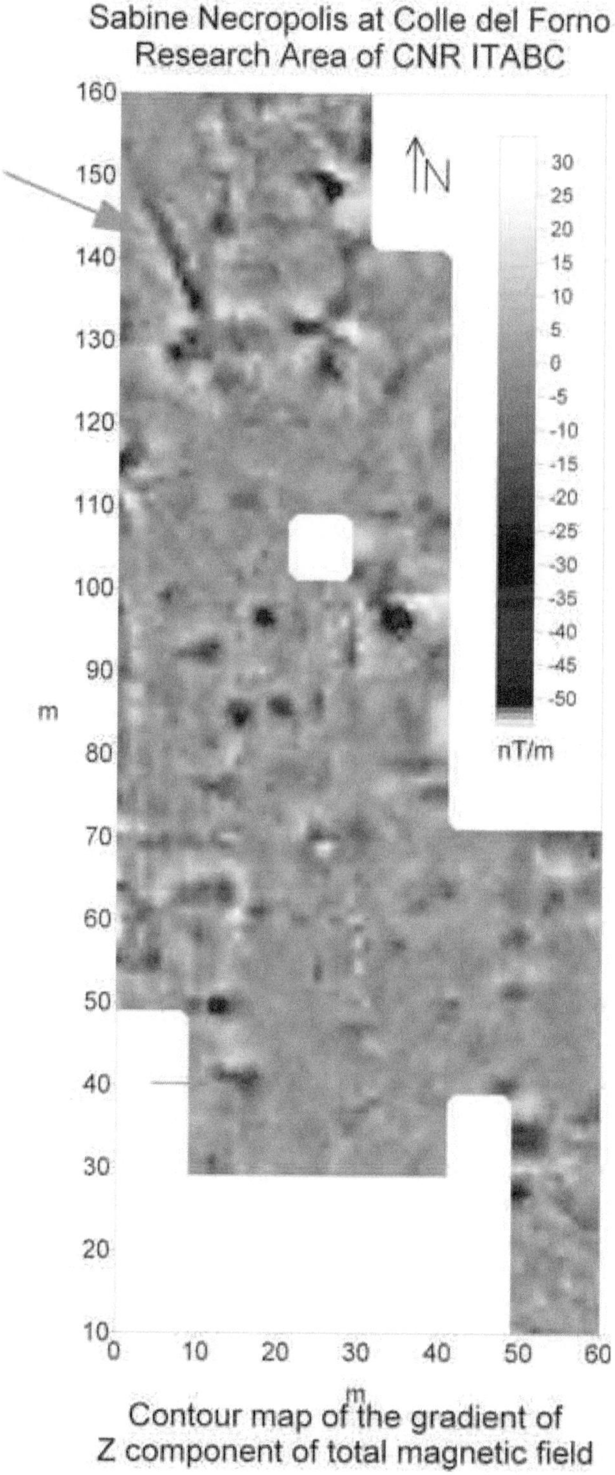

Sabine Necropolis at Colle del Forno
Research Area of CNR ITABC

Contour map of the gradient of
Z component of total magnetic field

Figure 4.   Sabine Necropolis at Colle del Forno. Contour map of the gradient of Z component (of TMF) for the total investigated surface.

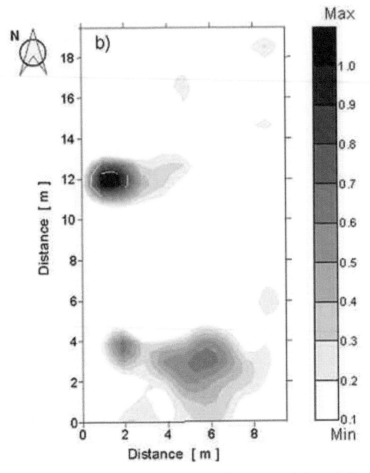

Figure 5a.   Sabine Necropolis at Colle del Forno. Results of the cross-correlation analysis of the vertical gradient of the Z component for Test_area1 and Area2.

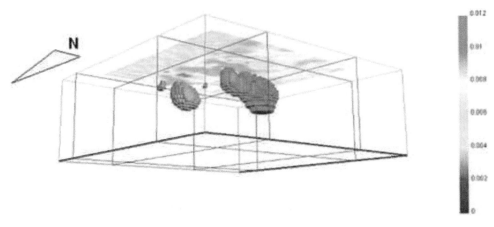

Figure 5b.   Sabine Necropolis at Colle del Forno. Rendering of the results of 3D inversion; the main structures, for Area1 and 2, are well pointed out.

3   THE DIRECT CURRENT GEOELETRIC METHOD

The direct current geoelectric method (DG) consists in the experimental determination of the resistivity distribution, closely characterizing the electrical structure underground. The principle of the method is as follows: a constant electrical current is sent into the ground, via two point electrodes (metal rods A and B) and the resulting distribution of potentials is mapped by means of a couple of electrodes connected to a voltmeter. By doing this, a potential distribution is established in the subsurface depending on the resistivity distribution in the earth. Two potential electrodes (M and N) can be used to measure the potential difference between two selected points on the surface or in boreholes (Fig. 6).

Knowing this voltage, the introduced current and the position of all four electrodes an apparent resistivity can be calculated.

Inhomogeneities, like electrically better or worse conducting bodies, are inferred from the fact that they deflect the current and distort the normal potential behaviour (Fig. 7).

The hypothesis of stationarity implies the divergence-free condition for the electrical current density vector, i.e.

$$\nabla \cdot \mathbf{J} = 0, \tag{3.1}$$

where J is the current density. Equation 3.1 states that all the current lines going into an arbitrary volume of material must leave the same volume, unless there is a source or sink inside. Taking into account Ohm's law:

$$E = \rho J, \tag{3.2}$$

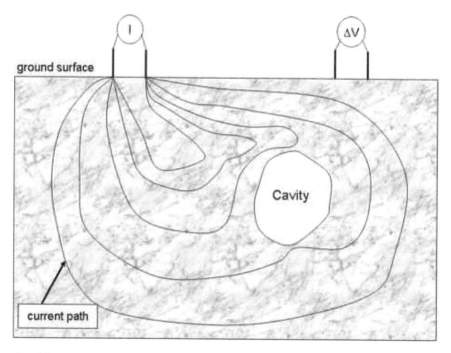

Figure 6.   Schematic representation of the electric current path in presence of the cavity.

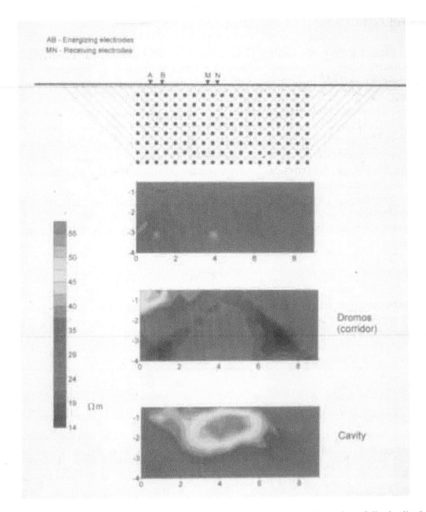

Figure 7. Field procedure sketch for the dipole-dipole geoelectrical array. Examples of dipole-dipole pseudosections with different anomalies.

where E is the electrical field and ρ is the resistivity, from equation 3.1 we obtain:

$$\nabla \cdot \frac{\mathbf{E}}{\rho} = \frac{\nabla \cdot \mathbf{E}}{\rho} + \mathbf{E} \cdot \nabla \frac{1}{\rho} = 0 \qquad (3.3)$$

Equation 3.3 leads to Poisson's differential equation

$$\frac{\nabla \cdot \nabla U}{\rho} + \nabla U \cdot \nabla \frac{1}{\rho} = 0 \qquad (3.4)$$

where U is the electric potential, related to E by

$$\mathbf{E} = -\nabla U \qquad (3.5)$$

Equation 3.5 is of pratical importance in the determination of the resistivity. In fact, it can be demonstrated that, if the subsoil is uniform, by measuring the potential drops across two arbitrary points on the ground surface, we are able to determine in a simple way the true resistivity as follows:

$$\rho = k\frac{\Delta U}{I} \qquad (3.6)$$

where I is the current flowing between two energizing electrodes and k is a geometrical factor depending on the disposition of the four (energizing plus receiving) electrodes.

If the subsoil is inhomogeneous, equation 3.6 may be again used to define the so called apparent resistivity ρ_a. As long as the electrode arrangement is varied, the ratio in equation 3.6 will generally change. This results in a different value of the calculated resistivity for each measurement. So, each measured resistivity is influenced by the true resistivity distribution in the volume interested by the current circulation, related to the location of the electrodes.

The apparent resitivity will equal the true resistivity of the ground only in case of a homogeneous half-space. In general, the apparent resitivity is an integral measurement, mapping a complicated three-dimensional resistivity distribution onto a single value. A single resistivity measurement is not sensitive to one certain location in the ground but will be affected by the whole distribution. A single measurement can be regarded as a "projection" just like a "flat" shadow of a complicated three-dimensional object.

Any combination of four electrodes shows a typical pattern of sensitivity, i.e. some parts of the ground will affect the measurement more than others. For instance, if the spacing between the electrodes is increased, the arrangement will be more sensitive to the resistivity in greater depths. Another effect of the sensitivity distribution is that a decrease in resistivity in certain parts of the ground may even cause an increase in the measured apparent resistivity.

Several electrode arrangements are used in geoelectrical prospecting.

In the Wenner array, the four electrodes are equally spaced along a straight line. If the distance between any two adjacent electrodes is a, the geometric factor k in equation 3.6 is explicitated as follows:

$$k = 2a \qquad (3.7)$$

In the Schumberger array, two very closely spaced measuring electrodes are placed mid-way between the two current electrodes. If a and b are the spacing between the energizing electrodes and the receiving ones respectively, the ratio:

$$-\Delta U/b \qquad (3.8)$$

gives an estimate of the electrical field in the midpoint of the array and an approximation of the geometric factor is given by:

$$k = \pi\frac{a^2}{b} \qquad (3.9)$$

In general, varying the relative position of the four electrodes, we obtain the distribution of the apparent resistivity values. In archaeological prospecting, the DG method is adopted especially for the detection of buried ditches, tombs, cavities and stone foundations. Moreover, the study of DG microsurveys with non-invasive electrodes over floors, walls, columns and any other limited pieces of historical materials is a new reality, which deserves particular importance.

Due to the generally limited dimensions of the targets ultra-high resolution techniques are needed. In particular, the dipolar geoelectrical tomography along a selected profile can lead to an extremely detailed picture of the electric resistivity behaviour across the vertical plane passing through the profile. In Fig. 7 the field procedure is sketched (Cammarano et al. 1998). An adequate representation of the acquired data consists in the attribution of each apparent resistivity value to the intersection between two lines starting from the midpoints of the dipoles and with an inclination of 45° with respect to the ground surface. The dipolar apparent resistivity values are then calculated by means of:

$$\rho_a = \left[\pi\delta(n-1)n(n+1) \right] \frac{\Delta U}{I} \tag{3.10}$$

where δ is the energizing or receiving dipole length, n is the ratio between the inter-electrodic distance and ΔU is the potential drop across the receiving dipole and I is the current sent by the energizing dipole.

The obtained distribution of the apparent resistivities is then submitted to a comparative analysis in order to determine a set of isolines.

The instrumental components for the execution of all DG measurements concern the energizing and receiving apparatus. The first one is essentially a constant voltage generator coupled with an automatic polarity inverter, which allows to obtain a square current wave, whose period is selected on the basis of the signal-to-noise ratio characterizing the measuring site. In fact, by inverting the polarity of the input signal, we are able to eliminate the effect of spontaneous potentials which can contaminate the acquisitions. The receiving apparatus consists of a digital millivoltmeter connected to a computer, which stores the potential drops across the receiving electrodes and calculates in real time the mean of all the data acquired during the energizing cycle. This technique permits to lower stochastic noises.

In general the rock resistivity depends on many factors, say water content in fissures and fractures, porosity, degree of saturation and nature of pore electrolytes. In the dry state most rocks are non-conducting, i.e. they have extremely high resistivities, which decrease rapidly with the existence of fluids, usually containing various ions to form the electrolytic solution.

In archaelogical prospecting, the presence of a high apparent resistivity anomaly is usually an indicator of some resistive structure, such as a cavity, a stone wall or a foundation, hosted within a less resistive material. A very interesting situation regards the study of historical buildings, where for capillaric ascent of humidity and ingression of more or less aggressive waters, internal alteration nucleuses, tipically characterized by very low resistivities, become the sources of degradation and even disgregation of the structure. In such cases, to readily ascertain the presence of the alteration sources, again micro-tomographies have to be programmed, using dipole spacing of the order of a few centimetres.

3.1 *Resistivity Imaging—Electrical Resistivity Tomography (ERT)*

If a whole set of resistivity measurements is combined, it is possible to reconstruct a model of the resistivity distribution in the ground. The model will be closer to the true resistivity distribution the more distinct measurements are combined. Sticking to the "shadow analogy" it is obvious that every projection from a different angle will add some more information to our knowledge about the shape of the unknown object. Many sets of measurements consisting of only a few different "projections" have been invented to produce simple, one-dimensional models of the subsurface, describing only the variation of resistivity vertically but not laterally.

However, advanced field technology now allows a much greater number of measurements in a shorter time. Sophisticated inversion strategies making use of increased computing performance have been invented to obtain high-resolution spatial images of subsurface resistivity distributions

Geoelectrical Tomographies (x,y) (Dipole-Dipole array)

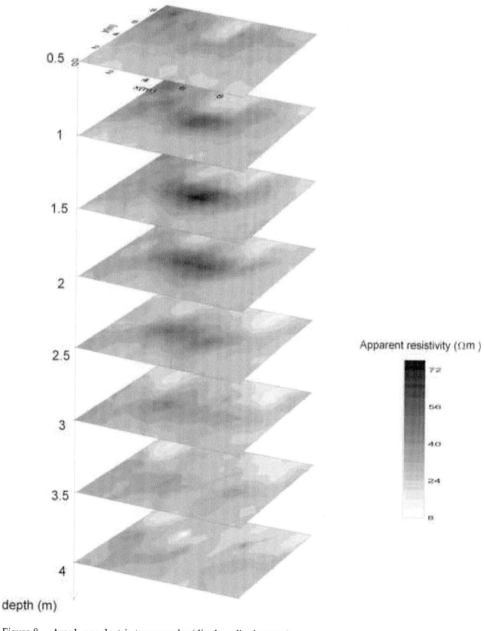

Figure 8. Area1, geoelectric tomography (dipole—dipole array).

(Tsokas et al. 1995b, 1997a; Walker, 2000). Analogous to medical imaging technologies these applications are commonly called Electrical Resistivity Tomography (ERT).

Electrical resistivity tomography (ERT) or electrical resistivity imaging (ERI) is a geophysical technique for imaging sub-surfaces structures from electrical measurements made at the surface, or by electrodes in one or more boreholes. It is closely related to the medical imaging technique electrical impedance tomography and mathematically is the same inverse problem. The technique evolved from techniques of electrical prospecting that predate digital computers, where layers or anomalies were sought rather than images. Early work on the mathematical problem assumed a layered medium (see for example Langer (1933), Slichter (1933)). who is best known for his work on regularization of inverse problems also worked on this problem. He explains in detail how to solve the ERT problem in a simple case of 2-layered medium.

When adequate computers became widely available the inverse problem of ERT could be solved numerically, and the work of Loke and Barker (1996a,b) at Birmingham University was among the first such solution, and their approach is still widely used and enhanced (Tsourlous et al. 1997; Cardarelli and Fishanger, 2006a; Papadopoulous et al. 2006). Applications of ERT include mineral prospecting, monitoring of ground water flow and archaeology.

In Figs. 8 and 9 the examples of geoelectrical tomographies (x,y), obtained from data curried out using dipole-dipole array, are shown and visualized as 2D contour map at different depths.

4 THE INDUCTIVE ELECTROMAGNETIC METHODS

The large category of the Inductive Electro-Magnetic (IEM) methods can give a notable contribution to the knowledge of the resistivity distributions in the subsoil, through the measurements of the components of a suitable IEM field on the ground surface. The scope of the IEM methods in archaeology is to give a representation of the electric geometry of any buried man-made ambient, by mapping variations in the electrical conductivity. This is obtained by a very detailed reconstruction of the pattern of the electric current paths induced by natural or artificial time-varying EM sources.

4.1 *IEM method with natural source*

The basic hypothesis of the method is that a natural time-varying electromagnetic field is normally incident over the ground surface. It diffuses underground and then gives rise to a secondary electromagnetic field related to the induction of currents and charges in the subsurface resistivity anomalies and on electrical discontinuities, respectively. So, we are able to measure on the ground surface the whole set of the components of the total electromagnetic field, i.e. the primary plus the secondary field.

It can be demonstrated that (Kaufmann & Keller, 1981), for each frequency in a suitable range, the components of the measured electromagnetic field are related by tensorial relations as follows:

$$E_x = Z_{xx}H_x + Z_{xy}H_y \tag{4.1}$$

$$E_y = Z_{yx}H_x + Z_{yy}H_y \tag{4.2}$$

where the tensor Z, given by:

$$\mathbf{Z} = \begin{vmatrix} Z_{xx} & Z_{xy} \\ Z_{yx} & Z_{yy} \end{vmatrix} \tag{4.3}$$

Geoelectrical Tomographies (x,y) (Dipole-Dipole array)

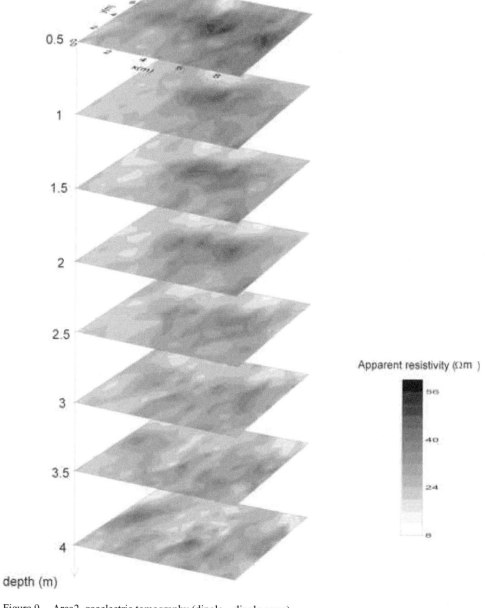

Figure 9.   Area2, geoelectric tomography (dipole—dipole array).

is the impedance tensor. From this tensor the apparent resistivities are usually extracted as follows:

$$\rho_{ij}^a = \frac{1}{\omega\mu}\left|Z_{ij}^2\right|, \qquad \text{where } i, j = x, y \tag{4.4}$$

where ω is the angular frequency and μ is the magnetic permeability of the medium, supposed equal to that of the free space.

The functions (4.4) are a representation of the electric structures of the subsoil, because the deepening of electromagnetic waves increases as frequency lowers. Therefore, by analysing the behaviour of the apparent resistivities as a function of the frequency and measuring site, we can obtain information on the resistivity distribution underground. Obviously, the frequency range involved in this study is related to the particular archaeological target and to the mean resistivity of the hosting material. In fact, the depth of penetration d of the studied electromagnetic field is approximatively given as

$$\delta = 500\sqrt{\frac{2\pi\rho}{\omega}}\,(m) \tag{4.5}$$

Moreover, the vertical component of the magnetic field may help gain a greater resolution of the electric geometry underground. In fact, it is the most sensitive electromagnetic component to any lateral variation of resistivity and its sources are the oscillating electrical dipoles induced in the subsurface conductors.

The ultimate scope of the IEM methods is to recover the distribution of such dipoles, for they are fine indicators of resistivity contrasts in the subsoil and hence of the presence of a buried anthropic ambient.

After this consideration, it follows that the major methodological goal in the study of the IEM method, is the realization of a high resolution 3D tomography for the most correct signature of the boundary lineaments of any finite 3D body.

The main disadvantage of the method is the possibility of negative influence from electromagnetic noise. It can be demonstrated (Berdichevsky & Zhadnov, 1984) that the above linear relations (4.1) and (4.2) yield only in a frequency range in which the primary field varies with distance not faster than a linear trend. In the case of strong electromagnetic noise, the IEM method with artificial source is recommended.

4.2 IEM method with artificial source

Let us now suppose that a time-varying electromagnetic field (Tx) is generated on the ground surface. Again, it gives rise to induction currents within buried conductors. On the other hand, these currents are the sources of a secondary electromagnetic field, which can be measured on the ground surface (Rx) together with the primary incident field, in order to obtain information about the existing resistivity distribution underground (Fig. 10).

The basis of the IEM method with artificial source is the law of induction, i.e. Neumann-Lenz law. If the magnetic flux normal to a closed circuit changes with time, an electromotive force is induced in the circuit so that an electrical current will flow with the result of generating a magnetic field opposing the time-varying flux. This latter magnetic field is the secondary field which can be measured at the ground level. In general, it will differ in phase from the primary field. For the sake of simplicity, we can admit that the total physical effect of the electromagnetic induction is the formation underground of numerous magnetic dipoles, oscillating at the same frequency of the primary field. Now, supposing that each magnetic dipole is situated in a homogeneous isotropic

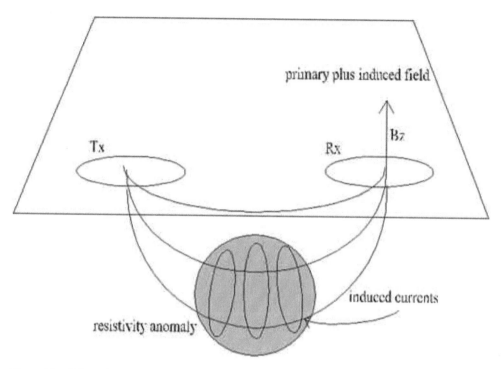

Figure 10. Schematic representation of the IEM method. Induction of a secondary field in presence of a resistivity anomaly.

medium characterized by a resistivity ρ, a dielectric constant ε and a magnetic permeability μ, the magnetic field relative to each dipole of strength m_o (Parasnis, 1986) is (Fig. 11):

$$B_r = \frac{\mu m_o}{2\pi r^3}\left(1 - ikr\right)e^{-i(\omega t - kr)}\cos\vartheta \qquad (4.6)$$

$$B_\vartheta = \frac{\mu m_o}{4\pi r^3}\left(1 - ikr - k^2 r^2\right)e^{-i(\omega t - kr)}\sin\vartheta \qquad (4.7)$$

Where:

$$k = \sqrt{\omega^2 \varepsilon\mu + i\omega\mu/\rho} \qquad (4.8)$$

If the frequency of the primary field is sufficiently low and the interested volume is spatially limited, as is often the case we may encounter in archaeological prospecting, we have that:

$$k|r| \to 0 \qquad (4.9)$$

This is the so called near-field zone, and the magnetic field components take the form:

$$B_r \approx \frac{\mu m_o}{2\pi r^3}e^{-i\omega t}\cos\vartheta \qquad (4.10)$$

45

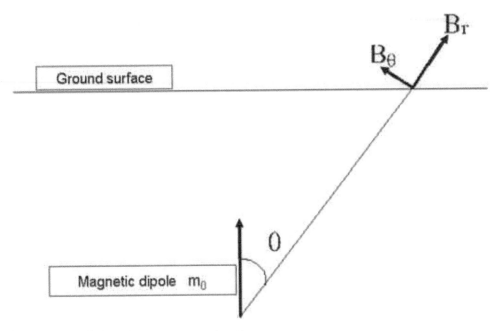

Figure 11.   Magnetic field generated by a magnetic dipole.

$$B_{\vartheta} \approx \frac{\mu m_{o}}{4\pi r^{3}} e^{-i\omega t} \sin \vartheta \qquad (4.11)$$

Under this condition, the total field of the dipole is in phase with the primary field. Also in this case, the scope of the IEM method is to recover the distribution of the oscillating magnetic dipoles, knowing their effects on the ground surface and taking into account that the greater is the strength of the dipoles the higher will be the conductivity of the excited medium.

5   THE GROUND PENETRATING RADAR (GPR)

The Ground Penetrating Radar is an electromagnetic impulsive method much suited for shallow depth investigations, as it can supply subsurface profiles grouped in vertical radar sections. The transmitter-receiver antenna is pulled along the surface of a site, signals are sent with a highly directive radiation pattern into the ground and echoes are returned from targets in the ground within a few meters (see Fig. 12a).

The emitted radar signal is a pulse of electromagnetic radiation with nominal frequency value in the range 15–2500 MHz (1 MHz = 106 Hz). The velocity of an electromagnetic wave in air is 30 cm/ns (1 ns = 10–9 s). In soils the velocity is less, as typical values are in the range 5–15 cm/ns.

According to the media impedance and to their heterogeneity, the radar pulse can be reflected towards the surface or else it can be attenuated and diffused, hence quickly totally dissipated. When reflected echoes emerge, the received signal can be correlated with the transmitted one and the delay of arrivals, i.e. the travel time in the ground, is a function of velocity (Fig. 12).

The vertical scale in radargrams is proportional to the two-way travel time (twt): it can be transformed in a depth scale if the wave propagation velocity is known.

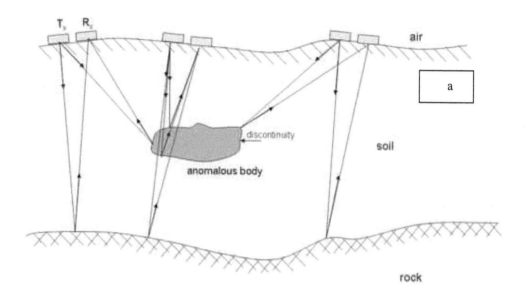

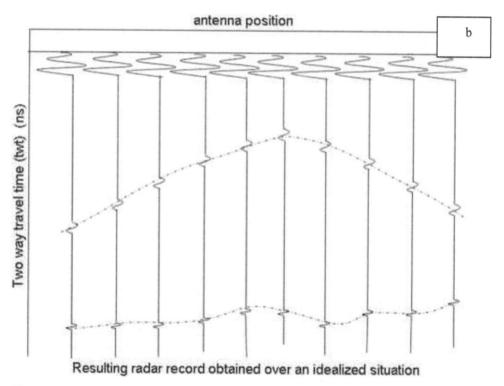

Figure 12.   a) Schematic view of the physical mechanism of the GPR source/target coupling effect. b) Example of GPR profile.

In general, it can be stated that the wave propagation velocity is affected by the dielectric constant (er) and by the magnetic susceptibility (m) of the media. The electric conductivity (s) contributes to the wave attenuation and to its reflection.

The propagation of electromagnetic waves is described by the wave equations:

$$\nabla^2 E = \gamma^2 E \tag{5.1}$$

$$\nabla^2 H = \gamma^2 H \tag{5.2}$$

where

$\gamma^2 = i\omega\mu(\sigma + i\omega\varepsilon)$, $i = \sqrt{-1}$, $\omega = 2\pi f$ in rad/s, f is the frequency in Hz, s is the conductivity in mho/m, $\mu = \mu_0\mu_r$ is the magnetic susceptibility and $\varepsilon = \varepsilon_0\varepsilon_r$ is the electric permittivity.

The propagation factor or wave number $\gamma = \alpha + j\beta$ is that square root of γ^2 whose real and imaginary parts are positive. The attenuation factor α and the space shift constant b are:

$$\alpha = \omega\left\{\frac{\mu\varepsilon}{2}\left[1 + \left(\frac{\sigma}{\omega\varepsilon}\right)^2\right]^{1/2} - 1\right\}^{1/2} \tag{5.3}$$

And

$$\beta = \omega\left\{\frac{\mu\varepsilon}{2}\left[1 + \left(\frac{\sigma}{\omega\varepsilon}\right)^2\right]^{1/2} + 1\right\}^{1/2} \tag{5.4}$$

respectively. Moreover, the wave length λ and phase velocity v are respectively defined as

$$\lambda = \frac{2\pi}{\beta}, \tag{5.5}$$

And

$$v = lf = \frac{\omega}{\beta} \tag{5.6}$$

The electromagnetic field E0 originating at $z = 0$, $t = 0$, at a distance z and time t will be described by $\mathbf{E}(z, t)$ as follows

$$\mathbf{E}(z, t) = \mathbf{E}_0 e^{-\alpha z} \cdot e^{i(\omega t - \beta z)}. \tag{5.7}$$

The first exponential function is the attenuation term and the second is the propagation term. From the first exponential function it is seen that the attenuation is 1/e at a distance z = 1/a, which is also called the skin depth d.

The intrinsic impedance Z is the relation between the electric field and the magnetic field, i.e.

$$Z = \frac{\mathbf{E}}{\mathbf{H}}. \tag{5.8}$$

Z is a complex quantity which is calculated according to

$$Z = \sqrt{\frac{i\omega\mu}{\sigma + i\omega\varepsilon}}$$ (5.9)

The following simple formulas can be used in practical field measurements

$$v = \frac{c}{\sqrt{\varepsilon_r}}$$ (5.10)

which is the wave propagation velocity, where c is the velocity of the light in free space (30 cm/ns),

$$s = \frac{vt}{2}$$ (5.11)

which is the reflector depth, where t is the travel time t (ns),

$$K = \frac{\sqrt{\varepsilon_{r_1}} - \sqrt{\varepsilon_{r_2}}}{\sqrt{\varepsilon_{r_1}} + \sqrt{\varepsilon_{r_2}}}$$ (5.12)

which is the reflection coefficient (R = 1–K is the penetration coefficient), and

$$A = 12.863 \cdot 10^{-8} f \sqrt{\varepsilon_r} \left[\left(tg^2\delta + 1 \right)^{1/2} - 1 \right]^{1/2}$$ (5.13)

which is the attenuation in the medium.

A limiting factor for the GPR survey is the attenuation, or signal power loss, Fig. 13.

This is expressed in decibels/meter (dB/m), the number of decibels being 10 times the logarithm of the ratio of the signal power to the initial power. Considering antenna efficiencies and reflection losses, a typical system can operate with a 50 dB loss (Weymouth, 1986b).

Another important attribute of radar is the resolving power, or the ability to locate small objects. The wavelength affects the ability of the Georadar to identify thin layers or isolated features. Resolution is more than 1/2 and the depth of horizontal interfaces can be determined to about 1/10 (Weymouth, 1986b). In order to get a better resolution, a higher frequency antenna could be used, but this would increase attenuation, while low frequency antennas have a coarser resolution but their penetration depth is remarkably better.

Strong attenuations occur in electrically conductive media. If the conductivity is low, but the number of electrical interfaces is high, multiple reflections could reduce penetration depth, while poor conductivity combined with a small number of reflecting interfaces will cause the wave to be attenuated as a function of the distance between the antenna and the reflecting interface (Finzi & Piro, 1991, 2000).

The strength of reflections of radar pulses depends primarily on the magnitude of change in the dielectric coefficient or conductivity at a discontinuity, and not on the bulk magnetic susceptibility or on the resistivity contrast. The dielectric constant is a measure of how easily charges polarize or separate in a target excited by electromagnetic radiation.

Metals have essentially an infinite dielectric coefficient and thus produce very strong reflections. A pit with a well-defined boundary will produce a better reflection. Walls and foundations

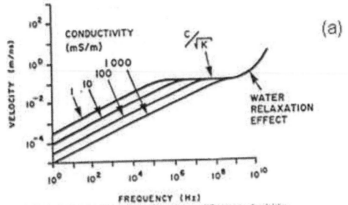

(a)

The relation between velocity and frequency at different conductivities.

-Typical dielectric constant, electrical conductivity, velocity and attenuation observed in common geological materials at 100 MHz

(b)

Material	K	σ (mS/m)	V (m/ns)	α (dB/m)
Air	1	0	0.30	0
Distilled water	80	0.01	0.033	2×10^{-3}
Fresh water	80	0.5	0.033	0.1
Sea water	80	3×10^{4}	0.01	10^{3}
Dry sand	3–5	0.01	0.15	0.01
Saturated sand	20–30	0.1–1.0	0.06	0.03–0.3
Limestone	4–8	0.5–2	0.12	0.4–1
Shales	5–15	1–100	0.09	1–100
Silts	5–30	1–100	0.07	1–100
Clays	5–40	2–1000	0.06	1–300
Granite	4–6	0.01–1	0.13	0.01–1
Dry salt	5–6	0.01–1	0.13	0.01–1
Ice	3–4	0.01	0.16	0.01

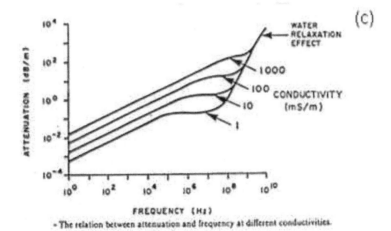

(c)

- The relation between attenuation and frequency at different conductivities.

Figure 13. Values of velocity (a–b) and attenuation (b–c) for different frequencies of an electromagnetic wave and for different conductivities (after Davis & Annan, 1989).

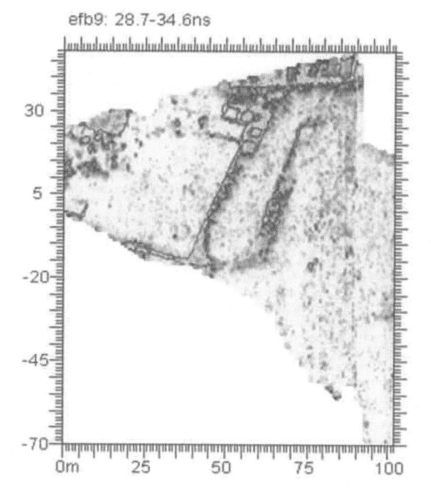

Altopiani di Arcinazzo (Roma) – Emperor Traiano's Villa

efb9: 28.7-34.6ns

GPR survey (500 MHz) – May 2002
GPR time-slice in the time-window 29-35 ns
(estimated depth 1.50 m from the surface).

Plate 2.   Traiano's Villa, GPR time slice in the time window 29–35 ns (twt). The map shows clear anomalies due to the presence of walls and rooms. (See colour plate section)

are also good reflectors. Typical archaeological applications are the search for graves, buildings and the identification of anthropic soil transformations.

The increasing necessity for detailed three-dimensional resolution of the shallow depth structures makes the 3D GPR acquisition one of the most important remote sensing tools. The advantages of 3D surveyineg are documented for the case of mapping geological features (Grasmueck 1996; Sigurdsson et al. 1998); as well as archaeological investigations, where the higher horizontal and vertical resolution is required (Conyers et al. 1997; Leckbush 2000; Malagodi et al.

1996). High-resolution acquisition techniques, using a sub-meter profile spacing interval have been successfully applied in locating subsurface archaeological structures (Goodman et al. 1995; Malagodi et al. 1996; Pipan et al. 1996, 99, 2001; Basile et al. 2000), and also to image large scale archaeological features (Edwards et al. 2000; Nishimura et al. 2000; Neubauer et al. 2002; Piro et al. 2001, 2003).

One of the most useful representation of the GPR data sets collected along closely spaced parallel profiles is to display the data in horizontal maps of recorded reflection amplitudes measured across the survey grid. These maps, referred to as amplitude time slices, allow easy visualisation of the location, depth, size and shape of radar anomalies buried in the ground. The maps can be created at various reflection time levels within a data set to show radar structures at a specified time (depth) across a surveyed site. Mapping the energy in the reflected radar returns across a survey grid can help to create useful information that can sometimes mirror the general archaeological site plan result obtained from invasive excavation.

The raw reflection data acquired by GPR is nothing more than a collection of many individual traces along 2-D transepts within a grid. Each of those reflection traces contains a series of waves that vary in amplitude depending on the amount and intensity of energy reflection that occurred at buried interfaces. When these traces are plotted sequentially in standard 2-D profiles, the specific amplitudes within individual traces that contain important reflection information are usually difficult to visualise and interpret. In areas where the stratigraphy is complex and buried features are difficult to discern, amplitude time slice analysis is one of the most efficient post processes which can be applied to the raw data to extract the 3-D shapes of buried remains (Malagodi et al. 1996; Conyers and Goodman, 1997; Goodman et al. 1995, 2004a,b; Piro et al. 2000, 2001).

Due to velocity changes across the area and with depth, a slice map made across a constant level time window, will not represent a level slice in terms of depth in the ground. Horizontal time slices must therefore be considered only approximate depth slices. Without very detailed velocity control throughout a grid, it is impossible to construct perfectly horizontal depth slices (Leckbusch, 2000).

To compute horizontal time slices, the employed software compares amplitude variations within traces that were recorded within a defined time window. When this is done, both positive and negative amplitudes of reflections are compared to the norm of all amplitudes within that window. No differentiation is made between positive or negative amplitudes in this analysis, only the magnitude of amplitude deviation from the norm. Low-amplitude variations within any one slice denote little sub-surface reflection, and therefore indicate the presence of fairly homogeneous material. High amplitudes indicate significant subsurface discontinuities and in many cases detect the presence of buried features. Finally data are interpolated and gridded on a regular mesh, (see Plate 2).

A high-to-low amplitude scale is normally presented as part of the legend of each map, but without specific units because, in GPR, reflected wave amplitudes are usually arbitrary.

6   INTEGRATED GEOPHYSICAL METHODS, CASE HISTORIES

6.1   *Search of superficial cavities (tombs)*

In this section the results of the integration of FDM (Fluxgate Differential Magnetic), DG (Dipolar Geoelectric) and GPR methods, applied to detect superficial cavities (dromos chamber tombs) in an archaeological test area, are shown (Piro et al. 2000, 2001a,b).

6.1.1   *The archaeological test site*

Geologically, the test area is characterized by a sequence of lithoid tuffs, with resistivity values of about 30–80 Ωm, about 10 m thick, lying on Pleistocene-Quaternary sandy-clayey sediments. At the surface the tuffs are covered by a layer of top soil of 20–30 cm of thickness.

The area of the Sabine Necropolis at Colle del Forno (Rome, Italy) is characterized by the presence of numerous unexplored tombs. From the results of archaeological studies (Santoro,

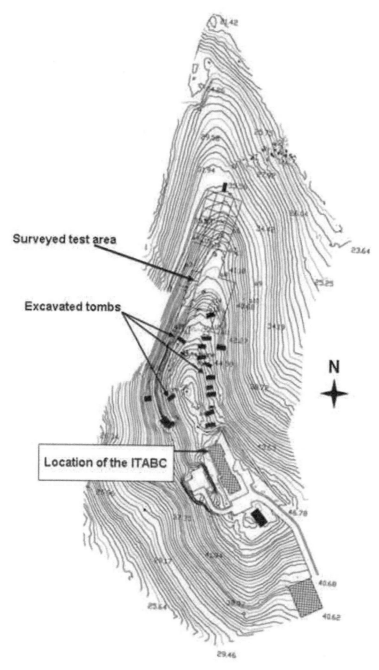

Figure 14. Topographic map of the Sabine Necropolis area at Colle del Forno (Montelibretti, Rome, Italy).

1977), these tombs can be considered as cavities with dimensions of 2 m × 2 m × 1.5 m. The roof is about 0.80 ÷ 1.0 m below the surface of the ground. The tombs present a corridor—dromos with dimensions of 6 m in length and 1 m in width. We have chosen two still unexplored zones, indicated as Test-Area1 and Test-Area2, where two tombs are believed to exist. Both areas have size of 9×9 m^2 and the expected tombs should be located according to the ancient excavation technique.

Fig. 14 shows the topographic map of the Sabine Necropolis at Colle del Forno, where all the known tombs are indicated as well as the area investigated by the geophysical studies.

6.1.2 The magnetic survey

For the magnetic survey the measurements were carried out using a Geoscan FM 36 fluxgate gradiometer. This instrument measures the vertical gradient of the magnetic component Z, with a fixed distance between the sensors of 0.5 m. In each area, 21 parallel S-N profiles, spaced 0.5 m apart, each 10 m long were carried out. With the aim to perfom a survey with a sampling spatial interval less than the body's dimensions, the measurements were collected with a sampling interval of 0.5 m along each profile. The bottom sensor of the gradiometer was taken at the constant height of 0.3 m from the soil.

After the common pre-elaboration techniques (despiking, filtering and re-ranging), the results were finally represented as a coloured contour map of the residual values of the gradient of the Z component (see Fig. 3). The analysis of these figures show the presence of two clear dipolar anomalies characterised by a very high negative value that can be associated to a structure which generates a negative susceptibility contrast. In the southern area (Test-1) the negative value of the anomaly corresponds to the same position of the anomalies obtained with the other methods in previous investigation. In the northern area (Test-2) the anomaly is positioned over a little area and it would delineate the position of the corridor, because the cavity, as it has been obtained with the other methods, is filled with etherogeneous materials.

To enhance the S/N ratio and to better define the spatial location and orientation of the possible targets, the bidimensional cross-correlation technique was applied (Piro et al. 1998). The cross-correlation function is a measure of the similarity between two sets of data. Each time there is an anomaly present in the raw field data, which is similar to the shape of the theoretical one, we obtain by applying the cross-correlation technique, a signal with a shape similar to that of the auto-correlation of the theoretical anomaly. The maximum output from the autocorrelation is converted to the more easily quantity of unity, in its normalised condition, when the data set is matched exactly. To apply this technique it is necessary to calculate the synthetic magnetic anomalies for three-dimensional anomalous bodies to be used as operators (Piro et al. 1998). The theoretical magnetic anomalies for a body with dimensions of $1 \times 1 \times 1$ (grid units) were computed using the relations proposed by Talwani (1965). The values of normalised crosscorrelograms are shown in Figs. 5a,b. The maps illustrate the presence of a nucleus of high value centred with the position of the expected body.

6.1.3 DG method

In each 9 m $\times 9$ m test area at Colle del Forno 10 parallel dipole-dipole profiles were carried out with a constant separation of 1 m from each other. For the measurements the BRGM SYSCAL/R2 resistivity-meter, which can operate in the stacking signal enhancement mode was used. Each apparent resistivity value was obtained with dipole-distance a of 0.5 m in length and attributed at a pseudodepth equal to half the spacing between the centres of the emitting and receiving dipoles along the median vertical axis through the line joining them. The measured values of the apparent resistivities in the ten sections of each area were then pulled together in order to obtain a three-dimensional matrix. In this way we were at last able to represent the horizontal apparent resistivity slices at different pseudodepths, which are displayed in Figs. 8 and 9, respectively for Test-1 and Test-2 area.

The analysis of the map of Fig. 8 shows an high value of apparent resistivity due to the presence of a cavity, and a shallower resistivity low, which can be due to the presence of the corridor of the tomb, filled with etherogeneous sediments. The same results were obtained in the Test-2 area, Fig. 9 where it is evident the high resistivity value, related to the presence of a cavity, with associated a low resistivity value due to the presence of the corridor.

6.1.4 *GPR method*

In a different area of the Sabine Necropolis, a previous GPR survey, made in 1994, allowed to single out an anomaly likely referable to a tomb (Malagodi et al. 1996).

For a better localization of the anomalous body (a tomb, in our case), a new 3D acquisition technique with sampling spatial interval of 20 cm, i.e. less than the expected dimension of the target, was used. In order to obtain a regular grid over the investigated area, a series of 46, 9 m long profiles, once again parallel to the previously defined x-axis, were made.

The radar traces were collected within a 0–100 ns time window, and in true amplitude mode, i.e. without including any kind of filtering and/or gain.

The velocity of 7.5 cm/ns for the electromagnetic waves in the selected ground was obtained from previous investigations.

Taking into account the geoenvironmental conditions of the investigated area and the results of the comparison between different GPR antennas operating at 100 MHz, 300 MHz and 500 MHz, made in the previous work (Malagodi et al. 1996), the radar system GSSI SIR-10 equipped with two antennas operating at 100 MHz (bistatic mode with a spacing of 95 cm) has been employed.

All the collected traces presented a low-frequency noise, probably caused by the ground-antenna coupling, and consequently a low signal-to-noise ratio (S/N) (Malagodi et al. 1996; Tillard, 1994). To attenuate the presence of low- and high-frequency noises and to enhance the S/N ratio, a pass-band filter was applied. With the aim of obtaining a planimetric vision of all possible anomalous bodies in the investigated area, the time-slice representation technique was applied, using all processed profiles.

A time-slice represents a cut at a specified time-value across the radar scan, as a function of x and y. Time slices are thus 2D horizontal contour maps of radar intensities across parallel profiles.

Plate 3 and Plate 4 show the results of the time-slice representation corresponding to different twt (the two-way travel time) on processed data for the 100 MHz antenna.

The GPR results obtained in Test-Area1 show a clear anomaly ascribable to the presence of a tomb. The twt of 25 ns would correspond to the top, approximately 90 cm deep. The twt of 30, 35 and 40 ns, approximately corresponding to the depths of 1.5, 2 and 2.5 m respectively, are likely to match the inner part of the cavity. Finally, the twt of 45 ns would correspond to the bottom of the tomb, approximately 3 m deep. No evidence of the corridor emerged.

On the contrary, the time-slices in Test-Area2 which are probably affected by a very high level of scattering disturbances, do not show any evident anomaly, Fig. 17a. This can be due either to lack of a tomb, or to its existence but in bad state of conservation, due to the collapse of the roof and accumulation of lithoid blocks and sediments inside.

6.1.5 *Integrated analysis*

By comparing all the results shown by each single method, the following integrated interpretations can be made for the two areas.

In Test_1 area all methods concur to give the global image F(x,y) of an intact chamber tomb with the dromos filled however with fine sediments, Plate 3d. In fact, GPR, dipole geoelectric and gradiometric data show quite matching anomalies corresponding to the inner chamber. The standard dimensions of the cavity are then confirmed by the 3D tomographic representation. Concerning the entrance corridor, the same methods show such contrasting effects as to be justified by an elongated body filled with conductive fine sediments of no magnetic relevance and vanishing em wave reflectivity. Indeed, a reversed resistivity contrast, compared with that of the nearby empty chamber, would be more than sufficient to explain both the resistivity low narrow zone.

In Test_2 area the situation appears somewhat more problematic, Plate 4d. Absence of GPR signals and of a gradiometric minimum in the zone where a resistivity small high occurs, would be indicative of a chamber body very likely filled with collapsed tuff blocks from the top. Such a filling would be enough to obliterate any GPR and magnetic contrast, though leaving sufficient residual voids as to induce some geoelectric response. For the dromos, the same considerations as above apply equally well.

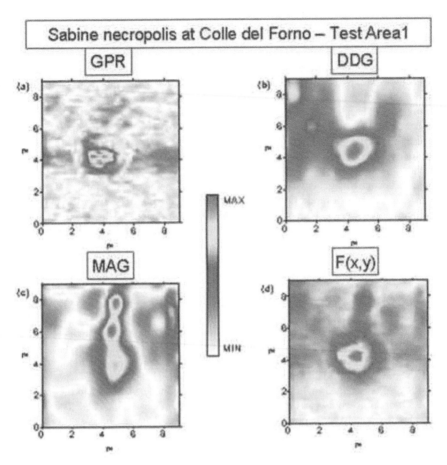

Sabine necropolis at Colle del Forno – Test Area1

GPR   DDG

MAG   F(x,y)

Plate 3. Test_Area1. Comparison of normalised results for the different employed methods: (a) GPR data, (b) dipolegeoelectric data, (c) cross-correlated magnetic data, (d) integration of all results. (See colour plate section)

In conclusion, the integrated analysis of the results obtained with the selected methods, has allowed us to obtain an exhaustive interpretation model of the field data. Therefore, with the above approach, it seems really possible to draw the maximum information out of each methodological contribution as well as of the multi-parametric cross-comparison.

6.2 Search of archaeological structures

In this section the results of the integration of FDM (Fluxgate Differential Magnetic) and GPR methods, applied to detect archaeological structures in the archaeological area of Emperor Traiano's Villa, are shown (Piro et al. 2003).

The Traiano's Villa, near the Affilani Mountains (Lazio region—Central Italy), is the most inland of all the imperial villas located near Rome. The site was ascribed to be the Villa of Emperor Marco Ulpio Traiano (a.d. 98–117) after earlier archaeological excavations made during the 18th–19th century. The buildings of the villa are located on flat plateaus with dimensions of about 4–5 hectares, and are supported laterally by walls with counterforts and niches. The lower terrace to the villa has a rectangular dimension 105 m × 35 m, and it is supported in the southern by the walls with counterforts and in the north by the walls with niches. The central area of this floor was probably occupied by a garden with an external portico. On the west corner of this lower terrace there are some buildings open to the public at the entrance to the villa (the triclinium,

56

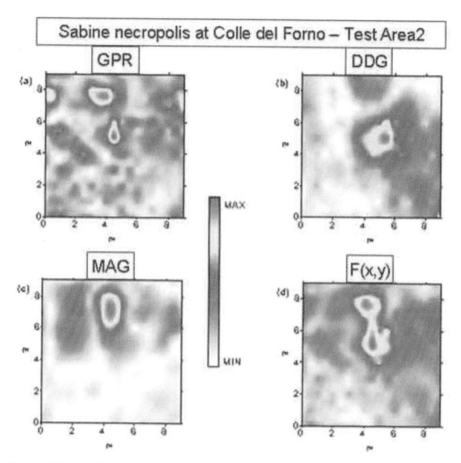

Plate 4. Test_Area2. Comparison of normalised results for the different employed methods. (See colour plate section). For other description see Plate 3.

atrium and nymphaeum structures), which have undergone excavations from 1955–85. The upper level of the villa is supported by walls over 200 m in length.

6.2.1 *Geological and morphological outline of the area*
Geologically the site is characterised by limestone formations of Miocene-age with a thickness of about 230 m. This formation is subject to karstic erosion processes which is exemplified by the strong fracturation. The morphology of the area is strongly influenced by the tectonics, which have created a consistently NW-SE trending fracture and fault system across the region.

The most important elements of the landscape associated with the karst-erosion are the dolina, lapiez, and karren stratigraphic formations. Permeation of the area has been facilitated by the difference in height between the Altopiani area (high-plateau) and the sources of the Aniene river.

In the North-East section of the Villa a spring-source that is present is fed by the Aniene river and arrives at the villa through fractures of Miocene-limestone basement. In this area a big rectangular cistern built in the Roman period is still present.

6.2.2 *The GPR survey*
GPR surveys were performed in November 1999, in May 2000 and in May 2001, in seven different area. In this section the result of GPR profiles collected in the first investigated area are shown.

For the measurements a GSSI SIR 10A+, equipped with a 500 MHz bistatic antenna with constant offset, has been applied.

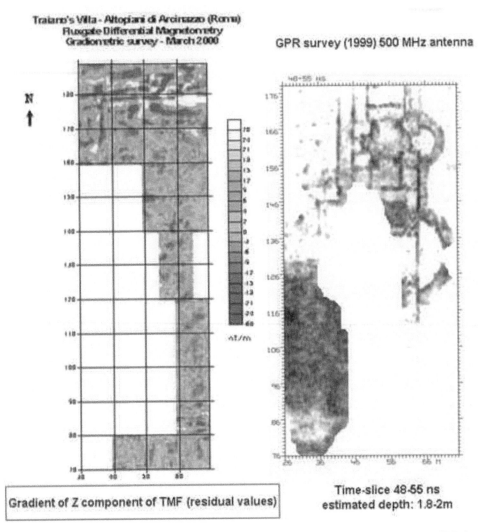

Figure 15. Traiano's Villa, comparison between the gradiometric contour map (on the left) and the GPR time slice in the time window 48–55 ns (twt) (on the right).

Single-fold exploratory profiles were first carried out at the site with the following objectives:

- preliminary identification of the targets;
- calibration of the instrument;
- selection of the optimum frequency antenna;
- analysis of the subsurface response as a function of the orientation of the profiles.

The first GPR survey was concentrated on the upper terraces of the site, in the east part of the area. Adjacent profiles at the site were collected alternatively in reversed and unreversed directions across the survey grids. The horizontal spacing between parallel profiles at the site was 0.5 m. Radar reflections along the transects were recorded continuously across the ground at 80 scan s-1, with a stack = 4. The gain control was manually adjusted to be more effective. Along each profile, markers were spaced every 1 m to provide spatial reference. The data were later corrected for a variation in speed to constant 30 scans per metre (or 1 scan per approximately 0.03 m).

All radar reflections within 75 ns (two-way-travel time) time window were recorded digitally in the field as 8 bit data and 512 samples per radar scan.

We have obtained a nominal microwave velocity of about 6 to 7 cm ns-1 using an experimental profile carried out in correspondence of a site where the archaeologists know the depth of a wall.

The survey was carried out within a block measuring 60 m (west-east) by 220 m (south-north). This area was investigated using the mentioned antenna (500 MHz).

Time slice analysis was applied to all the surveyed grids at the Villa of Traiano. For the site A (first) time slices were generated at 5 ns intervals. The time slice data sets were generated by spatially averaging the squared amplitude of radar reflections in the horizontal as well as the vertical. Horizontal averaging included creating spatial averages every 0.5 m along the radar transects. The data were gridded using a Krigging algorithm that included a search of all data within a 1.0 m radius of the desired point to be interpolated on the grid. Thresholding and data transforms were used to enhance various features detected on the time-scale maps.

In Fig. 15 (on the right) the clearest results were obtained at the 48–55 ns (1.8 to 2.0 m) slices, in which the location of many walls, having different shapes, size and orientation could be clearly imaged. This area is characterised by the presence of many rooms, halls, corridors, exedrae and bathes associated with the private areas of the villa. Several mushroom shaped anomalies are seen which are believed to be dipping pools in the bathhouse.

6.2.3 *The Magnetic survey*

The measurements were carried out, using the Geoscan FM36 fluxgate gradiometer, in the area named A, in the upper terrace of the Villa. This instrument measures the vertical gradient of the Z magnetic component with a fixed intersensors vertical spacing of 0.5 m. During the survey the bottom sensor was used at a constant height (0.3 m) from the soil. A total of 10000 measurements, in a 0.5×0.5 m regular grid, were taken in this area.

After the usual pre-processing, such as despiking, filtering and re-ranging the results have been represented as a coloured contour map of the residual values of the gradient of the Z component for the 25 assembled squares (Fig. 18). The analysis of this map shows that the area is characterised by many dipolar anomalies in a range of $[-30, +24]$ nT m^{-1}.

The gradiometric contour map in Fig. 15 (on the left) clearly shows the existence of three groups of dipolar anomalies spatially organised as pseudolinear structures or semicircular structures. From the comparison of the gradiometric map and GPR time-slice (Fig. 15) it is possible to verify the good matching between these two methods (see the location and extension of the anomalies (a)); for the two mushroom shaped anomalies, b1 and b2, there are some difference. The magnetic anomaly b1 overlaps only a half part of the semicircular body, individuated with the GPR survey. This could be due to the presence of the facing of the wall made with bricks, only for a half part of the body.

The comparison of the results of the two employed methods shows that each single method concurs to give the overall picture of the individuated archaeological structures.

From the analysis of the Fig. 18, the most significant result of the magnetic survey was that the anomalies obtained in correspondence of the mushroom shaped GPR anomalies present some differences. The larger structure (b2), individuated with the FDM method, corresponds to the same anomaly located with the GPR method. This could be interpreted as walls covered by bricks, which generate a high susceptibility contrast with the ground. The smallest structure (b1) individuated with the GPR method doesn't correspond with the obtained magnetic anomaly. This could be interpreted with walls covered with bricks only for a half portion.

REFERENCES

Aitken, M.J., Webster, G. &Rees, A. (1958): Magnetic Prospecting. Antiquity, 32, 270–271.
Aitken, M.J. (1974): Physics and archaeology, 2nd edition. Oxford: Claredon Press, 286 pp.
Aspinall, J.C. & Lynam, J.T. (1968): Induced polarization as a technique for archaeological surveying. Prospezioni Archeologiche, 3, 91–93.

Aspinall, J.C. & Lynam, J.T. (1970): An induced polarization instrument for detection of near surface features. Prospezioni Archeologiche, 5, 67–76.

Basile, V., Carrozzo, M.T., Negri, S., Nuzzo, L., Quarta, T. & Villani, A.V. 2000. A ground-penetrating radar survey for archaeological investigations in an urban area (Lecce, Italy). Journal of Applied Geophysics, 44: 15–32.

Becker H., 1995. From nanotesla to picotesla. A new window for magnetic prospecting in archaeology. Archaeological Prospection 2: 217–228.

Becker H. & Fassbinder J.W.M., 2001. Magnetic prospecting in archaeological sites. ICOMOS ed: Paris.

Becker H., Criminale M., Gallo D., 2005. Aerial photography and high-resolution magnetic surveys at the Celone River Valley (Southern Italy). Proceedings of 6th International Archaeological Prospection Conference, Ed. Piro S. (ISBN88-902028), Roma, pp. 21–24.

Benech C., 2007. New approach to the study of city planning and domestic dwellings in the ancient near East. Archaeological Prospection 14: 87–103.

Berdichevsky, M.N. & Zhdanov, M.S. (1984): Advanced theory of deep geomagnetic sounding. Amsterdam, Elsevier, 408 pp.

Bernabini, M., Brizzolari, E., Monna, D., Padula, G., Piro, S. & Versino, L. (1985): Individuazione di cavità sepolte mediante prospezione geoelettrica. Esempio di applicazione: ricerca di tombe nella necropoli di Colle del Forno nei pressi di Montelibretti (Roma). Bollettino del Servizio Geologico d'Italia, CIII, 67–79.

Bernabini, M., Brizzolari, E. & Piro, S. (1988). Improvement of signal—to—noise ratio in resistivity profiles. Geophysical Prospecting, 36, 559–570.

Bevan, B. (1983): Electromagnetics for mapping buried earth features. Journal of Field Arcaheology, 10, 47–54.

Black, W.E. & Corwin, R.F. (1984): Application of self-potential measurements to the delineation of groundwater seepage in earth fill embankments. 54th Annual International Meeting, SEG, Expanded abstracts, 162–164.

Bozzo E., Merlanti F., Ranieri G., Sambuelli L., Finzi E., 1991. EM-VLF soundings on the eastern hill of the archaeological site of Selinunte. Bollettino di Geofisica Teorica ed Applicata 34: 132–140.

Bozzo E., Merlandi F., 1992. Magnetic and geoelectrical measurements on the eastern hill of the archaeological site of Selinunte. Bollettino di Geofisica Teorica ed Applicata 34: 145–156.

Bozzo E., Lombardo S., Merlanti F., Pavan M., 1994. Geoelectric and electromagnetic measurements within an organised archaeological framework: the Marzabotto example. Annali di Geofisica 37 (suppl. 5): 1199–1213.

Bozzo E., Lombardo S., Merlanti F., 1995. Geophysical surveys at the Poliochni archaeological site (Lemnos Island, Greece): preliminary results. Archaeological Prospection 2: 1–13.

Breiner, S., (1973): Applications manual for portable magnetometers. Geometrics, Sunnyvale U.S.A., 57 pp.

Brizzolati E., Ermolli F., Orlando L., Piro S., Versino L., 1992a. Integrated geophysical methods in archaeological surveys. Journal of Applied Geophysics 29: 47–55.

Brizzolari E., Piro S., Versino L., 1992b. Monograph on the Geophysical Exploration of the Selinunte Archaeological Park. Foreword. Bollettino di Geofisica Teorica ed Applicata, 34.

Brizzolari, E., Cardarelli, E., Feroci, M., Piro, S. & Versino, L. (1992c): Vertical electric soundings and inductive electromagnetism used to investigate the calcarenitic layer in the Selinunte Archaeological Park. Bollettino di Geofisica Teorica ed Applicata, 34: 109–119.

Brizzolari, E., Orlando, L., Piro, S. & Versino, L., (1992d): Ground Probing Radar in the Selinunte Archaeological Park. Bollettino di Geofisica Teorica ed Applicata, 34: 181–192.

Brizzolari, E., Cardarelli, E., Feroci, M., Piro, S. & Versino, L. (1992e): Magnetic survey in the Selinunte Archaeological Park. Bollettino di Geofisica Teorica ed Applicata, 34: 157–168.

Brizzolari, E., Cardarelli, E., Piro, S. & Versino, L. (1993): Detection of subsurface magnetic anomalies of archaeological interest: computation of tridimensional magnetic anomalies and interpretation using bidimensional cross-correlation. In "Geophysical Exploration of Archaeological Sites, Series Theory and practice of Applied Geophysics", Vol.7, Ed. A.Vogel. Wiesbaden: Vieweg Publishing, 3–16.

Cammarano F., Mauriello P., Patella D., Piro S., 1997. Geophysical methods for archaeological prospecting: a review. Science and Technology for Cultural Heritage, 6 (2): 151–173.

Cammarano F., Mauriello P., Patella D., Piro S., Rosso F., Versino L., 1998. Integration of high resolution geophysical methods. Detection of shallow depth bodies of archaeological interest. Annali di Geofisica, Vol. 41, n.3, 359–368.

Carabelli, E. (1966): A new tool for archaeological prospecting: the sonic spectroscope for the detection of cavities. Prospezioni Archeologiche, 1, 25–35.

Carabelli, E. (1967): Ricerca sperimentale dei dispositivi più adatti alla prospezione elettrica di cavità sotterranee. Prospezioni Archeologiche, 2, 9–21.

Cardarelli E., Fischanger F., 2006a. 2D data modeling by electrical resistivity tomography for complex surface geology. Geophysical Prospecting Vol. 54 pp. 121–133

Cardarelli E., DI Filippo G., Tuccinardi E. 2006b. Electrical resistivity tomography to detect buried cavities in Rome: a case study. Near Surface Geophysics Vol. 4, pp. 387–392.

Cardarelli E., Fishanger F., Piro S., 2007. Integrated geophysical survey to detect buried structures for archaeological prospecting. A case-history at Sabine Necropolis (Rome—Italy). Near Surface Geophysics, (in press).

Carson, H.H. (1962): A seismic survey at Harpers Ferry. Archaeometry, 5, 119–122.

Ciminale M., Loddo M., 2001. Aspects of magnetic data processing. Archaeological prospection, 8: n.4, 239–246.

Clark, A.J., 1986. Archaeological geophysics in Britain. Geophysics, 51, 1404–1413.

Clark A.J., 1990. Seeing beneath the soil: prospecting methods in archaeology. London: Batsford.

Conyers L.B., Goodman D., 1997. Ground Penetrating Radar. An introduction for archaeologists. AltaMira Press, Walnut Creek, California, (ISBN 0-7619-8927-7).

Corwin, R.F. & Hoover, D.B. (1979): The self-potential method in geothermal exploration. Geophysics, 44, 226–245.

Dabas M., Hesse A., Tabbagh A., 2000. Experimental resistivity survey at the Roman Town of Wroxeter. Archaeological Prospection 7: 107–118.

Dalan, R.A (1991): Defining archaeological features with electromagnetic surveys at the Cahokia Mounds State Historic Site. Geophysics, 56, 8, 1280–1287.

Davis, J.L. & Annan, A.P. (1989): Ground-penetrating radar for high-resolution mapping of soil and rock stratigraphy. Geophysical Prospecting, 37, 531–551.

Di Filippo M., DI Nezza M., Marchetti M., Urbini S., Toro A., Toro B., 2005a. Geophysical research on Via Appia: the so-called "Monte di Terra" funeral monument. Proceedings of 6th International Archaeological Prospection Conference, Ed. Piro S. (ISBN88-902028), Roma, pp. 292–294.

Di Filippo M., Di Nezza M., Piro S., Santoro S., Toro B., 2005b. Integrated geophysical and archaeological investigations in the "Domus del Centenario", Pompei IX, 8 (Italy). Proceedings of 6th International Archaeological Prospection Conference Ed. Piro S. (ISBN88-902028), Roma, pp. 295–299.

Eder-Hinterleitner A., Neubauer W., Melichar P., 1996. Restoring magnetic anomalies. Archaeological prospection, 3: 185–197.

Fassbinder J.W.E & Reindel M., 2005. Magnetometer prospection as research for pre-Spanish cultures at Nasca and Palpa, Perù. Proceedings of 6th International Archaeological Prospection Conference, Ed. Piro S. (ISBN88-902028), Roma, pp. 6–10.

Finzi, E. & Piro, S. (1991). Metodo per impulsi elettromagnetici. Georadar. Atti del Seminario "Geofisica per l'Archeologia", Quaderni ITABC, 1, 53–70.

Finzi E., Piro S., 2000. Radar (GPR) methods for historical and archaeological surveys. In "Non-destructive techniques applied to landscape archaeology", The Archaeology of Mediterranean Landscape, Vol. 4, pp. 125–135.

Foster, E.J. (1968): Further developments of the pulsed induction metal detector. Prospezioni Archeologiche, 3, 95–99.

Francovich R., Campana S., Felici C., 2005. Large scale survey and archaeological mapping. The grosseto and Siena Projects. Proceedings of 6th International Archaeological Prospection Conference Ed. Piro S. (ISBN88-902028), Roma, pp. 25–30.

Frohlich, B. & Lancaster, W.J. (1986): Electromagnetic surveying in current Middle Eastern archaeology: Application and evaluation. Geophysics, 51, 1414–1425.

Gaffney V., Patterson H., Piro S., Goodman D., Nishimura Y., 2004. Multimethodological approach to study and characterise Forum Novum (Vescovio, Central Italy). Archaeological Prospection, Vol. 11, pp. 201–212.

Gibson, T.H. (1986): Magnetic prospection on prehistoric sites in western Canada. Geophysics, 51, 553–560.

Godio A., S. Piro, 2005. Integrated data processing for archaeological magnetic surveys. THE LEADING EDGE (SEG), Vol. 24, N. 11, 1138–1144.

Goodman, D. & Nishimura, Y. (1993): A Ground-radar view of Japanese burial mounds. Antiquity, 67, 349–354.

Goodman D., Piro S., Nishimura Y., Patterson H., Gaffney V., 2004a. Discovery of a 1st century Roman Amphitheater and Town by GPR. Journal of Environmental and Engineering Geophysics, Vol. 9, issue 1, pp. 35–41.

Goodman D., Schneider K., Barner M., Bergstrom V., Piro S., Nishimura Y., 2004b. Implementation of GPS navigation and 3D volume imaging of ground penetrating radar for identification of subsurface archaeology. Proceedings of the Symposium on the Application of Geophysics to Engineering and Environmental Problems, pp. 806–813. Environmental and Engineering Geophysical Society, Colorado Springs, Colorado.

Goodman D., Schneider K., Piro S., Nishimura Y. Pantel A.G., 2007. Ground Penetrating Radar Advances in Subsurfaces Imaging for Archaeology. In "Remote Sensing in Archaeology", Ed. J. Wiseman and F. El-Baz. Chapter 15, pp. 367–386.

Grasmueck, M. 1996. 3-D ground penetrating radar applied to fracture imaging in gneiss. Geophysics, 61 (4): 1050–1064.

Hay S., Keay S., Millett M., Piro S., 2005. Otricoli: an integrated geophysical and topographical survey. Proceedings of 6th International Archaeological Prospection Conference, Ed. Piro S. (ISBN88-902028), Roma, pp. 1–5.

Herbich T., 2005. Geophysical surveying in Egypt: recent results. Proceedings of 6th International Archaeological Prospection Conference, Ed. Piro S. (ISBN88-902028), Roma, pp. 421–426.

Hesse A., 1966. Prospections Geophysiques a faible profondeur. Applications a l'Archeologie. Dunod, paris.

Hesse A., Jolivet A., Tabbagh A., 1986. New prospects in shallow depth electrical surveying for archaeological and pedological applications. Geophysics 541(3): 585–594.

Hesse A., 1991. Les methods de prospection electromagnetique appliqués aux sites arcaeologiques. Atti del Seminario Geofisica per l'Archeologia. Quaderno n. 1 ITABC-CNR, 41–52.

Hesse A., 1992. A comprehensive archaeological and geophysical survey of a Knidian pottery workshop of amphorae in Resadiye, Datca Peninsula, Turkey. Proceedings of 28th International Symposium on Archaeometry (Archaeometry '92), Los Angeles, California, USA.

Hesse A., Doger E., 1993. Atelier d'amphores Rhodiennes et constructions en Pierre à Hisaronu (Turquie): un cas original de prospection electro-magnetique. Revue d'Archeometrie 17: 5–10.

Hesse A., Barba L., Link K., Oritz A., 1997. A magnetic and electrical study of archaeological structures at Loma Alta, Michoacan, Mexico. Archaeological Prospection 4: 53–67.

Huggins, R. (1984): Some design considerations for undertaking a magnetic survey for archaeological sources. Proceedings of the Society of Exploration Geophysicists Annual Meeting, Atlanta, Georgia, 209–212.

Kaufman, A.A. & Keller, G.V. (1981): The Magnetotelluric sounding method. Amsterdam: Elsevier, 595 pp.

Kuzma L. & Tirpak J., 2005. Proceedings of 6th International Archaeological Prospection Conference, Ed. Piro S. (ISBN88-902028), Roma, pp. 13–16.

Kvamme K.L., 2006. Integrating Multidimensional Geophysical data. Archaeological Prospection 13: 57–72.

Langer R.E., 1933. On an inverse problem in differential equations, Bull Am Math Soc, 39: 814–820.

Lapenna, V., Mastrantuono, M., Patella, D. & Di Bello, G. (1992): Magnetic and geoelectric prospecting in the archaeological area of Selinunte (Sicily, Italy). Bollettino di Geofisica Teorica ed Applicata, XXXIV, 133–143.

Leckebusch, J. 2000. Two- and Three-dimensional Ground-penetrating Radar Survey across a medieval chair: a case study in Archaeology. Archaeological Prospection, 7 (4): 189–200.

Linington, R.E. (1966): Test use of a gravimeter on Etruscan chamber tombs at Cerveteri. Prospezioni Archeologiche, 1, 37–41.

Loke, M.H. & Barker R.D., 1996a. Rapid least-squares inversion of apparent resistivity pseudo-sections using quasi-Newton method: Geophysical Prospecting, 48: 181–152.

Loke M.H. & Barker R.D., 1996b. Practical techniques for 3D resistivity surveys and data inversion: Geophysical prospecting, 44: 499–523.

Malagodi, S., Orlando, L., Piro, S. & Rosso, F. (1996): Location of archaeological structures using GPR method. 3-D data acquisition and radar signal processing. Archaeological Prospection, 3: 13–23.

Moffatt, D.L. (1974): Subsurface video pulse radar. Proceedings of an Engineering Foundation Conference on Subsurface Exploration for Underground Excavation and Heavy Construction, New England College. American Society of Civil Engineers, New York.

Morey, R.M. (1974): Continuous subsurface profiling by impulse radar. Proceedings of Engineering Foundation Conference on Subsurface Exploration for Underground Excavation and Heavy Construction, New England College. American Society of Civil Engineers, New York, 213–232.

Neubauer W., Eder-Hinterleitner A., 1997a. Resistivity and Magnetics of the Roman Town Carnuntum, Austria. An example of combined interpretation of prospection data. Archaeological Prospection 4: 179–189.

Neubauer W., Eder-Hinterleitner A., 1997b. 3D-interpretation of post-processed archaeological magnetic prospection data. Archaeological Prospection 4: 191–205.

Neubauer W., Eder-Hinterleitner A., Seren S., Melichar P., 2002. Georadar in the Roman Civil town Carnuntum, Austria. An approach for archaeological interpretation of GPR data. Archaeological Prospection, 9: 135–156.

Neubauer W., Locker K., Eder-Hinterleitner A., Melichar P., 2005. Geophysical prospection of middle neolithic circular ring ditch systems in lower Austria. Proceedings of 6th International Archaeological Prospection Conference, Ed. Piro S. (ISBN88-902028), Roma, pp. 43–47.

Nishimura, Y. & Goodman, D. 2000. Ground-penetrating radar survey at Wroxeter. Archaeological Prospection, 7 (2): 101–105.

Noel M., XU B., 1991. Archaeological investigation by electrical resistivity tomography: a preliminary study. Geophysical Journal International 107: 95–102.

Orlando, L., Piro, S. & Versino, L. (1987): Location of subsurface geoelectric anomalies for archaeological work: a comparison between experimental arrays and interpretation using numerical methods. Geoexploration, 24, 227–237.

Papadopoulos N.G., Tsourlos P., Tsokas G.N., Sarris A., 2006. Two-dimensional and Three-dimensional resistivity imaging in archaeological site investigation. Archaeological Prospection 13: 163–181.

Parasnis. D. S. (1986): Principles of applied geophysics. London: Chapman and Hall, 402 pp.

Patella, D. (1978): Application of geoelectric dipolar techniques to the study of an underground natural cavity of archaeological interest. Bollettino di Geofisica Teorica ed Applicata, XXI, 23–34.

Pipan, M., Finetti, I. & Ferigo, F. 1996. Multi-fold GPR techniques with applications to high-resolution studies: two case histories. European Journal of Environmental and Engineering Geophysics, 1: 83–103.

Pipan, M., Baradello, L., Forte, E., Prizzon, A. & Finetti, I. 1999. 2-D and 3-D processing and interpretation of multi-fold ground penetrating radar data: a case history from an archaeological site. Journal of Applied geophysics, 41: 271–292.

Pipan, M., Baradello, L., Forte, E. & Finetti, I. 2001. Ground Penetrating radar study of Iron Age tombs in South Eastern Kazakhstan. Archaeological Prospection, 8 (3): 141–155.

Piro S., Samir A., Versino L., 1998. Position and spatial orientation of magnetic bodies from archaeological magnetic surveys. Annali di Geofisica 41 (3): 343–358.

Piro S., Cammarano F., Mauriello P., 2000. Quantitative integration of geophysical methods for archaeological prospection. Archaeological Prospection, Vol. 7, n.4, pp 203–213.

Piro S., Santoro P., 2001a. Analisi del territorio di Colle del Forno (Montelibretti, Roma) e scavo nella necropoli sabina arcaica. In Orizzonti, II, pp. 1–16.

Piro S., Tsourlos P., Tsokas G.N., 2001b. Cavity detection employing advanced geophysical techniques: a case study. European Journal of Environmental and Engineering Geophysics Vol. 6, pp. 3–31.

Piro S., Goodman D., Nishimura D., 2001c. The location of Emperor Traiano's Villa (Altopiani di Arcinazzo—Roma) using high-resolution GPR surveys. Bollettino di Geofisica Teorica ed Applicata Vol. 43, n.1–2, pp. 143–155.

Piro S., Goodman D., Nishimura Y., 2003. The study and characterisation of Emperor Traiano's Villa (Altopiani di Arcinazzo—Roma) using high-resolution integrated geophysical surveys. Archaeological Prospection, 10, pp. 1–25.

Santoro, P. (1977). Colle del Forno, loc. Montelibretti (Roma). Relazione di scavo sulle campagne 1971–1974 nella necropoli.Atti dell'Accademia Nazionale dei Lincei, XXXI, 211–298.

Sarris A., Ball S., Georgica K., Kokkinou E., Karimali E., Mantzourani E., 1998. A geophysical campaign at the settlement of Kandou Kouphovounos (Cyprus). 31th International Symposium on Archaeometry, Budapest, Hungary.

Scollar, I. (1962): Electromagnetic prospecting methods in archaeology. Archaeometry, 5, 146–153.

Scollar I., Tabbagh A., Hesse A., Herzog I., 1990. Archaeological prospecting and Remote Sensing. Cambridge University Press: Cambridge.

Sigurdsson, T. & Overgaard, T. 1998. Application of GPR for 3-D visualization of geological and structural variation in a limestone formation. Journal of Applied Geophysics, 40: 29–36.

Slichter L.B., 1933. The interpretation of the resistivity prospecting method for horizontal structures, J. Appl. Phys., 4: 307–322.

Stright, M.J. (1986): Evaluation of archaeological site potential on the outer continental shelf using high-resolution seismic data. Geophysics, 51, 605–622.

Tabbagh, A. (1974): Méthodes de prospection électromagnétique applicables aux problèmes archéologiques. Archaeophysika, 5, 350–437.

Tabbagh, A. (1986). Applications and advantages of the slingram EM method for archaeological prospecting. Geophysics, 51, 576–584.

Tillard, S. (1994). Radar experiments in isotropic and anisotropic geological formation (granite and schists). Geophysical Prospecting 42, 615–636.

Tite, M.S. & Mullins, C. (1970): Electromagnetic prospecting on archaeological sites using a soil conductivity meter. Archaeometry, 12, 97–104.

Tsokas G.N., Giannopoulos A., Tsourlos P. Vargemezis G., Tealby J.M. Sarris A., Papazachos C.B., Savopoulou T., 1994. A large scale geophysical survey in the archaeological site of Europos (northern Greece). Journal of Applied Geophysics 32: 85–98.

Tsokas G.N., Papazachos C.B., Vafidis A., Loukoyiannakis M.Z., Vargemezis G., Tzimeas K., 1995a. The detection of monumental tombs buried in tumuli by seismic refraction. Geophysics, 60 (6), 1735–1742.

Tsokas G.N. & Hansen R.O., 1995b. A comparison of inverse filtering and multiple source Wenner deconvolution for model archaeological problems. Archaeometry, 37: 185–193.

Tsokas G.N. & Tsourlous P., 1997a. Transformation of the resistivity anomalies from archaeological sites by inversion filtering. Geophysics 62(1): 36–43.

Tsokas G.N., Sarris A., Pappa M., Bessios M., Papazachos C.B., Tsourlos P., Giannopoulos A., 1997b. A large scale magnetic survey in Makrygialos (Pieria), Greece. Archaeological Prospection 4: 123–137.

Tsourlous P., Szymanski J., Tsokas G.N., 1997. A fast smoothness constrained algorithm for the 2-D inversion of earth resistivity data. J. Balkan. Geophys. S., 1: 2–11.

Ulriksen, C.P.F. (1982): Application of the impulse radar to civil engineering. PhD Dissertation, Lund University of Technology, Lund, Sweden, 179 pp.

Vaughan, C.J. (1986): Ground Penetrating Radar surveys used in archaeological investigations. Geophysics, 51, 595–604.

Walker A.R., 2000. Multiplexed resistivity survey at the Roman Town of Wroxeter. Archaeological Prospection 7: 119–132.

Weymouth, J.W. (1986) a: Archaeological site surveying program at the University of Nebraska. Geophysics, 51, 538–552.

Weymouth, J.W. (1986) b: Geophysical methods of archaeological site surveying. In "Advances in Archaeological Method and Theory", Vol. 9, 311–395.

Wynn, J.C. & Sherwood, S.I. (1984): The Self-Potential (SP) method: an inexpensive reconnaissance and archaeological mapping tool. Journal of Field Archaeology, 11(2), 195–204.

Wynn, J.C. (1986): Archaeological prospection: An introduction to the special issue. Geophysics, vol. 51, pp. 533–537.

Worthington M.H., 1984. An introduction to geophysical tomography. First Break 2: 20–26.

Electrical methods

Seeing the Unseen – Campana & Piro (eds)
© *2009 Taylor & Francis Group, London, ISBN 978-0-415-44721-8*

Electrical and magnetic methods in archaeological prospection

A. Schmidt

Department of Archaeological Sciences, University of Bradford, UK

ABSTRACT: The relationship between archaeological features in the ground and their manifestation in geophysical surveys is complex. The archaeological interpretation of resulting data hence requires some understanding of both the geophysical principles and the archaeological setting. In this overview the concepts of electrical resistivity surveying are introduced with particular emphasis on soil properties (e.g. moisture content) and how these can be related to the contrast between soil matrix and archaeological features. Considering current flow in the ground these properties are linked to four-probe DC earth resistance measurements on the surface, with particular emphasis on the properties of different electrode arrays. Taking the simple example of a localised buried archaeological feature, responses from features at different depth measured with varying electrode configurations are presented. Pseudosections (or resistivity imaging) are vertical data sections through the ground that are compiled by systematically varying the dimensions of electrode arrays. Examples of different measurements techniques are presented. Archaeological features often also exhibit a contrast in magnetic properties, which can be used for their magnetic mapping, either directly through magnetic susceptibility measurements or indirectly through magnetometer surveys. Various mechanisms are discussed that can lead to magnetic susceptibility enhancement for archaeological features (e.g. Le Borgne effect, fermentation, magnetotactic bacteria). The use of magnetic susceptibility surveys in archaeological evaluation is demonstrated with several case studies.

1 INTRODUCTION

Geophysical methods are an essential tool for archaeological prospection on all scales of investigation: whether for detailed analysis of a single archaeological feature, to provide an overview of all features on an archaeological site, or for the assessment of a whole landscape. The relationship between geophysical measurements at the surface and buried archaeological features is complex and the interpretation of resulting data requires geophysical and archaeological insight. This chapter is a brief introduction to the two main geophysical techniques used in landscape archaeology, namely earth resistance and magnetic surveying. More detailed discussions have been published elsewhere (Clark 1990; Gaffney & Gater 2003; Schmidt 2007; Scollar et al. 1990) and current research is mainly made available through the journal Archaeological Prospection.

2 EARTH RESISTANCE SURVEYING

2.1 *Archaeology and earth resistance*

The general idea underpinning earth resistance surveying is fairly simple: an electrical current that is injected into a homogeneous ground spreads evenly (Fig. 1); but where it encounters obstacles in the form of archaeological features it has to change its course leading to measurable electrical effects at the surface. A map of the lateral surface variations will hence be a representation of buried archaeological remains.

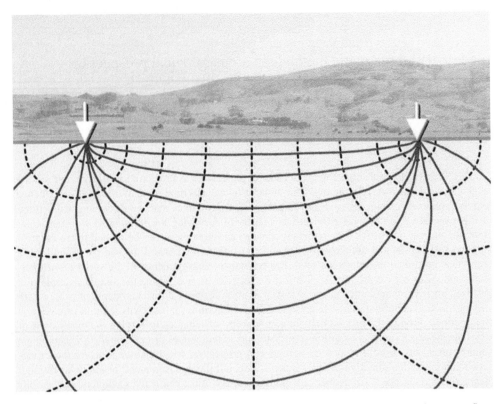

Figure 1. The spread of electrical current through homogeneous ground. Solid lines show the current flow, broken lines the resulting equipotential lines.

Electrical currents are carried by moving charged particles. In a metallic wire the current consists of electrons that freely move through the cable, its connectors and a battery or power supply. Such closed circuit will never show depletion or accumulation of electrons, since they can continue to flow around this loop. In contrast, a current through soil or sediments is entirely carried by ions, which are large charged particles. They are created when salt crystals in the ground (e.g. NaCl) dissociate in the presence of soil water (e.g. to form Na^+ and Cl^- ions). Since ions cannot leave the soil, their movement, and hence any current, would stop once they had all arrived at the surface. To avoid this, the polarity of a current used for earth resistance measurements has to be reversed continuously, forcing ions to alternatively move forward and backward.

There may be various obstacles to the movement of these ions in the ground and the associated weakening of any current is described by a soil's 'electrical resistivity'. Firstly, this is influenced by the initial abundance of salts. While there are some salts in all soils their concentration varies considerably between different soil types. Secondly, and more importantly, electrical resistivity depends on the availability of water. Water is needed to dissolve the salts into their constituent ions and also to facilitate their transport. Soil resistivity is hence mainly governed by the moisture content of the ground. The major factors influencing soil moisture are the sizes of individual soil particles (grains), the space between them (pores) and the availability of water. In addition, resistivity also depends on the mobility of ions in the water, which decreases with temperature and ceases when the water is frozen to ice (Scollar et al. 1990).

A typical example of a buried feature left by past human occupation is a ditch (Fig. 2).

After the abandonment of a settlement it may have gradually filled with sediments and is possibly no longer visible from the ground. However, it will still affect the flow of current, as its fill is

68

Figure 2.   A buried ditch shows contrast to the surrounding material in several physical parameters.

normally loosely packed, allowing the pores to retain water and the ditch will hence have a lower resistivity than the surrounding soil.

2.2   *The contrast of archaeological features*

It is clear from this example that it is not the absolute value of low electrical resistivity that allows to reveal the presence of such ditches, but the fact that this soil property is different from the surrounding material. It is this 'contrast', which makes them detectable. In this respect the geophysical measurement is no different from an archaeological excavation, where features can only be identified through their contrast to the surrounding soil or sediment matrix, either in their colour or texture. For example, mud-brick walls were not identified in Mesopotamian archaeology until archaeologists realised in the late 19th century the subtle contrast that this building material exhibits (Matthews 2003: 12). Geophysical prospection extends this concept and allows to look for archaeological features that may exhibit a contrast to their surrounding matrix in one or more physical properties that are not normally detectable by an excavator, for example electrical resistivity, magnetic susceptibility, remanent magnetisation etc. The geophysical technique to use for the detection of the buried features hence depends on the properties in which a contrast exists. Unfortunately it is often difficult to predict which property shows a pronounced contrast and in many case a number of trial surveys have to be undertaken with different methods to identify those that most suitable for the particular archaeological features, site and environmental conditions.

2.3 *Influence of climate on results from earth resistance surveys*

For a ditch that retains more moisture than the surrounding soil matrix the resistivity contrast is often referred to as being 'negative', since the resistivity of the feature is lower than that of the matrix in which it is embedded. However, there are situations in which this may change.

The moisture content of soil varies with external environmental factors (such as temperature, rain, wind and sunshine), which therefore also affect the electrical resistivity contrast. This can again be illustrated with the example of a buried ditch.

- In a warm and dry British summer, the soil matrix may have dried out considerably and only the ditch retains some moisture. This will lead to a pronounced negative contrast.
- If the dry weather continues, the ditch will also loose its moisture and the contrast will gradually disappear.
- Then it starts to rain heavily and the large pores of the ditch soak up the water very quickly, probably even quicker than the surrounding soil with its smaller pore size. This leads again to a significant negative contrast.
- Heavy rain continues and after several days all ground is thoroughly wet. By then the contrast will have been almost entirely lost.
- Next, sun and strong winds appear and the large pores of the ditch give off their moisture more easily than the surrounding soil, which therefore can lead to a positive resistivity contrast.

These exaggerated and idealised weather conditions help to understand the possible variations of soil moisture and resistivity contrast. An additionally complicating factor is the subsoil geology. Depending on the underlying drainage (e.g. good for chalk, poor for clay) further avenues for water loss or retention are available.

For a stone foundation, the resistivity cannot normally become lower than the surrounding soil and the contrast will hence always be positive (wet surrounding soil) or nearly zero (very dry surrounding soil).

2.4 *Measurement of earth resistance*

To measure electrical resistance of the ground, an electrical current is injected through two steel electrodes. As the current flows through soils and sediments an electrical potential ('voltage') develops. It can be visualised through equipotential lines that connect points with the same value of electrical potential (Fig. 1), similar to contour lines in a topographical survey. These hypothetical lines are distorted by any buried feature with a resistivity contrast. For example, current will tend to flow preferably through wet soil (e.g. within a ditch), but will be diverted around dry features, like walls and stones. However, these often pronounced changes in the ground only have a small effect at the surface, where the altered equipotential lines can be measured with two 'potential electrodes' in an earth resistance survey, providing indirect evidence for the presence of features in the ground.

Consequently, four electrodes are needed for an earth resistance survey (two for current injection and two for potential measurement) and there are many ways in which they can be arranged. Some of the possible 'four-electrode arrays' are more common than others. In archaeological prospection the most commonly used arrangement is the 'twin-probe array' (Fig. 3) in which one current and one potential electrode are mounted on a frame together with an earth resistance meter and this unit is moved across the survey area.

The other two electrodes are located at a distance and are connected with the measuring device through a long cable. The small spacing of the mobile electrodes on the frame leads to good spatial resolution and the arrangement is compact enough to make detailed mapping possible, for

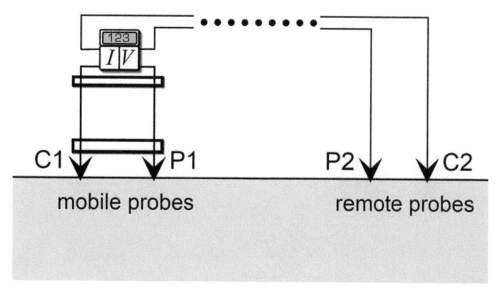

Figure 3.   Twin probe electrode array.

example in a raster with 1 m × 1 m resolution. Data collected systematically can subsequently be plotted so that the resulting map of earth resistance measurements provides clues about the resistivity contrast, and hence the archaeological features, in the ground.

2.5   *Resistivity of soil features*

The electrical resistance (R) is calculated and displayed by the earth resistance meter as the ratio of the electrical potential measured at the surface to the current injected into the ground (R = V / I) and is expressed in Ohms (symbol Ω). This earth resistance depends on two parameters: the resistivity of the ground and the arrangement of the electrodes. The latter dependency becomes clearer if one considers that the location of potential electrodes determines which voltage is sampled, even if the current electrodes are left in the same position.

The electrical resistivity of the buried feature (ρ) is measured in Ohm-metres (Ωm). As the current travels through the ground it will encounter areas of different resistivities, and the single value of earth resistance measured at the surface will be a complicated average of all these resistivites in the ground (Fig. 4).

To describe this behaviour, the concept of 'apparent resistivity' has been introduced. Given an earth resistance measurement R (in Ω) made with a certain electrode array, it is possible to calculate an associated value for the apparent resistivity ρ_A (in Ωm), which accounts for the spatial arrangement of the electrodes and represents the measured value of earth resistance in terms of the material property (i.e. electrical resistivity). To be useful, this conversion has to be such that in the simple case of a homogeneous ground the apparent resistivity is identical to the true resistivity of this ground, ρ. In the heterogeneous case the apparent resistivity becomes 'some sort of average' of all the resistivities in the ground. For its calculation, the exact electrode positions are taken into account and for the most common electrode arrays simple mathematical expressions exist. It follows from this brief discussion that earth resistance measurements do not allow an exact determination of the ground's resistivity at a single point since some averaging along the current path is unavoidable.

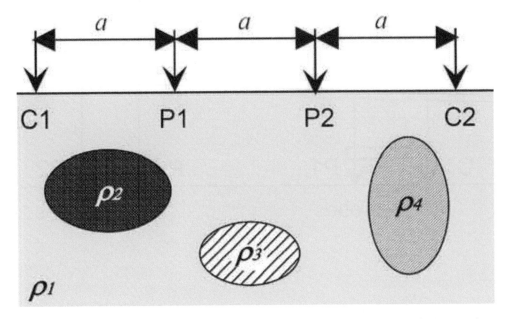

Figure 4.   A single earth resistance measurement will be influenced by all resistive bodies in the ground.

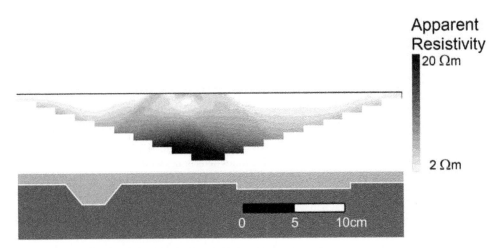

Figure 5.   A wider spacing of current electrodes forces the current to flow to greater depth and the measured earth resistance is hence influenced by deeper features.

2.6   *Depth of investigation*

By increasing the separation between the two current electrodes electrical current is able to penetrate deeper into the ground (Fig. 5) and the measured earth resistance is affected by features at greater depth. This relationship between electrode separation and depth of investigation can be used to probe the ground's resistivity at different depths.

For example in a 'pseudosection' the electrode separation of a chosen array configuration is systematically increased and the array then gradually moved forward along a defined line. The measured earth resistance is converted to apparent resistivity to make measurements with different

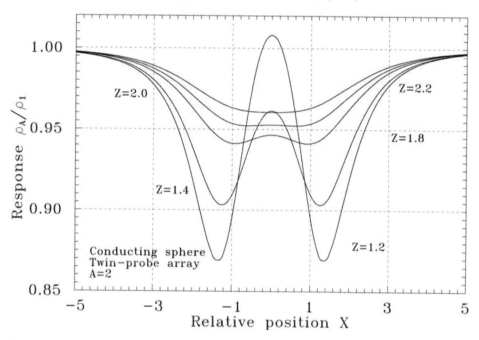

Figure 6.   Wenner pseudosection of two ditches in a water tank.

electrode separations comparable and the apparent resistivity of each measurement is then plotted at a depth calculated from the electrode spacing (e.g. half the spacing between current and potential electrodes (Griffiths & Barker 1994)).

As discussed above, earth resistance measurements are influenced by the average of the ground's resistivites and assigning one value of apparent resistivity to a particular depth is hence wrong. Nevertheless such pseudosections provide useful first insights into the vertical distribution of resistivities in the ground. Figure 6 shows a pseudosection recorded over two model ditches created in a water tank, illustrating clearly the potential of this technique.

Mathematical methods are available to further process the collected data and to calculate resistivity distributions that would result in exactly the measured earth resistance values. This process is referred to as 'inversion' (Loke & Barker 1996), but unfortunately it has no unique solution since several different resistivity distributions can be calculated that would all lead to the same measured earth resistance values. Some of these solutions produce overly smooth shapes for the buried features and are hence not always appropriate in an archaeological context. Results should therefore "be considered low-resolution (i.e., blurry and blunted) versions of reality" (Day-Lewis et al. 2006).

2.7   Earth resistance anomalies

When measuring earth resistance, either in a grid or along a profile, it would be best if recorded data reflected the shape of the buried features. For example, it would be convenient if a profile measured over a buried ditch would show a single dip over the centre of the ditch. However, due to the complicated paths that the electrical current will take around buried features, the resulting distribution of voltages on the surface can be very complex and plotting the apparent resistivity values along a profile may hence exhibit unexpected results. Figure 7 shows calculated traces over a localised archaeological feature (approximated as a sphere) with the same size as the twin-probe

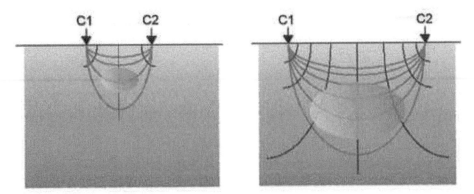

Figure 7. Earth resistance anomaly for a buried spherical conductor (radius r), measured with a twin-probe array. All length measurements are relative to the radius of the sphere: the electrode separation is a = A • r = 2r, the depth to the centre is z = Z • r, the lateral distance from the centre is x = X • r. The resistivity of the matrix in which the feature is embedded is ρ_l.

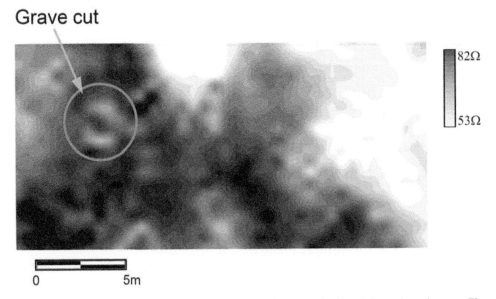

Figure 8. Earth resistance data over a Medieval graveyard, measured with a 0.5 m twin-probe array. The anomaly produced by a shallow and narrow grave cut is comparable to the theoretical results from Figure 7.

array that is used to measure it (e.g. a 0.5 m wide grave cut) for various depths of the feature. It is important to realise that the sequence of 'low-high-low' data in this profile is caused by one single feature and not by three separate entities. Data in Figure 8 show results from an earth resistance survey over suspected Mediaeval graves and the theoretically predicted effect can clearly be observed in the outlined area which is hence interpreted as one single grave cut. It is important to realise that such diagrams are not images of the subsurface features but are representations of the collected data, with all the inherent complications of geophysical signatures. It is therefore useful to clearly distinguish between the geophysical anomalies as manifest in the data and the causative archaeological features buried in the ground. Knowledge of geophysical signatures has to be combined with the relevant historical context for a successful interpretation of results. In the above example it was known that narrow graves were suspected in the investigated area and measured anomalies could hence be interpreted with greater confidence.

3 MAGNETIC METHODS OF SURVEY

3.1 *Magnetic field of the earth*

Magnetic methods of archaeological prospection have proven to be immensely successful because many archaeological features show a contrast in magnetic properties compared to the surrounding material. Underpinning the detection of such a contrast is the earth's magnetic field. Research on the causes of this field are still ongoing but it is most likely created by the movement of ions and electrons at the interface between the liquid core and the solid mantle deep inside the earth. In a first approximation, the resulting field can be portrayed as if it were produced by a large magnet situated at the earth's centre, its magnetic south pole pointing towards the northern hemisphere and hence attracting the northern tip of compass needles.

3.1.2 *Magnetism and archaeology*

Magnetism is usually described by 'magnetic fields', which at each point indicate how strongly a compass needle would be pulled and in which direction. All magnetism is caused by the movement of electrical charges. On the macroscopic scale this can be in the form of electric currents flowing through a coil while on a microscopic scale it is due to spinning and orbiting electrons and protons. Every atom therefore has a 'magnetic moment' that can be visualised as a small compass needle, its strength depending on the particular material.

3.1.3 *Induced magnetism*

By applying an external magnetic field (e.g. the earth's field), the elementary magnets of a feature become partly aligned with the external field and will therefore enhance it. The ease of alignment determines the strength of this enhancement and is described by the 'magnetic susceptibility' of a material. The higher the magnetic susceptibility, the bigger will be the magnetisation that is created by the overall alignment of the magnetic moments in a feature. A larger feature will create a bigger overall magnetic moment and to account for this, magnetic susceptibility is usually quoted with regards to the amount of the magnetic material measured, either its mass or its volume. In the SI system of units, mass specific susceptibility (χ) is quoted in m3 / kg . Volume specific susceptibility (κ) has no units in this system but to remind readers of this fact the expression '(SI)' is sometimes appended to a measurement (e.g. "$\kappa = 2 \times 10^{-5}$ (SI)").

Human habitation can lead to an increase of magnetic susceptibility, forming a contrast with the surrounding soil matrix, which is the reason why many archaeological features can be detected with magnetic methods. There are five main pathways through which soil magnetic susceptibility can be enhanced.

1. Heating. Soils often contain weakly magnetic iron oxides (e.g. haematite) that can be converted to more magnetic forms (e.g. magnetite or maghaemite) through heating in reducing conditions, in the presence of organic matter. The temperature at which this process starts is not well defined and values between 150°C and 570°C have been reported, with lower temperatures requiring longer exposure (Linford & Canti 2001; Maki et al. 2006). This pathway was first discussed by Le Borgne (Le Borgne 1955) and is often attributed to him.
2. Microbially mediated. Microbes thriving in rich organic deposits can change soil conditions sufficiently to trigger the conversion of weakly magnetic iron oxides to more magnetic forms (Linford 2004). Historically, this is referred to as 'fermentation' although strictly speaking methanogenesis is not required for this biogenic pathway (Weston 2002).
3. Magnetotactic bacteria. Some bacteria actively create intra-cellular crystalline magnetite to navigate in the earth's magnetic field (Fassbinder et al. 1990). These magnetic crystals remain in the soil even when the magnetotactic bacteria die and lead to an enhanced magnetic susceptibility.

4. Incorporated magnetic material. Magnetic enhancement of topsoil is also caused by the addition of magnetic material, for example broken pottery or brick fragments (Weston 2002). Such material is often found as discard or rubbish in archaeological middens and has been spread on arable fields with other manure, mainly in Medieval times.
5. Pedogenesis. Enhancement of soil magnetic susceptibility also occurs during soil formation processes, even without human influence. Maher and Taylor (1988) reported the formation of ultra-fine grained magnetite in soil despite the absence of any microorganisms.

The first three pathways rely on the availability of organic matter, which is usually more abundant in the upper soil horizon than in the subsoil, hence creating a magnetic differentiation of topsoil and subsoil. In addition, anthropogenic input further enhances these conditions (either through fire or deposition of organic material, like middens), sometimes allowing the identification of settlement areas through magnetic susceptibility mapping, or the differentiation of buried land surfaces (e.g. covered by non-magnetic windblown or alluvia deposits) from the magnetic stratigraphy. Archaeological environments with rich organic deposits include, for example, middens and decayed wooden posts. Fassbinder demonstrated that magnetic anomalies of post holes that were apparent in high-sensitivity magnetometer surveys are attributable to magnetotactic bacteria (Fassbinder & Irlinger 1994; Fassbinder & Stanjek 1993). Metalworking remains, for example hammerscale and slag, also become incorporated into soil layers and can greatly increase the magnetisation. Unfortunately, iron and steel fragments broken or fallen from modern farming machinery can also create undesirable magnetic anomalies in survey data. Whenever a cut archaeological feature (e.g. a ditch or a pit) is filled with magnetically enhanced soil the magnetic susceptibility contrast of the feature with the surrounding soil or sediment matrix makes it magnetically detectable.

3.1.4 *Remanent magnetism*
Induced magnetisation would disappear if the earth's magnetic field ceased, and it will follow any slow changes in the direction of the earth's field. In contrast, remanent magnetisation is created once and stays fixed in a material afterwards. For example, thermoremanent magnetisation is caused by heating a sample to about 650°C so that all elementary magnets become very mobile and align easily with the ambient earth's magnetic field. On subsequent cooling, this state of alignment is 'frozen' and a strong magnet is created. Remanent magnetisation will not change even if the earth's magnetic field alters its direction, as it has done in the past. By comparing the 'frozen' remanent magnetisation with calibration curves for ancient directions of the earth's field, the date for the last heating event can be established. This forms the basis for 'archaeomagnetic dating'.

Most soil features that were exposed to high temperatures during heating (e.g. hearths, kilns, kiln-fired bricks) or burning (e.g. burnt walls or houses) have acquired remanent magnetisation and exhibit a magnetic contrast.

3.2 *Magnetic susceptibility surveys*

Since human habitation can enhance magnetic susceptibility, mapping this material property can provide useful archaeological information (Linford 1994). Collecting soil samples and measuring their magnetic susceptibility in a laboratory provides accurate data but is time consuming. More convenient are measurements directly from the surface, using appropriate field instruments. The most commonly used instrument, the "MS2 Field Coil", has a penetration depth of only about 0.1 m (Lecoanet et al. 1999) but allows the rapid assessment of topsoil magnetic susceptibility. Areas of interest can either be mapped in detail (e.g. with sampling intervals of 1 m) to reveal individual archaeological features (e.g. charcoal burning areas (Schmidt 2007)), or with sparser sampling (e.g. 5–20 m) to obtain an overview of the magnetic susceptibility variation over a larger area and to identify 'hotspots' that can later be investigated with higher spatial resolution. Since the magnetic susceptibility of soil can vary considerably even over a short distance, it is advisable not to estimate (i.e. interpolate) values for areas between actual easurements. A display of the data as 'symbol plots' is often the most appropriate representation (Fig. 9).

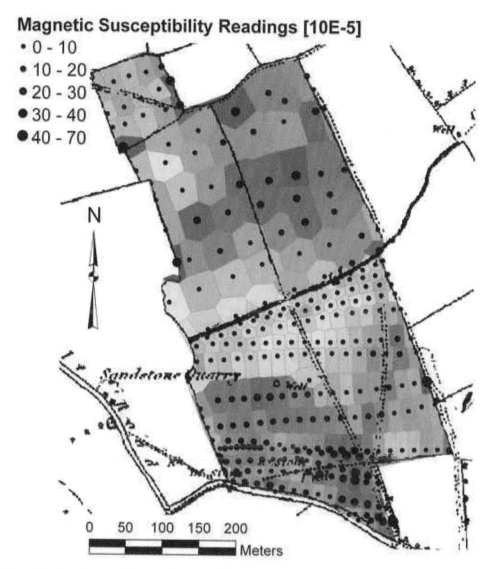

Magnetic Susceptibility Readings [10E-5]

- 0 - 10
- 10 - 20
- 20 - 30
- 30 - 40
- 40 - 70

N

0 50 100 150 200 Meters

Figure 9.   Sparsely sampled magnetic susceptibility survey. Each individual measurement is represented by a symbol of arying size ('symbol plot'), which is superimposed on the representation of the same data as Voronoi cells.

3.3 *Magnetic anomalies*

Buried archaeological features with a contrast in either induced or remanent magnetisation will act like bar magnets and create a magnetic field around them, the so-called 'anomaly field'. This anomaly combines with the earth's magnetic field to form the 'total field' that can be measured at the surface with a magnetometer and is usually expressed in 'Tesla', or more conveniently in 'nano Tesla' (1 nT = 10^{-9} T). The strength of the earth's magnetic field is about 30,000–50,000 nT. Mapping the magnetic field and its anomalies hence produces data plots that can be used to identify buried archaeological features. As with earth resistance surveys, the data plots show particular characteristics and are not a direct image of the buried remains. A localised archaeological feature can often be represented by a 'magnetic dipole' (i.e. a very short bar magnet) for which the magnetic field can easily be calculated. Figure 10 shows the magnetic anomaly that would be measured with a fluxgate gradiometer at 70° latitude, carried from south to north over the centre of a localised feature (the signal recorded by a caesium gradiometer at this latitude would look similar).

This anomaly has two important characteristics:

1. The positive peak is slightly shifted to the south of the buried feature.
2. To the north of the feature is a pronounced negative trough in the data.

The additional negative data are very characteristic of magnetic anomalies. Figure 11 shows the survey of an Iron Age enclosure where the positive signal of the ditch is accompanied by a fringe of negative data. To interpret these data correctly (i.e. as a single feature), it is important to take the signature of magnetic anomalies into account. The large circular pit in the SW of the enclosure also shows the effect of a halo of negative data, mainly to its north.

The intensity of the magnetic anomaly depends both on the strength of the magnetisation contrast and, very strongly, on the depth of the feature. It is hence impossible to use the signal amplitude for an estimation of a features' burial depths. However, the signal width is independent of the magnetisation contrast and can hence be used for its assessment (Schmidt & Marshall 1997): deep features (e.g. geological ore bodies) create broad anomalies, while shallow features (e.g. buried archaeological remains, modern ferrous parts fallen off a tractor) cause narrowly focussed anomalies.

Figure 10.   Magnetic anomaly over a localised archaeological feature.

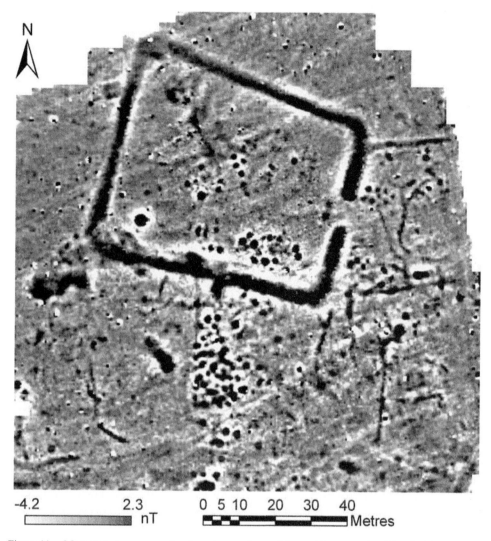

N

-4.2　　　　2.3　　　0 5 10　20　30　40
▭▬ nT　　▣▬▬▬▬▬▬ Metres

Figure 11.　Magnetometer survey of an Iron-Age enclosure. The positive anomaly of the ditch is accompa-nied by a fringe of negative data. Similarly, the large pit in the SW corner of the enclosure has a negative halo, most prominently to the north. Survey data courtesy of Dr Alistair Marshall

3.4　*Magnetometer measurements*

3.4.1　*Sensor types*

A magnetometer is a sensor that measures the 'total magnetic field', which is the field resulting from the combination of an anomaly with the ambient earth's field. The simplest sensor would be a compass needle suspended on a thread, but it is not sufficiently sensitive to measure typical archae-ological anomalies, which are often in the order of a few nT. Instead, a variety of sophisticated sensors are available and are discussed elsewhere (e.g. (Gaffney & Gater 2003)). As mentioned before, the magnetic field is characterised by its direction and strength and magnetometers are usually classified according to whether they measure the former or the latter. Fluxgate sensors, for example normally only measure the vertical component of the total field, while Caesium sensors measure its strength. The former therefore are directionally sensitive and have to be carried very consistently, whereas the latter have a great tolerance to changes in survey direction.

3.4.2 *Sensor arrangements*

The total magnetic field measured by a sensor is composed of the archaeological anomalies, fields created by underlying geological bodies and the earth's magnetic field. In a single-sensor survey, it is hence impossible to distinguish which of these contributions has caused a change in the recorded data. This is particularly problematic as the earth's field varies slightly throughout the day ('diurnal variations') and may even show strong and rapid changes ('magnetic storms'). These variations are caused by charged particles emitted by the sun, the 'solar wind', which interfere with the earth's magnetic field. Diurnal variations are simply caused by the greater proximity to the sun during daytime.

To determine the cause for a change in recorded results it is hence necessary to monitor the earth's field with a second sensor. This is most commonly achieved with a 'vertical gradiometer' arrangement in which two sensors are mounted on top of each other and the difference between them is recorded in the data logger. This eliminates all effects of the earth's magnetic field as the two sensors measure identical signals from the earth's field and the gradiometer reading is hence zero in the absence of an anomaly. Even geological anomalies are suppressed if their sources are far enough away. Gradiometers are hence sometimes referred to as 'intrinsic highpass filters'.

4 INTERPRETATION OF GEOPHYSICAL SURVEYS

It was shown for earth resistance and magnetometer measurements that recorded data are not simply an image of buried features but that resulting plots are strongly influenced by the geophysical signature of the measurement technique used. After data collection, interpretation of the results is necessary to relate geophysical measurements to possible archaeological features in the ground. This requires an understanding of the geophysical nature of the data as well as archaeological knowledge and an appreciation of the historical context. Combining different data sources (e.g. geophysical, remote sensing, aerial photography, old maps, historical texts) provides archaeological geophysicists with the plethora of data that is necessary to arrive at a meaningful archaeological interpretation of geophysical results. Similarly, geophysical prospection data alone cannot provide dating evidence for a feature. However, if geophysical data are combined with information on the morphology of detected archaeological structures (e.g. the typical trapezoidal shape of an Iron-Age enclosures, Fig. 11) or if a sequence of intersecting anomalies can be established (Schmidt & Fazeli 2007), broad dates may sometimes be estimated.

5 USEFUL RESOURCES

Archaeological Prospection Resources (www.bradford.ac.uk/archsci/subject/archpros.htm)
Journal Archaeological Prospection (www3.interscience.wiley.com/cgi-bin/jhome/15126)
International Society for Archaeological Prospection (www.archprospection.org)
MSc Archaeological Prospection (www.bradford.ac.uk/archsci/msc_ap.htm)

REFERENCES

Clark, A.J. 1990. Seeing Beneath the Soil, Prospecting methods in archaeology. London: Batsford.
Day-Lewis, F.D., E.A. White, C.D. Johnson, J.W. Lane (Jr.) & M. Belaval 2006. Continuous resistivity profiling to delineate submarine groundwater discharge—examples and limitations. The Leading Edge 25(6): 724–728.
Fassbinder, J.W.E. & W.E. Irlinger 1994. Aerial and magnetic prospection of an 11–13th century motte in Bavaria. Archaeological Prospection 1: 65–69.

Fassbinder, J.W.E. & H. Stanjek 1993. Occurrence of bacterial magnetite in soils from archaeological sites. Archaeologia Polona 31: 117–128.

Fassbinder, J.W.E., H. Stanjek & H. Vali 1990. Occurrence of magnetic bacteria in soil. Nature 343: 161–163.

Gaffney, C. & J. Gater 2003. Revealing the Buried Past: Geophysics for Archaeologists. Strout: Tempus Publishing Ltd.

Griffiths, D.H. & R.D. Barker 1994. Electrical Imaging in Archaeology. Journal of Archaeological Sciences 21: 153–158.

Le Borgne, E. 1955. Susceptibilité magnétique anormale du sol superficiel. Annales de Géophysique 11: 399–419.

Lecoanet, H., F. Leveque & S. Segura 1999. Magnetic susceptibility in environmental applications: comparison of field probes. Physics of the Earth And Planetary Interiors 115: 191–204.

Linford, N. 1994. Mineral magnetic profiling of archaeological sediments. Archaeological Prospection 1: 37–52.

Linford, N.T. 2004. Magnetic ghosts: mineral magnetic measurements on Roman and Anglo-Saxon graves. Archaeological Prospection 11(3): 167–180.

Linford, N.T. & M.G. Canti 2001. Geophysical evidence for fires in antiquity: preliminary results from an experimental study. Paper given at the EGS XXIV General Assembly in The Hague, April 1999. Archaeological Prospection 8(4): 211–225.

Loke, M.H. & R.D. Barker 1996. Rapid least-squares inversion of apparent resistivity pseudosections by a quasi-Newton method. Geophysical Prospecting 44: 131–152.

Maher, B.A. & R.M. Taylor 1988. Formation of ultrafine-grained magnetite in soils. Nature 336: 368–371.

Maki, D., J.A. Homburg & S.D. Brosowske 2006. Thermally activated mineralogical transformations in archaeological hearths: inversion from maghemite gamma-Fe2O4 phase to haematite alpha-Fe2O4 form. Archaeological Prospection 13(3): 207–227.

Matthews, R. 2003. The archaeology of Mesopotamia: theories and approaches. London, New York: Routledge.

Schmidt, A. 2007. Archaeology, magnetic methods. In D. Gubbins and E. Herrero-Bervera (eds) Encyclopedia of Geomagnetism and Paleomagnetism. Encyclopedia of Earth Sciences Series Heidelberg, New York: Springer.

Schmidt, A. & H. Fazeli 2007. Tepe Ghabristan: A Chalcolithic Tell Buried in Alluvium. Archaeological Prospection 14(1): 38–46.

Schmidt, A. & A. Marshall 1997. Impact of resolution on the interpretation of archaeological prospection data. In A. Sinclair, E. Slater and J. Gowlett (eds) Archaeological Sciences 1995: 343–348. Oxford: Oxbow Books.

Scollar, I., A. Tabbagh, A. Hesse & I. Herzog (eds) 1990. Archaeological Prospecting and Remote Sensing. Cambridge: Cambridge University Press.

Weston, D.G. 2002. Soil and susceptibility: aspects of thermally induced magnetism within the dynamic pedological system. Archaeological Prospection 9(4): 207–215.

Seeing the Unseen – Campana & Piro (eds)
© 2009 Taylor & Francis Group, London, ISBN 978-0-415-44721-8

Electrical resistivity tomography: A flexible technique in solving problems of archaeological research

G.N. Tsokas, P.I. Tsourlos & N. Papadopoulos
Exploration Geophysics Lab., Department of Geophysics, Aristotle University of Thessaloniki, Thessaloniki, Greece

ABSTRACT: The need for information about the depth extend of the buried targets was one of the main reasons for the development of the electrical resistivity tomography (ERT). The term implies that automated multiplexers are used for data acquisition which are subsequently inverted by some mathematical scheme. The present work presents the merits of resistance mapping and also its drawbacks which created the need to apply ERT in archeological prospection. An algorithm for inverting resistivity data is presented in brief both in its 2-D and 3-D versions. Examples of 2-D ERTs demonstrate the potential of the method and also its applicability and efficiency in treating problems of the archaeological research. The employment of the method for large scale surveys is discussed and a practical scheme to collect tomographic data along the resistance survey is presented. The scheme is based on the modification of a widely used commercial system. Its use reduces significantly the effort and time required to carry out a large scale ERT survey. The pseudo 3-D images obtained by combining parallel tomographic profiles are compared against the full 3-D inversion of the data. A practical guide is presented for the production of acceptable and reliable full 3-D tomographic images when 2-D data are used which have been collected along parallel traverses.

1 RESISTIVITY SURVEYING—THE NEED FOR 2D PROSPECTING TECHNIQUES

Resistivity surveying, in general, involves the introduction of DC electrical current into the ground via two metal probes (source and sink) and the measurement, at varying positions, of the potential differences (via two other probes), with each reading providing information about the interaction of the current with the three-dimensional variations of the resistivity of the subsurface. The method is commonly used in archaeological prospection due to several reasons:

- It tends to pick up clear responses from most of the targets that are commonly encountered in archaeological sites.
- The instrumentation is of low cost and relatively easy to use.
- Processing is relatively straight forward and nowadays easily performed.
- The main aim is to convert the resistance or the resistivity distribution into a form that is more easily interpretable and understood by non experts, in other words, to have a precise image that resembles the plan view of the concealed antiquities (Scollar et al., 1986; 1990). This approximates the plan view that would have been drawn if excavation had taken place and the plan view of the unearthed ruins was drawn. Therefore, interpretation can be carried out by all involved parts, i.e. geophysicists, archaeologists, land developers, etc.

There are various positional combinations of the four electrodes and several measuring configurations that can be used. In archaeological prospection the resistivity surveying is traditionally conducted by carrying out measurements at grid points to construct a two-dimensional anomaly "map" of the area—each line of data is referred to as a profile. During the survey the spacing of the probes remains stable and thus no variable depth information is available.

Systems involving the movement of only two electrodes offer an obvious operational advantage when used for profiling. Hence, systems like the Schlumberger array (in gradient mode) and the pole-pole electrode configurations have become very popular in archaeological geophysics (Clark, 1986). The square array (Habberjam and Watkins, 1967; Habberjam, 1979) employed also for archaeological exploration work. Clark (1986; 1990) claimed that some advantages could be achieved if small square spacings are used. The most popular configuration, though, is the so called "twin-probe" array which is a modified version of the pole-pole array (Aspinall and Lynam, 1970).

Nowadays, resistivity profiling in archaeological sites has become easy to implement by the introduction of light resistivity meters that are mounted on a frame which allow a relatively fast and automated probing procedure. Instruments such as the RM4TM–DL10TM or the more advanced RM15TM developed by Geoscan Research Ltd allow the application of several electrode configurations and the automatic logging of the measurements. In (Fig. 1) a sketch of the typical field procedure involving mobile resistivity instrument is depicted.

Further, special towed resistivity meters with automatic recording systems have been developed. The ancestor of these systems, the so called RATEAU (Hesse et al., 1986) developed as far back as the beginning of 80 ties. It consisted of four steel wheels, which act as a square frame and is towed by a vehicle providing one reading per 10 cm at a cruising speed of 15 miles/hour. Later on, more sophisticated and advanced systems were developed (Panissod et al., 1997; 1998a; 1998b, Walker, 2005).

A plethora of successful applications of the method exist in the literature. An example is presented in (Fig. 2) extracted from the work of Tsokas et al. (1994) at the archaeological site of Europos (Region of Macedonia, N. Greece). In this particular example, the twin probe arrangement was employed setting the mobile probes distance to 0.5 m and the profile spacing at 1 m and measuring stepwise at 1 m intervals. The produced image-like map reveals the ancient urban

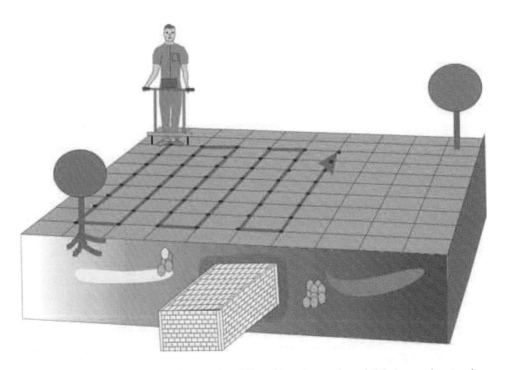

Figure 1. A sketch of the typical field procedure followed in resistance (or resistivity) surveying at archaeological sites.

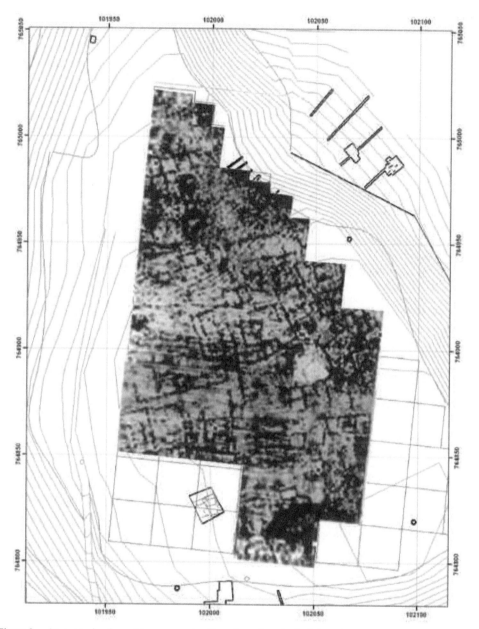

Figure 2. An example of a resistance mapping from the archaeological site of Europos (N. Greece). The produced image-like map reveals ruins of foundations at the Acropolis of the ancient city (Tsokas et al. 1994).

structure of the acropolis of the city. The high resistivity anomalies (dark) correspond to the foundations of buildings. Yet, in the map, there is a second network of less dark (resistive) anomalies which, by test excavations, was found out that it corresponds to a deeper Classical phase which underlies the Hellenistic and Roman ones. With this example, the limitations of resistance mapping are demonstrated. The main demerit is of the method is that it does not provide any direct or unambiguous depth information.

Attempts have been made to process resistivity profiling data and to extract the exact extend of the existing features. These are based on inverse filtering (Tsokas and Tsourlos, 1997), on filtering using the Fourier transform (Bernabini et al., 1988; Brizzolari et al., 1986; 1992, or on pattern recognition schemes (Schollar et al., 1986).

Overall, resistivity profiling can provide useful site information, but there are a number of major deficiencies (Szymanski and Tsourlos, 1993).

- The field data does not provide any direct depth information.
- Quantitatively different maps will be obtained with different array configurations.
- The image-like nature of the data can be misleading, and careful interpretation is required.
- The data represent only an indirect effect of the earth's resistivity variations.

To overcome some of the limitations associated to the resistivity profiling, a 2-D surveying approach was suggested for archaeology (Noel and Walker, 1991), which produces data containing information about both the lateral and vertical variations of the earth's electrical properties. This type of approaches have evolved from the restricted requirements of larger-scale survey applications: a series of electrodes is laid out on to the ground and by using a multi-core cable and a multiplexing unit a series of profiles with varying electrode spacing over the same area is obtained automatically (Griffiths et al., 1990; Dahlin, 1992).

This type of data is difficult to interpret directly. In fact, the interpretation method falls into a class of approaches known as 'inverse imaging'. The aim is to extract information from the surface measurements about structural variations some distance below the surface. Treatment of the data-sets with the so-called 2-D non-linear inversion schemes is necessary. Several non-linear resistivity inversion algorithms which can handle ill-conditioning have been reported in the literature, mainly based on Marquadt's method (Trip et al., 1984; Smith and Vozoff, 1984; Pelton et al., 1978) or Occam's inversion (Constable et al., 1987; Tsourlos, 1995; Loke and Barker, 1996).

The development of automatically multiplexed electrode arrangements and automatic measuring systems facilitated the acquisition of large numbers of measurements in a limited time. Further, the advent of fast computers allowed the development of the automated resistivity inversion schemes which aim to construct an estimate of a subsurface resistivity distribution which is consistent with the experimental data. Among others, the smoothness constraint inversion (Constable et al., 1987) has become the most popular for interpreting 2-D resistivity data since it produces a simplified subsurface resistivity model which is a reasonable representation of the subsurface and at the same time guarantees inversion stability. The combination of the techniques allowing swift data collection along with the development of inversion schemes was termed Electrical Resistivity Tomography (ERT). Nowadays, the ERT technique is considered as one of the most significant new geophysical methodologies to emerge in the last decade. ERT is now widely used for environmental, archaeological, engineering and groundwater resource mapping.

2   AN INVERSION SCHEME

All inversions performed in the framework of this paper are 2.5 or 3D. They follow the scheme to be described in this paragraph and they were implemented by custom built software developed in the Exploration Geophysics Lab. of the Aristotle University of Thessaloniki (Greece).

In 2.5-D modelling the change in resistivity is considered to be two dimensional but the current flow pattern is a three dimensional one. In other words, the measured values correspond to a three dimensional subsurface where the resistivity is allowed to vary in only two dimensions and remains constant in the strike direction (Tsourlos et al., 1999). To include the potential variability in the strike (y) direction a cosine Fourier transformation is applied.

The FEM treats the problem by discretizing the earth into homogeneous triangular (in this case) regions called elements. The potential within each element is approximated by a simple interpolation function (basis function). In order to minimize the error between the approximated and real potential, the Galerkin minimization criterion is applied. After applying the Galerkin minimization scheme to every element, the individual element equations can be assembled in to one global system which has the form

$$KA = F, \qquad (1)$$

where A is the unknown transformed nodal potential vector, F is the vector describing the sources and K is a matrix which is related to the nodal coordinates. By applying the homogeneous Dirichlet and Newman boundary conditions the system of equation (1) is being solved and the transformed nodal potential is obtained. After solving equation (1) for several wavenumbers the total potential is recovered by applying the inverse Fourier transform. Since the nodal potential is known, point to point potential differences and apparent resistivities are easily obtained.

The approach in the 3-D case is very similar to the 2.5-D one, however here the problem is treated in 3-D so there is no need to use a Fourier transformation. The solution of the differential equation that governs the flow of the electrical current in the ground (Poisson equation) is sought by subdividing the area into hexahedral elements. After applying the Galerkin minimization scheme to every element, the individual element equations are assembled to form the global system which has the form of equation 1.

The final step is to solve the system of equations: for the 3D case which in general involves large systems of equations an iterative technique is preferable. In this work the conjugate gradient method for solving large sparse linear systems is used.

The inversion core algorithm is the same for both the 2D and 3D case. A non-linear smoothness constrained inversion algorithm was used (Sasaki, 1992). The inversion is iterative and the resistivity x_{k+1} at the $k + 1$th iteration is given by

$$x_{k+1} = x_k + dx_k = x_k + [(W_d J_k)^T (W_d J_k) + \mu_\kappa (C_x^T C_x + C_z^T C_z)]^{-1} (W_d J_k)^T W_d dy_k \qquad (2)$$

where C_x, C_z are matrices which describe the smoothness pattern of the model in the x and z axes respectively (de Groot-Hedlin and Constable, 1990), dy_k is the vector of differences between the observed data dobs and the modeled data dkcalc (calculated using the forward modeling technique 2.5D or 3D), J_k and J_k and μ_κ is Jacobian matrix estimate and the Lagrangian multiplier respectively for the kth iteration, W_d is the diagonal matrix of the data variances, and T denotes the transpose. The adjoint equation approach (McGillivray & Oldenburg, 1990) was incorporated into the FEM scheme in order to calculate the Jacobian matrix J (Tsourlos, 1995). Depending on the dimensions of our problem the Jacobian matrix is calculated either by the 2.5D or the 3D forward solver.

3 EXAMPLES OF 2-D ERTS

(Fig. 3) shows the plane view of the ERTs conducted in Montelibreti near Rome (Piro et al., 2001), in an area where a Sabin necropolis exists. The underground tombs are chambers carved into volcanic tuff, usually associated with a corridor (dromos). Two of them are buried in the investigated area and their outline is also depicted in (Fig. 3). The dipole-dipole array was employed and the spacing between adjacent ERTs was set to 1 m. Each apparent resistivity reading was obtained using 0.5 m long dipoles.

The result of the inversion of the resistivity data for the tomography having the code Ty4 is shown in (Fig. 4). It clearly seen that the monument has created a pronounce high resistivity anomaly due to the high resistivity masonry material and empty space. This anomaly defines more or less both the vertical and lateral extent of the monument. Further, the dromos is also picked up and imaged. The advantage of carrying out parallel ERTs arranged in a regular grid is that the separate inversions, which is the outcome of the survey, can be combined to yield a quasi3-D image. However, this image is a pseudo 3-D depiction of the subsurface resistivity distribution

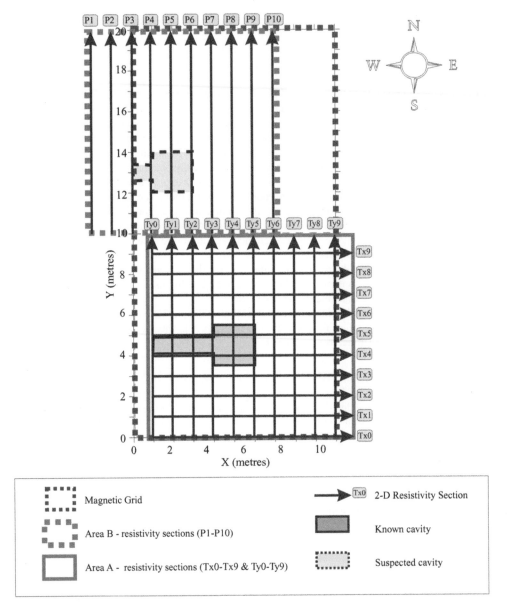

Figure 3. Lay out of the tomographies carried out at the Sabin necropolis of Montelibreti, near Rome (Piro et al., 2001). The outlines of two buried tombs are also shown. The area of the tombs is shaded.

since the inversion is not done in the full 3-D context. An example of such a combination is shown in Plate 5. It has been taken from the survey performed over a drain in Fountains Abbey (England) (Tsourlos, 1995; Coppack et al., 1992). The image is a very good reconstruction of the actual shape of the drain and shows also its dimensions very precisely. In fact, the 2-D images were combined to give a quasi3-D impression.

In (Fig. 5), another example of combined 2-D inverted subsurface images over the buried ruins of an ancient road is presented (Tsokas et al., 1999), in the same context as before. The example

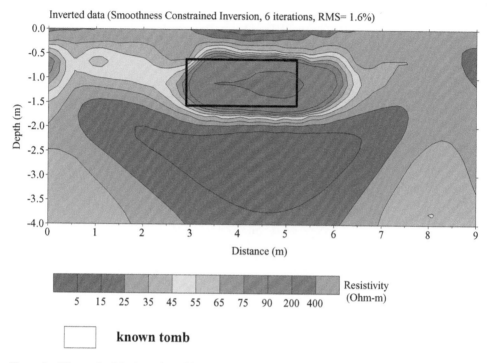

Figure 4.   The result of the inversion of the data for the ERT which has the code Ty4 in Figure 3. The outline of the buried monument has been also drawn on the figure (Piro et al., 2001).

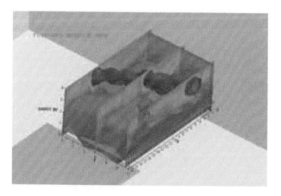

Plate 5.   An image produced by inverting resistivity data over a drain at Fountains Abbey (England) (Tsourlos, 1995; Coppack et al., 1992). (See colour plate section)

is from the site hosting the ruins of the first capital of the Macedonian kingdom in N. Greece (ancient Aegae; modern Vergina). Five tomographies were combined, having the lay out shown in (Fig. 6) superimposed on the result of the resistance mapping in the area. A very good image of the ancient road had been produced either by the tomographies or by resistance mapping. The cylindrical appearance of the ancient road is due to the contouring algorithm.

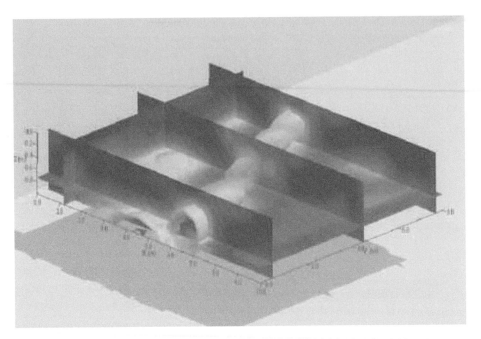

Figure 5.   An example of combined 2-D inverted subsurface images over a buried road in ancient Aegae (Vergina, Region of Macedonia, N.Greece): a pseudo 3-D subsurface image is created (Tsokas et al., 1999).

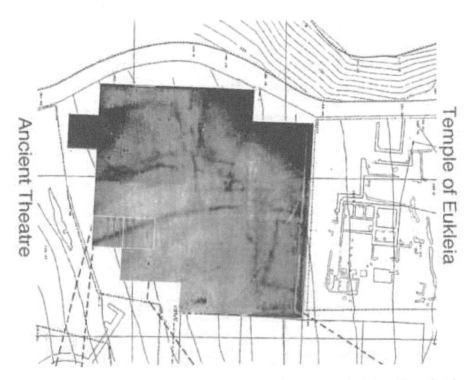

Figure 6.   An example of combined 2-D inverted subsurface images over a buried road in ancient Aegae (Vergina, Region of Macedonia, N.Greece): a pseudo 3-D subsurface image is created (Tsokas et al., 1999).

4   LARGE SCALE ERT SURVEYS FOR ARCHAEOLOGICAL PROSPECTION

For large areal coverage, the conduct of a survey by means of ERT is more time and effort consuming than carrying out resistance mapping. In this respect, the ERT technique is employed in selected limited areas that are pinpointed by the mapping to produce informative 2D/3D subsurface images of the "true" resistivity of the most important targets.

However, there are certain circumstances where either the resistance mapping is inadequate or the field conditions render the ERT technique as the proper solution. Of course the discussion concerns only the cases where a sufficient resistivity contrast exists between the target and the hosting medium but the prerequisite property contrast needs to hold for the successful application of any geophysical technique. Otherwise, perhaps some other method would be the best solution for exploring the subsurface.

During the preparations for the Athens 2004 Olympic games, the writers were asked to perform archaeological prospection in the place where the new rowing center was to be constructed. The request had an urgent character making necessary the conduct of the survey in mid summer of 2001. However, the ground of the site is clay, a fact that rendered the resistance mapping as impossible because of the high contact resistances and the hardness of the topsoil. Further, tests of magnetic prospection proved that the method could not be applied because no sufficient magnetic contrast existed between the targets and the surrounding medium. GPR also failed because of the low resistivity of the clay subsoil. Therefore, ERTs were carried out by hammering the electrodes into the ground and reducing the contact resistances by pouring salty water.

A large scale geophysical survey was carried out at the foothills of a topographic table hosting the ruins of the ancient city "Trikorinthos", in the area of Marathon north of Athens. The explored site assumed to host remnants of ancient constructions belonging to the ancient city. Although the explored area is associated with the final phase of the famous Marathon battle, it did not expected to bear any signs of that event.

A total area of about an hectare (10000 m^2) was covered by resistivity tomographies. The distance between the measured electrical profiles was 0.6 to 2.4 m, resulting into a large number of tomographic data. Such an extensive application of electrical tomographic imaging in archaeological site investigation rendered the survey as one of the largest appearing in literature. The Schlumberger and pole–dipole arrays were used, having the electrodes spaced at 0.6 m. The instrumentation allowed to use 42 channels in each spread and resulted to 144 measurements for each tomography. The total number of ERTs conducted was 186.

The interpretation of the tomographic data was carried out using the inversion scheme of Tsoulros (1995). The inverted data was combined to produce depth slices and quasi3-D images. The result for a particular bit of land north of the rowing channel is shown in (Fig. 7). The depth slice at 0,45 m shown along with two quasi 3-D views. The whole operation resulted in detecting and mapping various concealed structures, some of which were verified by the subsequent excavation. In the quasi 3D images in particular a high resistivity feature corresponding to an ancient road is shown. Therefore, the ancient use of the area was revealed by the combined result of the geophysical investigations and digging.

5   A SIMPLE SYSTEM TO PERFORM ERT ALONG WITH RESISTANCE MAPPING

A new instrumentation technique to collect tomographic resistivity data in a relatively small amount of time using the RM15TM of Geoscan Research Ltd has been proposed by Papadopoulos et al. (2006). The instrument was appropriately configured using some other units provided by the manufacturing company. Then it was programmed to conduct the survey with the pole-pole array, by performing a small resistivity sounding at the nodes of a grid. Usually, the grid is the common 20×20 m^2 cell which is used for conventional geophysical prospection.

Five probes were placed on the frame as shown in (Fig. 8). One of them was always used as the current electrode A, while the rest four probes (M1, M2, M3, and M4) were used to measure the

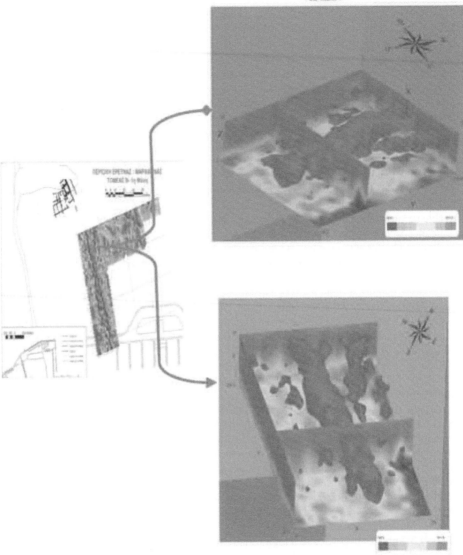

Figure 7.   (Left): Depth slice at 0.45 m obtained by inversion of the resistivity data of parallel tomographies. The particular bit of land comprises part of the area explored in Marathon, around the rowing center which was under construction when the survey carried out. The plan view of rowing center is shown at the lower right part of the figure. Revealed ancient ruins are depicted by black solid lines at the upper part. (Right): Two quasi 3-D images of the subsurface resistivity distribution corresponding to different view angles depicting the remnants of an ancient road. The part of the subsurface imaged corresponds to the area indicated by the arrows.

potential. The distance between the pair of electrodes A-M1, A-M2, A-M3 and A-M4 was 0.5 m, 1 m, 1.5 m and 2 m respectively. Extra cables were used to connect the system with the remote probes (B, N), which theoretically have to be at infinity and not close one to the other (Fig. 9).

The parallel to x-axis tomographies are performed by moving the system along the Y-axis in a parallel mode. Accordingly, tomographies parallel to Y-axis are carried out by moving the system

Figure 8. The RM15TM resistivity meter of Geoscan Research Ltd, complemented with various other switches and connectors, is mounted on a frame modified such that 4 readings are taken at every node of the grid. The readings reflect deeper resistivity changes as the distance of the potential prove (Mi) from the current one (A) increases (Papadopoulos et al., 2006).

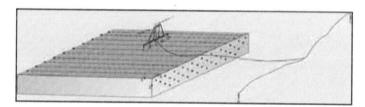

Figure 9. In the particular setting depicted, the system is moving along the Y axis and thus tomographies are performed along X axis. In this manner a small resistivity sounding is performed at every measuring position. The data are then combined to produce pseudosections and inverted to yield the distribution of the real subsurface resistivity (Papadopoulos et al., 2006).

along the X-axis (Fig. 9). The inter-profile spacing can be set to any desirable distance. However, because of the subsequent pseudo 3-D presentation, but also if 3-D processing is to be done, it is better to be 0.5 m, i.e. the length of the basic dipole.

The system described above is capable of conducting ERT survey along with the resistance (or resistivity) mapping. The extra effort needed is much less than conducting tomographies in the conventional manner. Yet, it remains considerable with respect to simple mapping. A demerit of the this measuring strategy is that it can not be applied to all terrains, in particular if they are very rough. However, its ability to collect tomographic data in a swift manner renders the technique as a convenient alternative to the conduction of conventional ERT surveys.

The example of Plate 6 is from the geophysical investigations in Sikyon (Papadopoulos et al., 2006), a site hosting the ruins of a big ancient urban center. It is at the north coast of Peloponnese in Southern Greece. A 10×15 m² grid was explored using the system described above. Tomographic data were collected along both the lateral dimensions, allowing thus comparisons. The data of each tomography were inverted using the scheme described in the second paragraph. Then, they were combined to produce the depth slice of Plate 6.

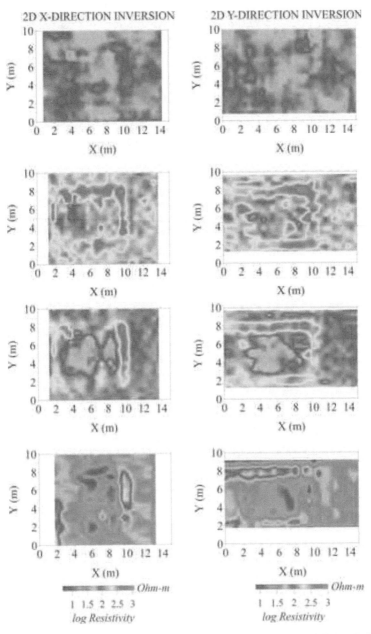

Plate 6. Example of depth slices obtained from tomographic data using the modified RM15™ resistivity meter. The slices are at 0.125, 0.375, 0.625 and 0.875 m depth from top to the bottom. Data from the archaeological site of Sikyon in Greece (Papadopoulos et al., 2006). (See colour plate section)

Essentially, the 2-D electrical resistivity tomography is the collection of a number of profiles in an area, with increasing inner-electrode spacing. Alternatively, it can be seen as a series of successive electrical soundings along a line. In practice, a series of electrodes are placed on the ground surface at equal distances one from the other. Then measurements are taken for varying electrode spacing. In this manner, the horizontal and vertical variations of the subsurface resistivity are recorded.

Nowadays, 2-D resistivity measurements are easily performed and they are of low-cost. Their acquisition is very fast since very efficient and fully automated instruments have been developed. Also, very efficient and reliable two dimensional inversion resistivity algorithms have been developed. As a net result, the 2D measuring mode produces reliable results in cases where the subsurface conditions approximate also a 2-D setting. In other words, wherever, the strike of the concealed structures extends to a practically infinite distance, perpendicular to the ERT profile. However, artefacts will be introduced if the subsurface setting is complex, constituting thus a 3-D resistivity variation. By no means the actual 3-D variation can be imaged precisely following a 2-D approach.

It is clear that correct imaging of the 3-D subsurface structure requires also full three dimensional measurements. Usually, the electrodes are arranged in a rectangular grid having the same spacing in the X and Y direction (Loke and Barker, 1996). In this lay out, current is inserted through a pair of electrode and the potentials differences can be measured in other pair. Due to the reciprocity theorem, if the current and electrode pairs are interchanged, the same apparent resistivity measurement is yielded. Therefore, each pair of current and potential dipoles it measured only once, i.e. when they interchange of electrodes occurs the reading is omitted.

The pole-pole configuration is commonly used for the 3-D surveys. Xu and Noel (1993) showed that the maximum number of independent measurements is nmax = P(P–1)/1, P. For example, if a rectangular grid of 24 electrodes is established on the ground surface, then, using the pole-pole configuration, 276 independent readings can be taken. It is obvious that it is rather time consuming to collect complete sets of three dimensional measurements using typical single channel resistivity meters.

The recently introduced multi-channel automated resistivity meters reduce substantially the overall necessary time to collect 3-D data. Yet, the time needed has not been reduced to a reasonable amount, such as to allow 3-D surveys to be conducted on a routine commercial basis. Therefore, the practice in collecting 3-D data is to record the apparent resistivity by conducting dense parallel 2-D profiles. Ideally, the inter-line spacing should be equal to the basic inter-electrode spacing. The data can be collected either conducting profiles parallel to X axis (X-survey) or parallel to Y-axis (Y-survey). Of course one can conduct profiles parallel to both axes (XY-survey), a fact that would increase the amount of information to be used for the subsequent inversion.

The parallel 2-D tomographies are easily and very quickly interpreted using 2-D inversion algorithms. The results are then combined to create a quasi-3D (x, y, z) image of the subsurface resistivity distribution. This approach is followed for the majority of the tomographic data collected worldwide. However, usually the results suffer from artefacts either due to the fact that 3-D responses are attributed to 2-D structures and/or due to the varying level of misfit that individual 2-D inversions may reach to. It is therefore reasonable to assume that the quality of the quasi-3D images depends on the degree of noise of the measurements plus the deviation of the subsurface conditions from the two dimensionality assumption.

Papadopoulos et al. (2006) investigated the application and the effectiveness of applying full 3-D inversion algorithms in interpreting dense 2-D data collected along parallel profiles. They also examined the differences between the quasi-3D and real-3D images. Their results in both synthetic and real data showed that quasi-3D images suffer from artifacts because of the three dimensionality of the archaeological targets. Only the combination of an XY-survey with 2D inversion schemes can give satisfactory results. However, if 3-D inversion algorithms are used for the data along parallel profiles, the result is satisfactory. In fact, they produced almost identical results for the 3D-X, Y and XY surveys. Further, each one of these results individually considered, proved superior in comparison with 2-D ones. In summary, they illustrated that dense parallel profiles along X or Y direction, interpreted using 3-D algorithms, are adequate to image the buried structures. Another outcome of their study is that the 3-D algorithm used functioned well, even in the presence of very noisy data.

The inversion algorithm used by Papadopoulos et al. (2006) is the one presented in the second paragraph of this paper. Also, they used data sets collected with conventional schemes as well as using the modified RM15TM system presented here.

7 FLAT BASE ELECTRODES

The modern expansion of the urban centres and the extended cover of the ground surface created the need to use geophysical techniques indoors, in paved surfaces, above roads, etc. Reasonably, the Ground penetrating Radar surveys are the most suitable for the case since they are non-destructive and simultaneously the offer survey speed. A decade ago, applying standard geoelectrical methods in such cases was out of any discussion because holes had to be drilled through the material to insert the metal probes. The case was rather unthinkable for archaeological sites and existing monuments since drilling would damage the structures.

However, the resistivity method can be also applied in such environments if special probes are used which enable electrical coupling with the ground. The capacitance electrodes (Kuras, 2002) provide a solution. Another option is the use of the so called "flat base" electrodes, which allow the application of resistivity techniques also in a non destructive manner (Moussa et al., 1977). Carrara et al. (2001) reported the use of copper flat base electrodes to explore the are beneath a

Figure 10.   The type of flat base electrode used by Athanassiou et al. (2007).

Roman mosaic floor. Other applications have been reported by Cosentino and Martorana (2001), Karastathis et al. (2002) and Tsokas et al. (2006).

Athanassiou et al. (2005; 2007) provided some practical guidance and studied the performance of flat base electrodes against the conventional metal probes. They also carried out field tests and investigated their effectiveness. The electrodes used consisted of a copper square flat base, 7×7 cm^2 wide and 1 cm thick. This plate was put on the surface of the surveying area. A thin (diameter of 1 cm) cylindrical copper segment (length of 7 cm) was attached to the flat base part to facilitate handling and provide the means for cable connections (Fig. 10).

A jell was used to establish electrical coupling with the surface of the surveying area. The gel consisted of water, salt and cellulose powder (industrial thickener). Athanassiou et al. (2007) claim that alternatively, conductive gel used in medical applications or even confectionery gel can be used. They carried out tests with these material and they proved equally good as the cellulose powder. Moreover, they proved that instead of flat base electrodes, bentonite mud can be used.

Figure 11.   The line of metal stakes define an ERT profile carried out west of the Parthenon on the Acropolis of Athens.

Extensive use of the later type of electrodes was done by the authors, at the geophysical investigations on top and on the walls of the Acropolis of Athens (Tsokas et al., 2006). An example extracted from this work is presented here. (Fig. 11) shows an ERT profile defined on the ground by the line of metal probes. The inverted data along the interpretation is shown in (Fig. 12).

The same profile was measured once more using simple contacts consisted of small quantity of bentonite mud flat base electrodes. In fact, the nuggets of the bentonite mud were placed at exactly the same positions where previously steel electrodes were inserted into the ground. The electrodes consisted of pure bentonite mud mixed with salty water. The line of the bentonite nuggets is shown in (Fig. 13) and the inferred tomography in (Fig. 14). Evidently, the images inferred using metal stakes and bentonite nuggets are almost identical. In fact, there is a discrepancy of less than 3% between the original data of the two tomographies. However, this is a very small error percentage and it could had been produced even if we had put off the metal stakes and then reinserted them same holes and repeat the measurements. Bentonite electrodes were used on the walls of Acropolis as well.

Tomographies were performed to investigate the condition of the walls and further to investigate the fill in the interior. A bentonite electrode is shown in (Fig. 15) which is stuck on the wall of the monument. Generally, the project had to face unusual technical (mounting the cable on the external wall) and processing challenges (wall-to-surface, cross-wall measuring modes), yet the final results suggest that ERT can be used successfully in this type of surveys.

According to Athanassiou et al. (2007), the flat base electrodes appear to perform well in most environments and surface materials except tarmac. However, the use of flat base or other type of contact electrodes, except metal stakes, certainly extends the range of applicability of ERT. Further, this kind of electrodes can be used in combination with conventional ones.

8 EXAMPLE OF FULL 3-D INVERSION

The 3-D inversion example of Plate 7 is coming from the survey in the centre of Piraeus (the port of Athens). The area explored is nowadays a park very close to the docks. The Wenner-Schlumberger array was employed and 37 parallel tomographies were carried out; each one of them consisting of 21 channels. The inter profile spacing was set to 1.2 m. However, because a pronounced target

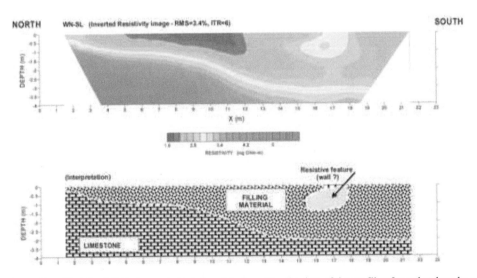

Figure 12. (Top) Resistivity tomography inferred by inverting the data of the profile of metal stakes shown in Figure 10. (Bottom) Interpretation of the distribution of resistivities along the profile.

Figure 13.   The same tomography as in Figure 11 but using bentonite instead of metal stakes.

was detected, 4 extra profiles were measured in the middle of the profiles of the initial grid. Thus, the overall number of tomographies became 41, some of which were spaced 0.6 m apart each from the other. Further, 13 tomographies carried out ranging perpendicular to those of the initial grid. These profiles had 41 channels each spaced at 0.6 m each from the other.

The pseudo 3-D image obtained combining the inversions of resistivity data along the profiles is shown in Plate 7, while the result of full 3-D inversion of all data is in (Fig. 16). Clearly, the artefacts seen in the pseudo 3-D image have been eliminated by the full 3-D inversion. Further, the 3-D inversion is more reliable.

9   CONCLUSIONS

In this work several examples of ERT applications in archaeological prospection demonstrating the flexibility of the technique are presented. ERT inverted images can provide valuable informa-

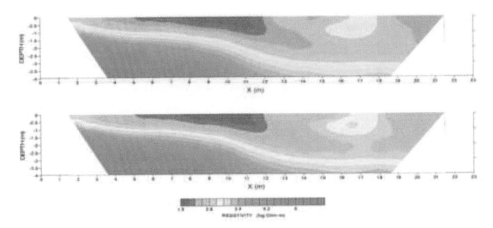

Figure 14. Resistivity tomography inferred by inverting the data of the profile of bentonite electrodes (top) compared to the inversion results obtained using conventional spike electrodes (bottom).

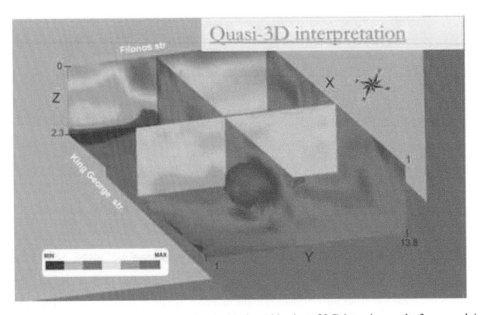

Plate 7. Pseudo 3-D image produced by the graphical combination of 2-D inversion results from a park in the centre Piraeus. A pronounced 3-D target is detected. (See colour plate section)

tion about the exact shape and burial extend of the archaeological targets. Typically ERT surveying is used in selected limited areas that are identified by the widely applied resistivity mapping in an attempt to produce informative 2D/3D subsurface images of the most important targets.

ERT can be performed either in isolated 2D sections or in a dense 2D grid. In the latter case the ERT data can be processed using 2D inversion algorithms and the 2D inversion results can be combined to produce quasi 3D geoelectrical images of the subsurface. Alternatively the dense 2D ERT sections can be processed using 3D inversion schemes to produce full 3D geoelectrical

100

Figure 15.   A bentonite electrode is in the red circle which is stuck on the wall of the Acropolis in Athens. The measuring tape seen was also stuck on the wall to facilitate the correct positioning of the electrodes. Part of the gear of the climbers who put the electrodes is also seen.

images. This approach although more computationally demanding is generally preferable since it produces more reliable subsurface images.

Extra flexibility and survey speed in performing ERT is gained by using alternative tomographic collection strategies by means of modified versions of standard geoelectrical instrumentation widely used in archaeological prospection. The presented configurations allow the collection of dense ERT lines making the full 3D processing of the collecting data an easier task.

Finally, the application of non-spike electrodes in performing geoelcetrical surveys gives the ERT technique the unprecedented ability to be applied in a fully non-destructive manner and thus extend the applications of geoelectrical techniques to environments that wouldn't normally be suitable for resistivity tomography. This is of particular interest to archaeological applications since in some cases standard spike electrodes can damage surface ancient structures.

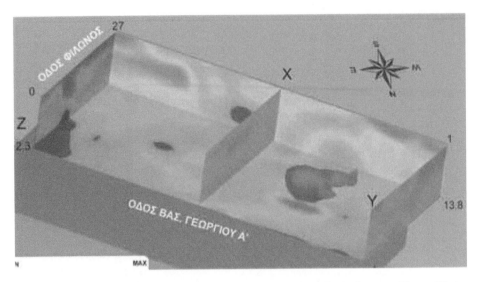

Figure 16.   Results of full 3-D inversion of the data of all measured progiles in the centre Piraeus. The image is compared against that of the pseudo 3-D view of Plate 7.

REFERENCES

Aspinall, A. & Lynam, J.T. 1970 An induced polarization instrument for the detection of near surface features. Prospezzioni Archaeologiche, 5, 67–75.

Athanasiou, E., Tsourlos, P., Tsokas, G.N., Papazachos, C. & Vargemezis, G. 2005 Nondestructive DC Resistivity Surveying Using Flat Base Electrodes. Extended Abstracts of NearSurface 2005, Palermo, Italy.

Athanasiou, E., Tsourlos, P., Vargemezis, G., Papazachos, C. & Tsokas, G.N. 2007 Nondestructive DC Resistivity Surveying Using Flat Base Electrodes. Near Surface Geophysics (in press).

Bernabini, M., Brizzolari, E. & Piro, S. 1988 Improvement of signal to noise ratio in resistivity profiles. Geophysical Prospecting, 36, 559–570.

Brizzolari, E., Orlando, L., Piro, S. & Samir, A. 1986 A frequency analysis and filtering of resistivity profiling of different resistivity arrays. Paper presented at the 48th E.A.E.G. meeting, Ostend, 3–6 June.

Brizzolari, E., Ermoli, F., Orlando, L, & Piro, S. 1992 Integraded methods in archaeological surveys. Journal of Applied Geophysics, 29, 1, 47–55.

Carrara, E., Carrozzo, M.T., Fedi, M., Florio, G., Negri, S., Paoletti, V., Paolillo, G., Quarta, T., Rapolla, A. & Roberti, N. 2001 Resistivity and Radar surveys at the Archaeological site of Ercolano. Journal of Environmental and Engineering Geophysics, 6 (3), 123–132.

Clark, A. 1986 Archaeological Geophysics in Britain. Geophysics, 51: 1404–1413.

Clark, A. 1990 Seeing beneath the soil: prospecting methods in archaeology. Batsford B.T. Ltd, London.

Constable, S., Parker, R. & Constable, C. 1987 Occam's inversion: A practical algorithm for generating smooth models from electromagnetic sounding data. Geophysics, 52, 289–300.

Coppack, P., Emerick, K., Wilson, K., Dittmer, J., Szymanski, J., Tsourlos, P.I. & Giannopoulos, A. 1992 Recent archaeological discoveries at the medieval site of Fountains abbey. Proceedings of MEDIEVAL EUROPE 1992, York, U.K. 21–24, September, York, UK, 201–206.

Cosentino, P. & Martorana, R. 2001 The resistivity grid applied to wall structures: first results. Proceedings of the 7th Meeting of the Environmental and Engineering Geophysical Society, European Section, Birmingham, UK.

Dahlin, T. 1992 On the automation of 2D resistivity surveying for engineering and environmental applications. Ph.D. Thesis, Lund University.

Degroot-Hedlin, C. & Constable, S. 1990 Occam's inversion to generate smooth, two-dimensional models from magnetotelluric data. Geophysics, 55, 1613–1624.

Griffiths, D., Turnbull, J. & Olayinka, A. 1990 Two-dimensional resistivity mapping with a computer-controlled Array. First Break, 8 (4), 121–129.

Habberjam, G.M. 1979 Apparent resistivity observations and the use of square array rechniques. Geoexploration Monographs, series 1, No. 9, Gerbrueder Borntraeger, Berlin.

Habberjam, G.M. & Watkins, G.E. 1967 The use of square configuration in resistivity prospecting. Geophysical Prospecting, 15, 445–467.

Hesse, A., Jolivet, A. & Tabbagh, A. 1986 New prospects in shallow depth electrical surveying for archaeological and pedological Applications. Geophysics, 51, 585–594.

Karastathis, V.K., Karmis, P.N., Drakatos, G. & Stavrakakis, G. 2002 Geophysical methods contributing to the testing of concrete dams. Application of the Marathon Dam. Journal of Applied Geophysics, 50, 247–260.

Kuras, O. 2002 The Capacitive Resistivity Technique for Electrical Imaging of the Shallow Subsurface. Ph.D. thesis, University of Nottingham.

Linford, N. 2006 The application of geophysical methods to archaeological prospection. Rep. Prog. Phys., 69, 2205–2257.

Loke, M.H. & Barker, R.D. 1996 Rapid least-squares inversion of apparent resistivity pseudosections using a quasi-Newton method. Geophysical Prospecting, 44, 131–152.

Moussa, A.H., Dolphin, L.T. & Mokhtar, G. 1977 Applications of Modern Sensing Techniques to Egyptology. A Report of the 1977 Field Experiments by a Joint Team. SRI International, Menlo Park, California.

Noel, M. & Walker, R. 1991 Imaging archaeology by electrical resistivity tomography: a preliminary study. In Archaeological sciences 89, Budd, P., Chapman, B., Jackson, C. Janaway, R. and Ottaway, B., Oxbow, 295–304.

Panissod, C., Dabas, M, Jolivet, A. & Tabbagh, A. 1997 A novel mobile multipole system (MUCEP) for shallow (0–3 m) geoelectrical investigation: the 'Vol-de-canards' array. Geophysical prospecting, 45, 983–1002.

Panissod, C., Dabas, M., Florsch, N., Hesse, A., Jolivet, Tabbagh, A. & Tabbagh, J. 1998a Archaeological prospecting using electric and electrostatic mobile arrays. Archaeological Prospection, 5, 239–251.

Panissod, C., Dabas, M., Hesse, A., Jolivet, A., Tabbagh, A. & Tabbagh, J. 1998b Recent developments in shallow-depth electrical and electrostatic prospecting using mobile arrays. Geophysics, 63 (5), 1542–1550.

Papadopoulos, N.G., Tsourlos, P.I., Tsokas, G.N. & Sarris, A. 2006 2D and 3D Resistivity Imaging in Archaeological Site Investigation. Archaeological Prospection,13, 3, 163–181.

Piro, S., Tsourlos, P.I. & Tsokas, G.N. 2001 Cavity detection employing advanced geophysical techniques: a case study. European Journal of Environmental and Engineering Geophysics, 6, 3–31.

Pelton, W., Ruo, L. & Swift, J. 1978 Inversion of two-dimensional resistivity and induced polarization Data. Geophysics, 43, 788–803.

Sasaki, Y. 1992 Resolution of resistivity tomography inferred from numerical simulation. Geophysical Prospecting, 40, 453–464.

Scollar, I., Weidner, B. & Segeth, K. 1986 Display of archaeological magnetic data. Geophysics, 38, 349–358.

Scollar, I., Tabbagh, A., Hesse, A. & Herzog, I. 1990 Archaeological prospecting and remote sensing. Cambridge University Press.

Smith, N. & Vozoff, K. 1984 Two-dimensional DC resistivity inversion for dipole-dipole data. IEEE Trans. Geosc., 22, 21–28.

Szymanski, J. & Tsourlos, P. 1993 The resistive tomography technique for archaeology: an introduction and review. Archaeologia Polona, 31, 5–31.

Tripp, A., Hohmann, G. & Swift, C. 1984 Two-dimensional resistivity inversion. Geophysics, 49, 1708–1717.

Tsokas, G.N., Giannopoulos, A., Tsourlos, P.I., Vargemezis, G., Tealby, J.M., Sarris, A., Papazachos, C.B. & Savopoulou, T. 1994 A large scale geophysical survey in the archaeological site of Europos (nothern Greece). Journal of Applied Geophysics, 32, 85–98.

Tsokas, G.N. & Tsourlos, P.I. 1997 Transformation of the resistivity anomalies from archaeological sites by inversion filtering. Geophysics, 62, 36–44.

Tsokas, G.N., Soupios, P., Tsourlos, P., Vargemezis, G., Savvaidis, A., Paliadeli-Saatsoglou, C.H. & Drougou, S. 1999 Geophysical investigations in the area between Eukleia's temple and the theater in ancient Aegae (Verghina) using various methods. Proceedings of the 1st Conference on Physics in Culture, Thessaloniki, October 28–30.

Tsokas, G.N., Tsourlos, P.I., Papadopoulos, N., Manidaki, V., Ioannidou, M. & Sarris, A. 2006 Non destructive ERT Survey at the South Wall of the Akropolis of Athens. Extended Abstracts of NearSurface 2006, Helsinki, Finland, B040.

Tsourlos, P.I. 1995 Modeling interpretation and inversion of multielectrode earth resistivity data-sets. Ph.D. Thesis. University of York.

Tsourlos, P.I., Dittmer, J. & Szymanski, J. 1995 A study of non-linear techniques for the 2-D inversion of earth resistivity data. Expanded abstracts of the 57th meeting of the EAEG, Glasgow, Scotland, 29 May–2 June.

Tsourlos, P., Szymanski, J. & Tsokas, G. 1999 The effect of terrain topography on commonly used resistivity arrays. Geophysics, 64, 1357–1363.

Walker, R., Gaffney, C., Gater, J. & Wood, E. 2005 Fluxgate gradiometry and square resistance survey at Drumlanrig, Dumfries and Galloway, Scotland. Archaeological Prospection, 12, 131–136.

Xu, B. & Noel, M. 1993 On the completeness of data sets with multielectrode systems. Geophysical Prospecting, 41, 791–801.

Seeing the Unseen – Campana & Piro (eds)
© 2009 Taylor & Francis Group, London, ISBN 978-0-415-44721-8

Theory and practice of the new fast electrical imaging system ARP©

M. Dabas

Département de Géophysique Appliquée, UMR Sisyphe, Université Pierre et Marie Curie—Paris VI, France

ABSTRACT: The aim of the presentation given during the International Summer School in Archaeology at Grosseto in July 2006 was to show the principle of a new towed system devoted to electrical mapping of soils: the ARP© system (*Automatic Resistivity Profiling*) and to give some examples obtained with this system in Archaeology. The principle of the ARP© is very simple because it relies upon the standard galvanic electrical method widespread for different applications since its discovery by Marcel and Conrad Schlumberger in the 30s. Effectively, the ARP© system was first designed for agricultural applications in 2001 (GEOCARTA company, spin-off from CNRS, France). It was not before 2004 that the system was released for archaeological surveying, due to the necessary increase in terms of positional accuracy and measurement accuracy. We will discuss first from a more theoretical point of view the design of such instruments through 1D (one dimensional) and 3D numerical simulations in order to compare the responses with other instruments in the market and in order to design the optimal geometry of the ARP system. Depth of investigation and spatial response of these instruments are inferred from these calculations. On the other hand, the instruments have practical limitations in terms of their design, calibration, use in the field, etc. and these points have to be taken also into consideration. For that purpose, a practical comparison between existing sensors was done during the European Conference on Precision Agriculture in 2003. This comparison was also performed to validate the previous theoretical results. Finally, the design of this new instrument, now used at a wide scale by Terra NovA, will be explained with some recent results on archaeological sites.

Part I   Theory, instrumentation and field validation

1   GENERALITIES, PRINCIPLE OF ELECTRICAL METHOD COMPARED TO EM METHODS

In electrical methods, an electrical current is injected in the soil by means of a pair of electrodes. This current is either a D.C. current or a slow alternating current (several Hertz) to avoid polarization effects and/or eddy currents. The current flow in the whole volume of the soil and sub-soil and its spatial distribution is a function of the spatial distribution of the electrical resistivities.

As the soil is rarely uniform, geophysicists use the term apparent electrical resistivity to name the 'average' resistivity of the volume where the current is able to flow. This spatial distribution is measured by two or more electrodes on the ground surface, which measure the resulting voltage. The ratio of the voltage to the current, multiplied by a constant (the geometrical factor which takes into account the orientation of the 4 electrodes) is the apparent electrical resistivity (ρ_a). Because subsurface materials have generally different resistivities, measurements at the surface of the soil can characterize the vertical and horizontal distribution of underlying structures. In archaeological prospecting, typical resistivities span from 10 Ohm.m (clay soils) to several hundreds ohm.m or even thousand ohm.m over crystalline or metamorphic areas. Generally, resistivity of subsurface materials is higher than soil resistivities in nearly all cases due to, broadly speaking, a lower amount of water. Would the soil be perfectly homogeneous and very deep, the apparent resistivity would be the true resistivity of the soil. Generally, the geophysical data are interpreted as 'anomalies'. This refers to the discrepancy between an 'undisturbed' environment with a uniform true resistivity and the observed apparent resistivity.

To a given distribution of structures corresponds a unique apparent resistivity. But the opposite is false: to a given set of data (apparent resistivity or conductivity, or any other geophysical parameter) can correspond different structures. This non-uniqueness of the inverse problem makes ER/EC geophysical data set difficult to interpret. Consequently some additional data are often needed to characterize the anomalies.

In EMI methods, a current is injected in the soil by means of low frequency variations of a magnetic field (H_p) originating from an oscillating current into a coil (antenna) above the soil. This magnetic field induces in the soil eddy currents (Faraday's law) which spatial distribution is a function of the electrical conductivity (σ) of the soil. The conductivity is the opposite of the electrical resistivity. Electrodes on the ground surface could measure these currents but, for EMI methods, they are measured by a second coil (Rx): the eddy currents generate a very small oscillating magnetic field (H_s). The Rx coils measures both the primary field (H_p) and the secondary field (H_s). What is measured on the Rx coil is a function of the different conductivities in the sub-soil but also of others factors like: orientation and distance between the two coils, operating frequency and magnetic susceptibility. Under specific constraints, named Low Induction Number (LIN), a simple direct expression can be found between apparent conductivities (σ_a) and the ratio of the quadrature out of phase primary to secondary field (H_p/H_s). In this case, the measurement can be translated directly to conductivity and these instruments are often named conductivity meters. For archaeological purposes, we could be interested by the use of the in-phase component (proportional to magnetic susceptibility) but none of the available instruments on the market have proved to our opinion useful (problems of drift, abrupt or random changes, etc...).

Using galvanic or electromagnetic induction instruments is a choice that has to be made by the geophysicist in the field. Our experience is that the best 'images' of the soil and its structures and especially for Archaeology is given by galvanic methods. EM is still a research topic for archaeological mapping.

We have found that the main drawbacks of using EMI instruments are:

– Time drift of the electronics that makes for example the merging of several maps acquired at different times very difficult,
– Mechanical drift of the instruments: any slight change of coil direction with respect one to each other results in a signal that can be as high as the one due to archaeological structures. Of course, when using towed systems, this problem becomes even more important.
– No absolute calibration: this is related to the first mentioned problem. Calibration of EM instruments is a very difficult problem. Normally obtaining a zero is done by raising the instrument (>2 m) or making the calibration over a very resistive ground. Moreover the in-phase response (magnetic response) has also to be zeroed because of problems of phase mixing,
– Depth of investigation can be changed by coil orientation or coil separation but this is very limited in available instruments. Moreover, definition of depth investigation is more difficult than in galvanic methods,
– Limited dynamic of measurements, especially for moderate and high resistivities (>200 Ohm.m).
– Limited time response: due to the low amplitude of the secondary magnetic field in the receiver coil, time integration has always to be done in order to higher the signal to noise ratio. This integration makes the time response of these instruments very long, typically more than one second (time between the measurement of the 'anomaly' and the output measured signal). This is not important for hand-made instruments but becomes clearly a drawback for continuously towed systems.

The advantages are:

– Simple instruments to operate in the field and that can be towed easily by an operator by hand and/or with a vehicle,

106

– Very good response over conductive soils (clay or salty soils for example).

To our experience, the drawbacks for using galvanic methods are:

– Contact resistance: when the soil is too dry, it is difficult and sometimes impossible to drive the current into the soil due to a too high contact resistance (>100 Kohm for example),
– No measure possible when the soil is frozen (contact resistance + high resistivity of the ice),
– Workload heavier and longer than when using EMI instruments.

The advantages of using galvanic methods are:

– Many depths of investigation possible which are controlled mainly by the distance between current and injection electrodes,
– Absolute measurements (no calibration and no drift in the field),
– Wide dynamic of measurement (a few ohm.m to several ten thousands Ohm.m),
– Quick time response (a few milliseconds).

Considering the drawbacks of galvanic methods, we have tried to:

– Design a resistivimeter which can cope high contact resistance, have the best time response and easy to operate in the field,
– Design a mechanics that can be used as rolling electrodes and towed by an all-terrain vehicle even in very harsh environments,
– Design a hardware and software to drive the instruments, help the operator while driving (auto-guidance) and also makes the real time check of quality parameters.

This system is named ARP© (Automated Electrical Profiling) and is patented (number 0101655, European extension pending).

Before explaining the design of such an instrument, some numerical simulations were performed for estimating for example the best position of electrodes in order to maximise the volume of investigation, but also to compare with other available instruments.

A complete explanation can be found in "Resistivity-conductivity mapping for Precision Agriculture, a global overview" (M. Dabas, 2007, in press).

2 NUMERICAL SIMULATIONS

Before testing an instrument in the field, it is important to predict and optimise the effect of a number of parameters related directly to the instrument (position of electrodes in galvanic surveying, position and orientation of coils in EMI surveying) or to the field where the instrument is operated (influence of resistivity, influence of any layering in the soil, influence of small changes of coil geometry, etc.).

1D and 3D simulations were performed in order to test for example the vertical response of instruments (depth sensitivity) in DC and EMI mapping. We have found that the control of the depth of investigation is something very important specifically during the archaeological interpretation of ARP maps.

2.1 *Depth sensitivity in DC and EMI mapping*

In order to calculate the contribution of structures in the soil for different geophysical instruments, it is useful to compute the effect of a very thin layer for different depth z: this is what is called to depth sensitivity that is different from the investigation depth (see next paragraph).

This computation was performed for 3 different instruments: the EM38 of Geonics, the Veris2000 of Veris Technologies (another continuous galvanic towed system) and the ARP of Geocarta. The software used for DC modelisation is SELQCQ and was designed by Jeannette Tabbagh (Paris VI University). For EM, SH1DBF was used. The effect of a horizontal layer of thickness 0,1 m was computed for depth ranging from 0 to 2 meters that accounts for most of the archaeological structures found in the field.

For the EM38 (f = 14600 Hz, 1 m between coils), the coil was considered to be at a height of 0,1 m above ground. The two orientations HCP (plane of the coils horizontal, vertical moment) and VCP (plane of the coils vertical) were studied. A contrast between resistivities of 4 was chosen both for a resistive layer (60 Ωm in a medium of 15 Ωm) and for a conductive layer (15 Ωm in a medium of 60 Ωm).

Similar computation was performed for the DC systems (Veris and ARP).

For the Veris, the geometrical configuration is the one of a simpleWenner quadripole with a = 0,20 m for the "Veris shallow" configuration and a = 0,61 m for "Veris deep" configuration.

For the ARP system, the 8 electrodes are not in line and form a 2D pattern with the following coordinates: A (0;–0,5), B(0,+0,5), M1(+0,6 ; –0.25) N2 (+0,6; +0.25) for channel 1; M2(+1,2 ; –0.5) N2 (+1,2; +0.5) for channel 2; M3(+2,2; –1) N3 (+2,2; +1) for channel 3.

The effect of a conductive and a resistive layer is shown on the following figures (Figs. 1–2).

Some conclusions can be drawn from these curves: the depth sensitivity of the ARP-Channel 1 is equivalent to the Veris Deep system. The Veris shallow has the most important sensitivity to the very superficial layers (20 cm), which in most cases is not important in Archaeology. For a resistive layer, the EM38-VCP is equivalent to the ARP-Channel 1 and the EM38-HCP to ARP-Channel 2. This is not valid for a conductive layer where the EMI instrument is more sensitive to deeper layers. The ARP-Channel 3 has the most widespread response versus depth and

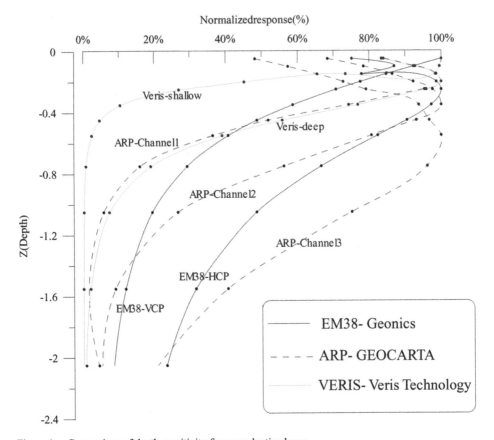

Figure 1.   Comparison of depth sensitivity for a conductive layer.

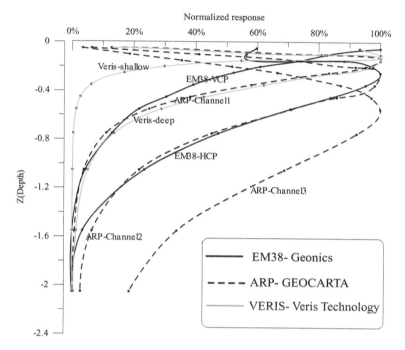

Figure 2. Comparison of depth sensitivity for a resistive layer.

there is still a contribution of 20% coming from layers at a depth of more than 2 meters. We see of course that there is an overlap of the sensitivities of the channels with depth. This theoretical problem is very difficult to address at the moment and it is not possible unfortunately to get the maximum of the sensitivity over a very restricted depth interval. Therefore, it is important to remember that even 'resistivity images' obtained for example for the deeper channel (3) have still a contribution from the superficial layers. Of course, this contribution has a tendency to disappear due to the more important contribution and volume of the deeper layers.

2.2 Investigation depth in DC and EMI mapping

There exist a lot of definitions for investigation depth in geophysics. We prefer for practical reasons in agriculture, the one that is defined as the depth where the effect of a layer is under the threshold of detectability for the particular instrument used. A threshold of 10% was chosen. Of course, the investigation depth is a function of the electrical contrast. For example, in order to find the investigation depth in the resistive case with a contrast of 4, we use a model with an upper layer of 15 Ωm of variable thickness and a substratum of 60 Ωm. The depth z_i of the substratum is lowered from 0 to Z and the apparent resistivity is computed for every z_i. For $z_i = 0$, the apparent resistivity is the one of the substratum (60 Ωm), and for very large z_i the apparent resistivity tends towards the one of the upper layer (15 Ωm). The depth of investigation is z_i where the apparent resistivity is equal to 54 Ωm.

In the next figure 3, we show the decrease of apparent resistivity with depth for the resistive case and a contrast of 10.

In the following Table 1, we have computed for three different contrasts, the investigation depth in the case of a resistive medium.

We can see that the investigation depth is not a linear function of the electrical contrast for both EMI and DC instruments. The ARP system has the maximum depth of investigation compared to

109

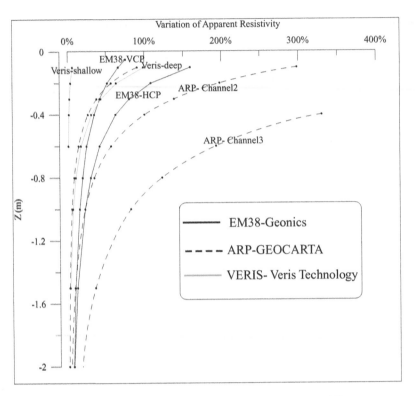

Figure 3. Comparison of investigation depth for a resistive substratum (contrast 10).

Table 1. Investigation depth (m) for a resistive medium.

	Contrast 2	Contrast 4	Contrast 12
EM38-HCP	1.7	2.4	3.0
EM38-VCP	1.2	1.7	2.0
Veris shallow	0.2	0.3	0.4
Veris deep	0.3	0.4	0.5
ARP-Channel 1	0.6	0.9	1.0
ARP-Channel 2	1.1	1.7	1.7
ARP-Channel 3	2.0	2.6	3.0

the other systems. We can also notice that the computed depths of investigation of EM38 are very different to what is quoted in literature, Table 2.

We can notice that investigation depth for DC methods (ARP and Veris) are very similar in the resistive and conductive case. This fact is important for surveying: we will not overestimate conductive layers vs. resistive layers. This is not the case for EM methods: they will have a tendency to overestimate the presence of conductive layers. Moreover, depths of investigation quoted in the conductive case are too high for archaeological studies (we can notice here an interesting linear relationship between electrical contrast and depth of investigation). Except for loessic soils, the depth of the soil is generally less than 2 m.

Finally, we have made other simulations with Wenner a = 0,5 1 and 2 m and found that depths of investigation are very close to ARP channel 1, 2 and 3 respectively.

Table 2. Investigation depth (m) for a conductive medium.

	Contrast 2	Contrast 4	Contrast 12
EM38-HCP	2.2	4.5	9.0
EM38-VCP	1.7	3.5	7.0
Veris shallow	0.2	0.3	0.4
Veris deep	0.2	0.3	0.4
ARP-Channel 1	0.6	0.8	0.9
ARP-Channel 2	1.1	1.4	1.6
ARP-Channel 3	1.9	2.4	2.8

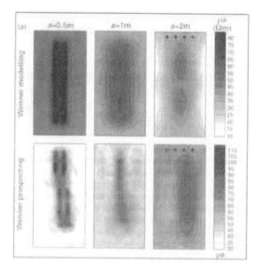

Figure 4. Top: result of simulation for three different spacing with a quadripole perpendicular to the structure. Bottom: measurements in the field. Depth range: from 0.5 m to 2 m.

2.3 *Horizontal response in DC mapping*

It is important that the response to a simple structure is simple: during the interpretation phase, the geophysicist and archaeologist have to get the best archaeological images of the structures and not complicated anomalies due to a bad choice in the acquisition parameters. For example we would like that the image of a wall looks like a wall and is centred exactly over that wall.

For that purpose, the effect of a simple wall close to the surface was simulated by 3D modelling and tested in the field over an artificial wall buried in the test site of Garchy, France (Panissod et al., 1998).

2.3.1 *Wenner configuration*

On the upper part of the figure 4, the result of simulation is shown for three different spacing with a quadripole perpendicular to the structure. On the bottom part of the figure 4, the measurements in the field are displayed. With 0,5 m spacing, the wall appears like a double structure because it is close to the surface. This well-known effect is clearly confirmed in the real case study (bottom part of figure). With an 'a' spacing close to the vertical extension of the wall, the shape of the wall is more simpler but tends to smooth. With a larger spacing, the resistivities on the border of the wall tend to be even lower than reality (trapping of current under the wall).

2.3.2 *ARP configuration*

Modelisation of the effect of the same structure using the ARP system gives a clearer anomaly than with the Wenner configuration (figure 5). Doubling of the anomaly still appears for larger spacing (a = 2 m) and this effect has to be taken into consideration during the interpretation process. Of course if the upper part of the wall was deeper buried, this effect would not happen. Another point that is important is the isotropy of the response: using Wenner configuration, the shape of the anomaly changes when the orientation of this in-line quadripole is changed relative to the strike of the wall (not shown on figure). For an ARP configuration, this shape does not change (see in the example of the fanum in part II). This is clearly an advantage when surveying because the elongation of archaeological features is generally not known in advance.

A more comprehensive discussion can be found in (Dabas, 2001).

3   VALIDATION OVER A TEST SITE (ATB, GERMANY)

In order to test these theoretical results and available instruments, we have initiated with the Agricultural Engineering Potsdam in Bornim (ATB) and Potsdam University a unique field demonstration during the fourth European Conference on Precision Agriculture. We have processed nearly all data coming from this experiment and also from later experiments on the same field in order to test the accuracy of the geophysical instruments used but also to measure the autocorrelation and intercorrelation between each instrument. Only results concerning EM38, Veris and ARP will be shown here and full details can be found in (Dabas, 2007 in press).

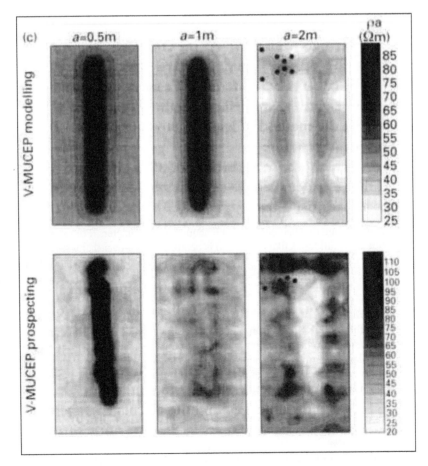

Figure 5.   Modelisation of the effect of the same structure of Fig. 4, using the ARP system.

The test site is situated in Germany near Potsdam (Bornim), 30 km west of Berlin and was surveyed in very dry conditions (summer 2003). Soil texture of the test site is dominated by loamy sand (glacial drift area inserted with fluvial plans and lowlands). The site has a history of intensive land use. Since 1927, an agricultural research station was established, nowadays known as the ATB. In the last decades, the test site was not used for plant production. In the sixties a water pipe was laid through the field. The site is covered with grass. Due to this history of land use, the soil is strongly compacted in parts of the site, we find rubble, and beside at least two pipelines in the underground, many other metal objects are present near the surface of the soil.

This site was chosen for its practicability during the fourth European Conference on Precision Agriculture. It is probable that most results obtained like sensitivity depth, conditions of operation, degree of autocorrelation between two periods of time, or intercorrelation between instruments are still valid whatever the type of soil encountered.

3.1 *Experimental design*

Measurements with the different geophysical instruments were conducted on this small field. The soil was very dry due to low precipitations at this time of the year. Measurements over the whole field were conducted along parallel lines. Some of them were done by hand and located by tape or dGPS (EM38), while other measurements were done continuously using an ATV or a four-wheel car and located by dGPS (Veris, ARP, EM38DD). In addition to this mapping, an intensive survey took place on a transect (158 m length) located within the field on the west boarder. This transect was used for instruments that cannot be for the moment operated continuously (multiplexed electrodes or electrical bore hole logging for example). A software was designed in order to project the nearest data over this transect for the instruments that were operated over the whole field.

Some of the geophysical instruments were used over the whole field:

– Two systems for continuous measurement of electrical resistivity: the ARP system (Geocarta, France) and the Veris system (Veristech, USA).
– Two systems for continuous measurement of electrical conductivity: the EM38 in VCP and EM38DD (Geonics, Mississauga).

Some methods were only applied over the transect (defined *a posteriori* before the mapping of the field), they include:

– A multi-electrode switching system (Geotom 200/100 RES/IP, GeoLog Germany). Wenner configuration was chosen with a = 0.5 to a = 4 m, an electrode spacing of 0.5 m (stack of 8, maximum noise: SD of 1 %).

Later on during winter 2003 and spring 2004 additional measurements were made:

– An EM38 vertical during winter over the profile (in order to test the temperature drift of this instrument and influence of the height of the instrument).
– A Veris 3100 (summer measurements were not possible for this instrument due to dryness of the soil).

3.2 *Data analysis*

The huge amount of data was processed mainly by R. Gebbers in Germany and in parallel by M. Dabas in France using different software on the shelf and hand made routines. The original WGS84 geographical data were transformed using a UTM projection (ellipsoid DHDN12/PD Bessel 1841, 7 Helmert parameters transformation). The accuracy of the transformation was tested to be better than 1 cm.

Multi-electrode data were processed by RES2DINV. All maps were produced using the same software and interpolation methods (Surfer v8).

3.2.1 *1D comparison between instruments over a transect*

Comparison of maps is a very difficult exercise. Lot of factors interfere, like the dimension of mesh, the type of mathematical functions used as interpolator or the way to handle duplicate data for example. We want in a first attempt to focus only on data that are the closest as possible to the 'raw' data. We have in a first step visualized and correlated the data over a single profile. In a second step, maps were produced and 2D local correlations were computed (not shown here).

The GEOTOM data are probably the most accurate electrical data obtained over the transect. Spatial resolution is the minimum distance between electrodes (0.5 m). Noise was estimated by stacking measurements and a maximum SD of 1% was imposed. Using 8 electrode spacing ('a' = 0.5 to 4 m) enables a depth resolution of 0 to 4 m approximately. For comparison with EMI methods, all apparent ER were transformed into apparent EC. Conductivities show a general trend with increasing conductivities over 200 m. Dynamic is from 1 to 75 mS/m. Intermediate wavelength (50 m) but also very short wavelengths are observed. Conductivities are increasing with depth all over the profile. On the following figure 6, only EC from five electrode spacing is presented.

The depth of investigation of a Wenner configuration is known to be of the order of 'a'. This was verified by 2D numerical simulations (code SELQCBBIS) following the same way as depicted in Dabas and Tabbagh, 2002. The profile was inverted in order to compute the true conductivities (code RES2DINV, Geotomo software, Malaysia). On the following figure is shown the result of this inversion together with the results of the direct measurements made in the 14 bore holes (1 m deep). We can see the good accordance of inverted conductivities and bore hole conductivities except for point 60 where a 3D effect is clearly present, figure 7.

The Veris 3100 was used all over the field at the same time and data were processed in order to extract the conductivities along the transect. Data are not presented here because clearly the instrument could not tolerate the high contact resistance (probably more than 50 KOhm) especially in the low conductivities area: apparent conductivities were systematically underestimated by a factor of more than 100%. This is clearly a limitation of that instrument. Consequently, the experiment was done a second time in 'good' conditions (May 2004) when the soil was wet enough. During processing of the data, a shift was noticed between Geotom anomalies and Veris anomalies, figure 8. This was further explained by the shift between dGPS antenna and the centre of the quadripole sensors. This error does not seem to be taken into account in the processing of the VERIS3100. Next figure shows the same profile made in both directions and this shift is clearly visible for all three anomalies. VERIS data were further corrected from this shift.

Next figure 9 pictures the results obtained for the two Wenner configurations (Veris shallow a = 0.2 and Veris deep a = 0.61 m) compared to Geotom data. We can see the good concordance of the data specially the Veris deep with the Geotom a = 0.5 m. Some little discrepancies exist at the end of the profile, but this can be due to very short wavelength that cannot be measured by the Veris system (integration of data before output and measurement every second).

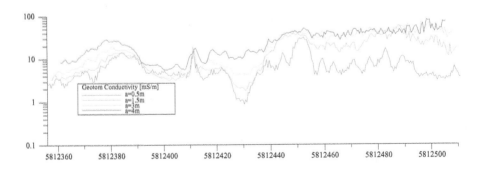

Figure 6.   Electrical conductivities measured by the Geotom multi-electrode system.

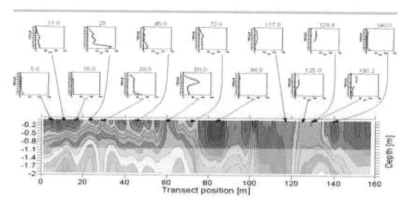

Figure 7.   Inversion of Geotom EC data and results of EC in bore holes.

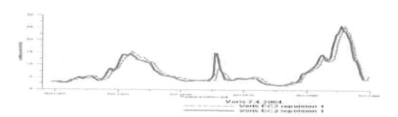

Figure 8.   Horizontal shift of EC measured in opposite direction by Veris3100.

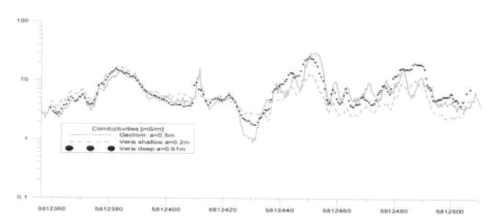

Figure 9.   Comparison of ECs measured by the Veris3100 system and Geotom ECs.

The same processing was done for the ARP data. The data were acquired over the whole field in July 2003. Despite the very difficult conditions already quoted before that prevent measurements with the Veris system, ARP data have shown a very good correlation with Geotom data in the Wenner 'a' = 0.5, 1 and 2 m inter-electrode configuration, figure 10. The absolute values follow clearly most of the details of this transect.

EM38-DD (two EM38 attached together in order to have at the same time measurements in vertical and horizontal mode) data were acquired over the whole field in July 2003. Next figure 11 pictures the variations of EC measured by the EM38-DD (absolute calibration with EMI instruments is always difficult and amplitudes were arbitrary shifted). Conductivities measured in the horizontal mode HCP (horizontal coils and vertical dipole) are higher than conductivities measured in the vertical mode VCP (vertical coils, horizontal dipoles). This is in accordance with the higher depth of investigation in the HCP mode. Despite the fact that the amplitudes follow the general trend of the transect, the dynamic is not the same as the one measured with Geotom. Data seem also to be much smoothed. Finally at some places, some low conductivities anomalies do not seem to have any correspondence with Geotom anomalies especially in VCP mode. For example at X = 5812372 the anomaly should originate from a buried metal object. At X = 5812414, a pipe is known to exist: the first high of the anomaly is probably due to the trench fill but the following low is due to the metallic response of the pipe. This effect is clearly a problem in EMI prospecting because the phase separation is not perfect and magnetic response can mix with the conductivity response.

In order to illustrate the temperature dependence of EM38, the following experiment was done: the instrument was moved from room temperature to outside temperature (21 to −8°C) and conductivities measured continuously while instrument remain at a fixed position. A global decrease of apparent conductivities was observed which results in a change of 15 mS/m in 1 hour and a half. Spikes were also observed, figure 12.

4 DESIGN OF THE NEW ARP SYSTEM

The ARP© system is the result of a long history, which can be split in three steps:

1. The original prototype dates from the eighties. This proto was developed for the mapping of archaeological targets. This experimental system (Dabas et al., 1989) was named RATEAU (*Résistivimètre Autotracté à Enregistrement Automatique*: Automatic recording and self-towed resistivity meter). It was made of four modules namely:

   – A square quadripole made with four rolling electrodes separated by 1 meter. Any type of all terrain vehicle (ATV) can tow this system (original speed was around 1 m/s).
   – A resistivimeter (RMCA4, CNRS©) that measures directly the electrical resistance. This specific resistivimeter tolerates very high contact resistance and is compatible with high-speed measurements (minimum 10 Hz).

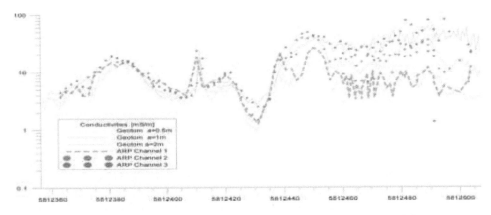

Figure 10.   Comparison of ARP ECs for 3 different depths of investigation and Geotom ECs.

116

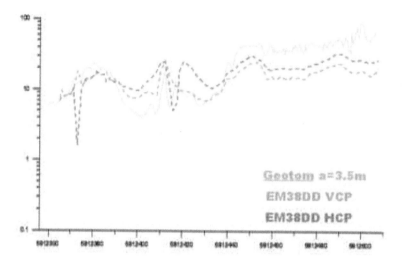

Figure 11.   Comparison of ECs between EM38-DD and Geotom instrument.

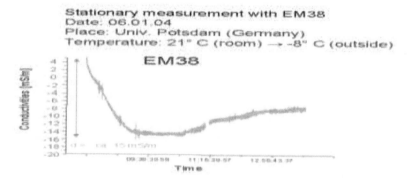

Figure 12.   Drift of measured EC (EM38) at a fixed point due to a change of air temperature.

- A computer that displays and stores the values of apparent electrical resistivities in real-time,
- A Doppler radar that triggers the ER measurements along the profiles at a fixed distance interval.

A tractor towed the system. The Center of Recherches Géophysiques de Garchy did the first application of RATEAU systems for Archaeology in the nineties with successful results published (Montbaron medieval mote, iron smelting sites in Burgundy for example).

2. This system was further enhanced to 3 depths of investigation. The geometry of this new system named Multi-depth Continuous Electrical Profiling: MuCEP (Panissod et al., 1997) was carefully designed by the above mentioned 3D simulations (forward modelling) in order to optimise the positioning of the electrodes on the ground surface versus depth information and also reduce the volume of the system which has to be packed in a "standard" van. The eight electrodes enable the ER to be measured for a depth of investigation approximately up to 0.5, 1 and 2 meters approximately.

Meanwhile, in 2000, Dabas et al. showed that it was possible to survey continuously with a man-towed frame but also with a simple pole-pole configuration (Wroxeter experiment). This experiment leads Geoscan Research (Bradford, England) to introduce recently a man-towed square quadripole using its standard RM15 resistivimeter (MSP 40 Mobile Sensor Platform).

3. The third system ARP© was developed specifically for agriculture by a spin-off from CNRS in 2001: Geocarta. Several improvements were made:

   – Absolute positioning by a dGPS or RTK GPS,
   – Possibility to acquire 3 measurements at a speed up to 6 m/s with a spatial resolution of 20 cm (area up to 1 ha by hour can now be surveyed in Archaeology),
   – Hydraulic rising of the system for a faster moving when changing direction,
   – Development of a new resistivimeter for optimised synchronous measurement of 3 channels with a quick time response (44 ms) and a high tolerance to contact resistance,
   – Development of a chip for interfacing the different electronic boards,
   – Developments of a real-time software on a ruggerized computer. Data are processed in real-time (both geographically and the 3 channel). Software evolved into a real-time GIS with auto-guiding possibilities and integration of other type of data (mainly satellite, orthophotos, project files),
   – Towing by a quad-bike.

In open-field areas, up to 10 ha can be surveyed in a day. In standard conditions, number of measurements in one hectare is around 150 000 (interprofile of 1 m, measurement along profile every 20 cm). The average velocity of data acquisition is around 4 m/s.

There are now seven generations of ARP systems working in the field. Most of them are built for specific applications: ARP01 (figure 13) to ARP03 for soil mapping in crop fields, ARP04 for wine yards, ARP05 for Archaeology, ARP06 for soil mapping in sugar cane fields.

Figure 13. ARP01.

The latest systems ARP07 and ARP08 are used for soil mapping in crop plots but can be used also for Archaeology.

Part II   Example over an archaeological site

Evaluation of archaeological potential in Rescue Archaeology in France is generally done by spot observations like trenches and field walking. This last method can be very successful but only if archaeological artefacts are near the surface and if the land has been ploughed recently. Some people also argue that the archaeological sites that can be discovered by this technique are so superficial that erosion processes are predominant and lower the interest of excavating such sites. Trenches are of course very destructive and can be opened over a very limited extent of the total area to be studied, generally between 5 and 10%. Moreover, this process implies that the land is already purchased which often implies a too short time period between evaluation and diggings.

Geophysical techniques can be planned in advance and have the advantage that the entire surface can be scanned. The main drawbacks from the archaeological point of view is that geophysical maps give no idea of the age and of the state of preservation of the structures.

Many geophysical techniques corresponding to different geophysical parameters can be used for mapping archaeological structures. Among the most well known methods are Electrical Resistivity (ER), Electromagnetic Induction (EMI), magnetometry, seismic and gravimetric techniques. But considering the spatial scale (meter) needed, the economic point of view, only two techniques are mostly used: electric (sometimes named as galvanic) and magnetic. They are generally operated manually and this limits the surface that can be covered in a day. Even using the most recent instruments with several sensors, 1 ha for magnetometry and 0,3 ha for resistivimetry can generally be surveyed with a resolution of 0,5 m in a single day. This limitation has to be overcome in Rescue Archaeology for example because the extent of the surface is of the order of tens to hundreds of hectares for a single site.

We have found that magnetic methods are invaluable for Rescue Archaeology but suffer from four drawbacks:

– A lack of depth of investigation using ordinary sensors due to the small distance between the two sensors in the mode of operation generally used (gradiometer),
– A lack of depth resolution (roughly speaking, inversion of data gives not a single solution for the depth of structures),
– Many sites cannot be detected due to a very low or even no contrast of the magnetic susceptibilities,
– Many sites have a very high magnetic "background" making the detection of archaeological impossible.

On the contrary, electrical resistivity is a very sensitive parameter in archaeological prospecting directly related to water content, with a wide dynamic and with a good control of depth of investigation. Typically the areas that could be surveyed manually were less than a few hectares. This had two consequences: the first one is that it was not possible to include at a reasonable cost a manual geophysical investigation in most of the projects in Rescue Archaeology; The second consequence is that the validity and coherence of the geophysical maps was weakened by looking only at restricted areas.

We have found that the main objectives in Archaeology of designing a system like the ARP or any other system which can survey let's say minimum 4 to 8 ha a day are the following:

– Optimum positioning of archaeological trial trenches,
– Improvement of the chance of detection and mapping of the geometry of archaeological sites,
– Gain of precision and time for scientist,
– Give affordable maps at a scale of 1/100 for archaeologists and developers,
– Manage the threat of urbanisation (Rescue Archaeology): give affordable solution for surfaces or more than 10 ha.

This was achieved by the design of the ARP system and has opened new possibilities in Archaeology. Hundreds of hectares have been surveyed since 2003 and it is now possible to draw some conclusions. The system has proved to work nearly over all sites except when the contact resistance were too high during summer and especially over clayey soils and also when the soil is frozen. In France these conditions happen only a few weeks per year. On another point of view, some sites did not show up because of the very small size of structures compared to their burial depth (post-holes mainly). This is why it is more useful to look also at other soil properties like magnetic susceptibility in order to find brick walls, postholes for example or fired structures. Consequently, we have recently designed a new magnetic towed system (AMP for *Automatic Magnetic Prospecting*). In the 'best world', it is of course better to have both type of information available (resistivity + mag. susceptibility) in order to higher the chance of not missing archaeological targets of interest but also to enhance the reliability during the interpretation process.

Compared to manual survey, the gain in terms of time of operation in the field was estimated to 20 at minimum. Since the first year where ARP surveys begun for Archaeology (2004) and 49 ha surveyed, the increase of the area surveyed every year is of 50%.

1 FIRST EXAMPLE: VALIDATION OVER A PREVIOUSLY SURVEYED SITE: THE ROMAN SITE OF VIEIL-EVREUX, FRANCE

Settled at the top of a plateau and at 6 km of *Mediolanum Aulercorum* (modern Evreux, a town in Normandy, West of Paris), the archaeological site of Vieil-Evreux: *Gisacum* is thought to be the main sanctuary of the tribe of *Aulerques Eburovices*. The name Gisacum, probably deriving from the local god *Gisacus* is still not certain. This site is known through archaeological diggings since more than 200 years and lot of aerial flights from Roger Agache and a local association: Archéo27. Following a first urbanisation in the first century a.d., a polygonal shape town emerges at the 2nd century covering an area of 250 ha. At the 3rd century, all dwellings seem to concentrate only at the border of this 'town-sanctuary' aligned along a 5600 m long portico. Since 1996, the Conseil General of Eure established a research program and geophysical surveys occurred since that time every year, thanks to the Mission Archeologique (L. Guyard). The main monumental structures have been surveyed manually (thermae, nymphea, fanum, theater) covering more than 8 ha and ARP is now operating over bigger areas (forum, polygon area).

We have used the fanum area as a test site covering approximately 1 ha. For the first time the ARP system was tested there for an archaeological application in December 2003.

The ER was first measured manually in June 1999 with a standard pole-pole configuration (a = 0,5 m), a distance between profiles of 50 cm and a distance along the profile of 50 cm. The time needed for field acquisition was nearly 10 days with two operators (34963 data with a = 0,5 m and 17477 data with a = 1 m). All the walls (in black: resistive), which are very superficial, can clearly be seen together with the central part of this building (cella) and very fine details (doors entrance, etc.). Darker lines parallel to the outer walls correspond to old trenches and lanes, figure 14.

The same area was re-surveyed in December 2003 with the ARP© system in less than two hours and one operator with profiles only every one meter and profiles in two orthogonal directions, figure 15–17.

The doubling of the structures explained in chapter 2–3–2 can be clearly seen on channel 3. The image obtained for channel 2 is nearly of the same quality as the one obtained "manually". Some slight differences can be seen but are related to positioning problems of the GPS used at that time: we noticed that the use of a GPS with subscription to a commercial service for the broadcast of real-time differential corrections (Omnistar-Landstar—Thalès) was not sufficient. The use of RTK technology or post-processing with a base station (code + phase) is necessary or the use of a virtual base station.

The isotropy of the response can be seen on the next figure 18 where acquisition was repeated two times with first profiles in the E-W directions (left) and in the North-South direction (right).

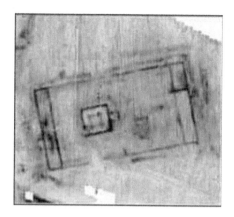

Figure 14.   Apparent electrical resistivities (a = 0,5 m) acquired manually.

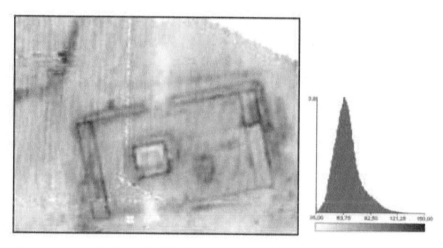

Figure 15.   Apparent resistivities with ARP, channel 1.

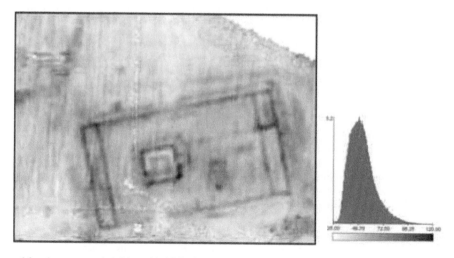

Figure 16.   Apparent resistivities with ARP, channel 2.

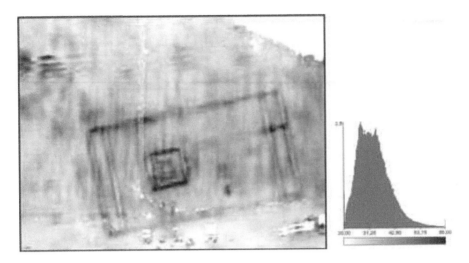

Figure 17. Apparent resistivities with ARP, channel 3.

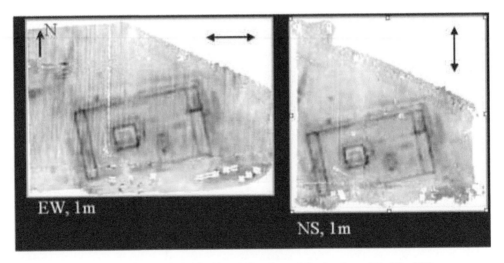

Figure 18. Comparsion of maps obtained with two directions of dipoles channel 2 of the ARP.

The slights ziz-zag observed in the E-W directions profiles is due to the roughness of the field which makes the GPS oscillates.

2 SECOND EXAMPLE: A SANCTUARY AND A THEATRE IN NORMANDY, FRANCE

In 2005, we were requested by the Ministry of Culture to undertake an ARP survey (6 ha) over the temple and the theatre of Bertouville (Eure, France). It is interesting to notice that all aerial photos have never given any information over this site, which is known by previous bronze discoveries (XIX century).

This survey was done in a single day and crops were already growing over the east part of the plot. The use of low-pressure tyres is very important in these conditions. The results clearly show the internal structure of the theatre and the triple fanum as resistive features (black), figure 19.

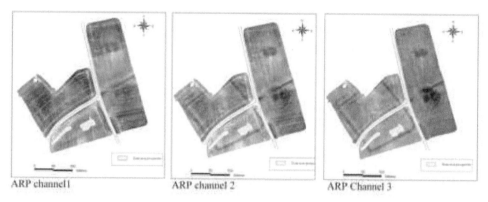

ARP channel 1   ARP channel 2   ARP Channel 3

Figure 19. ARP survey (6ha) over the temple and the theatre of Bertouville (Eure, France).

It is interesting to notice the usefulness of depth in that case: for example, the ditch all around the temple is better defined in map channel 3 whereas the walls of the north cella seem to disappear on that same channel. On channel 1 the white stripes on the west part and separated by 24 m corresponds to the effect of compaction of the tyres of tractors (soil compacted which retains more water). These structures disappear for channel 3. Compilation of the information given by the 3 channels gives clearly some ideas of the 3D extension of the structures.

Another important aspect is the information brought by the use of dGPS when acquiring the data. It is straightforward to derive a Digital Elevation Model of the area surveyed. In this example (see figure 20 below) the distance interval between height lines is 20 cm. The theatre for example corresponds to a local high in the DEM, which is hardly visible in the field.

3 LAST EXAMPLE: PROJECT OF AN INDUSTRIAL PARK AT VÉMARS, FRANCE

This last example is to show the possibilities of the ARP over bigger areas and in the context of Rescue Archaeology.

The industrial Park Prologis "Les Portes de Vémars" is located at Vémars, 30 km at the North-East of Paris and is close to A1 motorway. More than 50000 m² of buildings will be build over that area. Within this project, we are asked to survey a surface of 50 ha with a delay of less than one month in order to make an assessment of the archaeological potential, Plate 8. The geophysical phase was done in July 2006 just before a classical archaeological trial trenches phase.

The geographical context is simple: A plateau with slight undulations cut with two dry valleys (see figure). The red line pictures the limit of the surveyed area, Plate 8. This plateau is covered by quaternary silt with a vertical extension of minimum 8 meters. This formation overlaps the Saint Ouen limestone, which can be seen to the South. This context is indeed very favourable to the preservation of archaeological structures.

All data are interpolated on a square mesh of 0,5 × 0,5 m. Magnetic data were also acquired with AMP over the same area (not shown here). Of course, it is impossible to see the details of structures when printed on an A4 paper and all interpretation is done on screen using different GIS tools and filters.

As usual the horizontal stripes on the north part of this area are linked to the work of the soil by the farmer (compaction). At the scale of printing, only the geomorphology of this field clearly shows up: the two small valleys are filled with a more resistive material (colluviums around 70 Ohm.m as opposed to an average 30 Ohm.m over the plateau). An important distinction can be made between the North and the South: to the North these materials are near the surface (channel 1) and to the South their extension is clearly deeper (channel 3). For Archaeology, two sites were clearly located even if

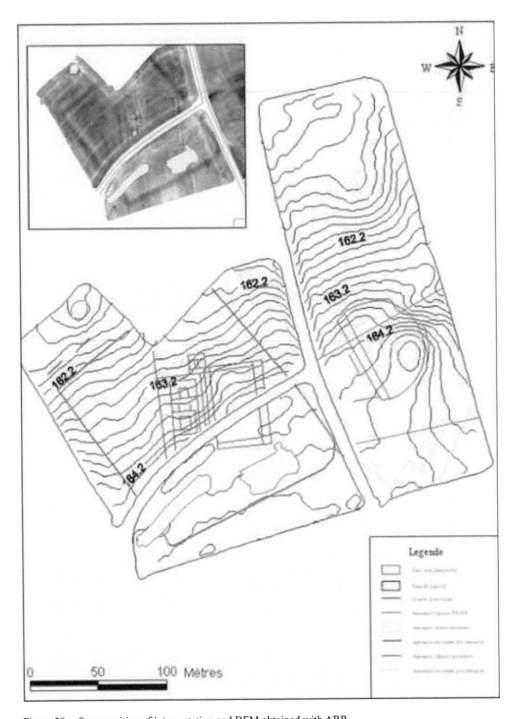

Figure 20.   Superposition of interpretation and DEM obtained with ARP.

124

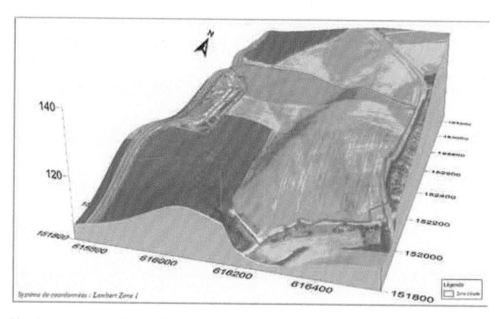

Plate 8.   Superposition of aerial photo over vertically exaggerated DEM. (See colour plate section)

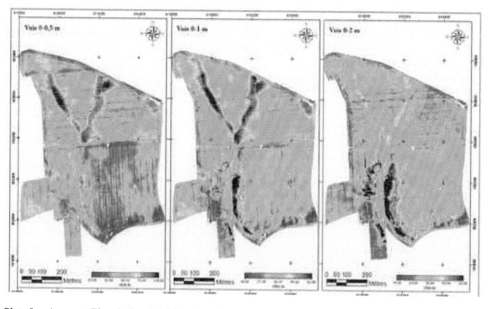

Plate 9.   Apparent Electrical resistivities ARP Channels 1, 2 and 3. (See colour plate section)

125

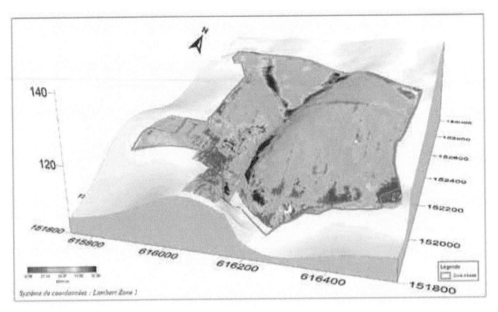

Plate 10. Superposition of resistivimetry (1m deep) over vertically exaggerated DEM. (See colour plate section)

their degree of erosion was very high. During the trial trenches phases, even a single stone of a few cubic meters was found coinciding with a geophysical anomaly, Plate 9 and Plate 10.

FURTHER READINGS

M. Dabas, A. Hesse, A. Jolivet, 1989: Prospection électrique de sub-surface automatisée, éd. CIRA (Centre Interdisciplinaire de Recherches Archéologiques), Bruxelles, pp. 73–81.

M. Dabas, G. Ducomet, A. Hesse, A. Jolivet, A. Tabbagh, 1989: Intérêt de la cartographie de la résistivité électrique pour la connaissance du sol à grande échelle, Science du sol, 27 (1), pp. 65–68.

M. Dabas, J.P. Decriaud, G. Ducomet, A. Hesse, A. Mounir, A. Tabbagh, 1994: Continuous recording of resistivity with towed arrays for systematic mapping of buried structures at shallow depth, Revue d'Archéométrie, 18, pp. 13–19.

M.Dabas, A. Hesse, J. Tabbagh, 2000: Experimental resistivity survey at Wroxeter archaeological site with a fast and light recording device, Archaeological prospection,7, 2, pp. 107–119.

M. Dabas, 2001: Du cas 1D au 3D: apport des simulations numériques à la notion de profondeur de détection et de réponse spatiale des principaux dispositifs employés en prospection électrique en Archéologie, in "Filtering, Optimisation and Modelling of Geophysical Data in Archaeological Prospecting", Fondazione Ing. Carlo Maurillo Lerici, Politecnico di Milano, ed. M. Cucarzi et P. Conti, Roma 2001, 202p., pp. 27–44.

M. Dabas, A.Tabbagh, 2003: A comparison of EMI and DC methods used in soil mapping-theoretical considerations for precision agriculture, pp. 121–129, in: Precision Agriculture, ed. J. Stafford and A. Werner, Wageningen Academic Publishers, Muencheberg, 783p.

M.Dabas, L. Guyard, T. Leppert, 2005: Gisacum revisité: croisement géophysique et archéologie, in Dossiers de l'Archéologie, no spécial «Géophysique et archéologie», no. 308, Nov. 2005, pp. 52–61, 79 p.

M. Dabas, O. Blin, C. Benard, 2005: Les nouvelles techniques de résistivité électrique employées dans la prospection de grandes surfaces en archéologie, Les Nouvelles de l'Archéologie, 101, 4, 24–32.

M. Dabas, 2006: La prospection géophysique, In "La prospection", Collection "Archéologiques" dirigée par A. Ferdière, éd. Errance, Paris, 2006, 167–216.

M. Dabas, 2007: resistivity-conductivity mapping for Precision Agriculture, a global overview, in: Near Surface Geophysics.

C. Panissod, M. Dabas, N. Florsch, A. Hesse, A. Jolivet, A. Tabbagh, J. Tabbagh, 1998: Archaeological prospecting using electric and electrostatic mobile arrays, Archaeological Prospection, 5, pp. 239–251.

Magnetic methods

Seeing the Unseen – Campana & Piro (eds)
© 2009 Taylor & Francis Group, London, ISBN 978-0-415-44721-8

Caesium-magnetometry for landscape-archaeology

H. Becker

Bavarian State Conservation Office, Munich, Germany

ABSTRACT: Using geophysical methods for landscape-archaeology special requirements on speed, sensitivity and spatial resolution are needed. Due to a large variety of magnetization processes, magnetometry especially in a multi-sensor set-up meets these requirements in an optimal manner. The comparison of total field caesium-magnetometry and fluxgate gradiometry shows clearly, that caesium-magnetometry should be preferred because of its ultra high sensitivity. The detection of deeply buried archaeological structures with low magnetization contrast is only possible by this method. The key for this application was the development "from nanotesla to picotesla" in the mid 1990s: A caesium-magnetometer with picotesla sensitivity, heretofore used only in aeromagnetics, was put to use on the ground for archaeological prospection. Also during these years J. Faßbinder from the Munich lab discovered a new biogenetic magnetization process: So-called magnetic bacteria, which have built-in magnetite single domain crystals, are involved in the rotting of organic materials and the subsequent formation of soil. This means, that a wooden post (non-magnetic!) becomes a magnetic structure. The next step towards high speed (and high spatial resolution) was done by applying the time mode sampling. This laid to the use of the two sensors of the gradiometer (single track) as a duo-sensor magnetometer (double track, which doubles the speed of the sampling in the field). By arranging the four sensors of two gradiometer systems horizontally one gets a quadro-sensor configuration. A first test of a quadro-sensor system was carried out in 1996 at Ostia Antica, the ancient harbour of Rome. Under smooth surface conditions the prospection of 1 hectare with 0.1/0.5 m spatial resolution can be done in 2 hours. Without these multi-sensor configurations it would have been quite impossible to measure huge sites like the Roman fort with vicus and necropolis near Ruffenhofen in Middle Franconia (c. one half square kilometre) in the short pauses available between the agricultural activities. But the largest area project was the prospecting at Qantir-Piramesse in Egypt from 1996 to 2004 reaching an over all area of nearly 200 hectares. Also the Celone Valley Project (2003—2006) in the Tavoliere (Apulia) with more than 100 hectares was only possible with the "magneto-scanner" with 4 sensors. Most of the projects also show, that geophysical prospection on the ground must be combined with other methods like aerial photography, laser scanning and high resolution satellite imagery for landscape archaeology. This was also demonstrated by the combination of magnetometry on the ground with Ikonos imagery at Uruk (Iraq) in 2001–02. The huge amount of data derived from high-speed magnetometry of large archaeological sites and for landscape archaeology every year arise new problems for data processing, visualization and archaeological interpretation. Still, there is a lot of work left in developing specialized software to transferring the information about the archaeological structures in the geophysical data to the human eye-brain system for interpretation, which cannot be done by computerized pattern recognition.

1 INTRODUCTION

Geophysical prospection applied to landscape archaeology includes some problems, which should be mentioned first. An archaeological landscape normally covers large areas. Very often they contain many different sites of different ages, but sometimes they consist just of one very big site. They may be situated in most different environments, on different geology, topography, morphology, climate, soils etc. In Egypt for example, the area from the boundaries of the agricultural land in the Nile valley at a extension to the desert of 5 to 10 km and over 1000 or 2000 km length may be considered as one archaeological landscape of burials, necropolises with pyramids, mastabas etc. for millions and millions of people being buried over many thousands of years. In the Nile valley on the other hand under completely different conditions there are the large cities, towns, villages, temple areas and palaces etc. situated and also forming large archaeological landscapes. Pi-Ramesse, the capital of Ramesses II. For example covers an area of about 30 square-kilometres and may be considered as the biggest city in the second millennium all over the world. When we are applying geophysics on the ground—and we must be very close to the surface and cannot do the job by airborne methods. Prospecting these very large sites means, that first of all we must be very fast. Secondly we need a high spatial resolution for mapping an archaeological site with their detailed structure. And finally we need an ultra high sensitivity for all the geophysical methods, because the contrast of the physical properties of the archaeological structures buried underneath the surface normally is very small. This means that for all the geophysical methods for prospecting in archaeology we are going near to the limits of the measurement in the field, the data evaluation, the visualization, and the interpretation. But trying to go over these limits makes the application of geophysics to archaeology so interesting. For geophysical prospection in archaeological sites we always must keep the three "s" in mind. They stay for speed, sensitivity and spatial resolution. These three "s" are the reason, why the application of geophysical methods for landscape-archaeology started rather late between 1970 and 1980, although geophysics in archaeology was used much earlier since 1940 or 1950, but only on rather small areas. In the course of the lectures you will see, that especially magnetometry has the potential to meeting the demands of the three "s" to the deepest extend. Over the last decades remarkable developments have been made in geophysical prospecting for archaeological purposes, beginning in the 1950s with the use of electricity to measure soil resistivity and of magnetics in the form of proton magnetometers. Martin Aitken was the very pioneer in magnetometry for archaeological purposes. When using a proton magnetometer in 1958 for the detection of ovens, he made the observation, that by this method it was also possible to detect pits by their contrast in the magnetization of the filling. This is described in his famous book "Physics and archaeology" (New York, 1961 and sec. Ed. Oxford, 1974 and also in his journal "Archaeomery", which marks the beginning of a new science in archaeology. He called this method "magnetic prospecting", not geomagnetic prospection, which is wrong, even when many authors are using it. In fact there must be applied quite sophisticated methods for the reduction of the geomagnetic field in the data to become able for the detection of the relevant signal from the archaeological structures. Of course it would be also nice to call the method archaeomagnetic prospection, as some other authors do so. But archaeomagnetism is also occupied by the magnetic effects of archaeological structures like ovens and other baked clay objects in situ, which can be used also for (archaeo-)magnetic dating (E. Thellier, 1962 and M. Aitken, 1974 etc.)).

Not many institutions in Europe or over the whole world have been developing magnetic prospecting for archaeological prospection, and even less for the needs of landscape archaeology. Even the very pioneer institute at Bonn (I. Scollar, Rheinisches Landesmuseum Bonn) stopped the very promising developments of combining geophysics and aerial archaeology in digital image processing techniques since the 1970s at the level of using proton-magnetometry, which is actually to slow and not sensitive enough for the application in landscape archaeology. The "Oxford Institute" (M. Aitken) concentrated on the refinement of TL-dating (thermoluminescence of baked clay), nearly all the institutions in Great Britain were using magnetic prospecting only based on fluxgate magnetometry with their limitations in gradiometry and sensitivity.

These critical remarks concerning the limits of fluxgate-gradiometry are not mend to neglect the profits of fluxgates for landscape archaeology. There are many examples for excellent magnetograms derived from fluxgate gradiometers. Nobody of the attendants of the Grosseto Summer school 2006 will forget the fantastic moment, when D. Powlesland (Landscape Research Centre, West Heslerton) enrolled his enormous print of a magnetogram of real landscapes archaeology taken with fluxgate magnetometers. Most of the fluxgate magnetograms look even more clear and sharper than a caesiummagnetometry. This is due to the higher depth of penetration of total field magnetometers, which also results in a summarizing effect of the magnetic anomalies over the deeper sources. For near surface structures with a high magnetic contrast fluxgate magnetometry always gives clearer images, but one should be aware, that it is not possible to detect the deeper structures. An example is given in figures 1a–c for a part of a very larges Old Kingdom necropolis in Dahshur in Egypt, where an bigger area had been measured both with fluxgate gradiometry and total field caesium magnetometry for comparison. Unfortunately the part with a rather deep mastaba could not be remeasured, because it had been excavated in the meantime. It consists only of one layer of sun-dried mud bricks (adobe) at a depth of 2 meters under sand.

In France at Garchy (A. Hesse) the development of techniques for archaeological prospection were concentrated mainly on resistivity and electromagnetic methods and only resistivity survey was atomised and adapted to the needs of large scale prospection. Even in the United States after the first application of high sensitive optical pumped magneometer (Rhubiduim magnetometry) in the 1960s (E. Ralph, "In search for Sybaris"), the idea of using high sensitive magnetometry had been not followed for landscape archaeology.

In Italy, where the application of technical methods in archaeology started also rather early in the 1950s by the "Fondazione Lerici" of Politecnico of Milano and developed by this company till the tragic death of Linington, have been followed by many scientific groups in the main Universities who have developed mainly the enhancement of inversion theory for magnetic, resistivity and GPR methods. In Spain, Portugal and the smaller European countries geophysical prospection for archaeology was never developed or institutionalised till today leaving all the very interesting projects in landscape archaeology to foreign (mainly commercial) groups. Very little research and application of geophysics in archaeology was done in the former Soviet Union and it was only after the opening of the "iron curtain" when some geophysical archaeology had started in Hungary, Czech and Slovakian Republics. So even at a global view there were only two institutions in Europe starting in the 1980s with the developments of caesium magnetometry for large scale archaeological prospection. First at the "Institute for Geodynamics and Meteorology" in Vienna (P. Melichar, later at the Archaeological Institute of the University by W. Neubauer), followed some months later by the "Institute for Geophysics" at the University of Munich by the author, later at the newly formed "Department for Geophysical Prospection and Aerial Archaeology" at the "Bavarian Authority for Monument Conservation" at Munich. Using almost the same instruments for high sensitive caesium magnetometry (Varian V-101) at Vienna and Munich, the field application at the two institutions varied mainly through more simple methods for the sampling in the field applied by the Munich group (H. Becker and J. Faßbinder), for being more flexible under difficult conditions at important projects abroad.

The application especially of caesium-magnetometry in Bavaria since 1981 had become quite unrivalled through continual improvements involving measuring techniques in the field and procedures for data evaluation, visualization and interpretation. Milestones included the construction of a differential proton magnetometer in the 1960s, automation of the digital sampling in the field, electronic processing of the data and finally digital imaging after the model of Irwin Scollar (Scollar et al. 1990). Building on his work at the Rhenish State Museum in Bonn, at the beginning of the 1980s the author was able to develop caesium magnetometry to the point where it could be used in archaeological prospecting. The author was very lucky being born just into an epoch with a very rapid development in electronics and hand-held computers and in taking part of the development of magnetometry for archaeological prospection since the middle of the 1960s from proton-magnetometers, differential protonmagnetometer and since the 1980s to the computerized systems for caesium-magnetometry.

The development "from nanotesla to picotesla" in the mid 1990s can be characterized as a "quantum leap" in a literal sense: a caesium magnetometer with picotesla sensitivity, heretofore used only in aeromagnetics, was put to use on the ground for archaeological prospection. The development of the caesium magnetometer, known as CS2/MEP720 (Scintrex/Picidas, Canada), became possible by the close collaboration with the engineer Bob Pavlic (Picodas) with the author working at the department "Archaeological Prospection and Aerial Archaeology" of the Bavarian

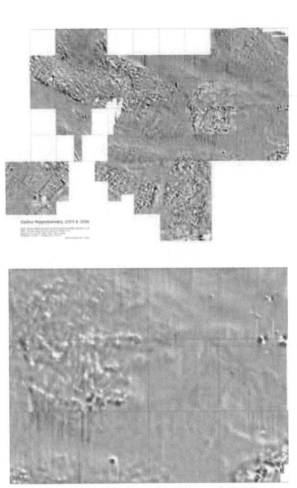

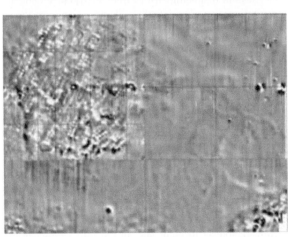

Figure 1. a–c Dahshur (Egypt). Caesiummagnetometry 2003 (total field), (Scintrex Smartmag SM4G in duo-sensor configuration) and fluxgate gradiometry 2005 (Geoscan FM36) of an Old Kingdom necropolis. The remeasurement of a group of shaft-graves with both instruments for comparison in b and c. Both measurements in the same grid, raster of caesiummagnetometry 0.1 × 0.5 m, raster of the fluxgate gradiometry 0.25 × 0.5 m (both interpolated to 0.25 and 0.25 m), dynamics ±10.0 nT and ±5.0 nT.

State Conservation Office. This was after the prospection of the city fortification of Homeric Troy in 1992 gained worldwide attention. Still in use today, this instrument was over one decade the most sensitive magnetometer that had ever been employed in archaeological prospecting. It was only during the last years, when a much faster and much more sensitive caesium magnetometer was developed—again by Bob Pavlik (Picoenvironment, Canada). The sensitivity of this instrument

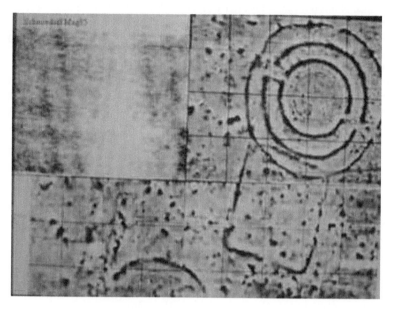

Figure 2a. Schmiedorf-Osterhofen 1985. Caesiummagnetometry of a Middle Neolithic ring ditch site with 2 palisades. Varian Caesiummagnetometer V101 in variometer mode. Raster 0.5×0.5 m, dynamics ± 7.0 nT, 20 m-grids.

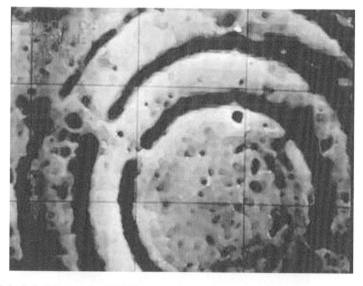

Figure 2b. Schmiedorf-Osterhofen 1994. Remeasurement of the interior of the ring ditch with Picodas CS2/ MEP720 in gradiometer mode. Raster 0.25×0.5 m, highpass filtering 10×10 pixel, dynamics ± 1.5 nT. Some posts of the palisade become visible.

is reaching a 10th of a picotesla at a cycle of 100 Hz (100 measurements per second). But we will see later, that the improvements of sensitivity and speed must be followed also by better methods for evaluation, visualization and interpretation. Still, there is a lot of work left in developing specialized software to transferring the information about the archaeological structures in the geophysical data to the human eye-brain system, which works much better in the detection of archaeology in geophysical data, than the highly sophisticated methods of pattern recognition. But the main advance of this ultra-high sensitivity of the CS2/MEP720 system in 1994 was a new procedure for time mode sampling of the data. Ten values per sensor can be measured and stored per second; at a fast walking tempo this accords with sample intervals of c. 10 cm. Secondly this time mode sampling of the data made available the socalled non-compensated (against time variations of the geomagnetic field) measurement (as opposed to the reduction of the time-based variations by a second sensor). This procedure allowed by an inbuilt band-pass filter the cancellation the high frequency portion of the time variation in the magnetic signal. This simultaneously laid to the use of the two sensors of the gradiometer (single track) as a duo-sensor magnetometer (double track, which doubles the speed of the sampling in the field).

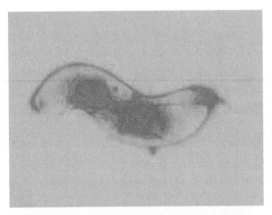

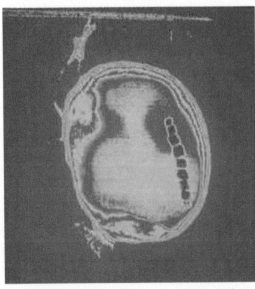

Figure 3a,b.   Magnetic bacteria with crystals of inbuilt chains of single domain magnetite (after J. Faßbinder 1994).

Also at this time Jörg Faßbinder from the Munich lab completed his dissertation on the magnetic properties and genesis of ferromagnetic minerals in the ground, as related to magnetic prospecting of archaeological sites (Faßbinder 1994). He had discovered a new biogenetic magnetization process: so-called magnetic bacteria, that have built-in magnetite single domain crystals are involved in the rotting of organic materials and the subsequent formation of soil. When soil formation is complete the bacteria die, leaving the magnetite crystals in the formally organic structure. Traces of a (non-magnetic) wooden post or a human burial e.g., thus become magnetic and can be identified even from above ground with the ultra-sensitive magnetometers through the anomaly of the geomagnetic field. This effect makes magnetic prospecting possible for a broad field of the wood/earth archaeology at the part of the world with humid climatic conditions.

2 DUO- AND QUADRO-SENSOR CONFIGURATION FOR HIGH-SPEED/ HIGH-RESOLUTION MAGNETIC PROSPECTING WITH CAESIUM MAGNETOMETRY

A new triumph in the use of magnetics for archaeological prospection occurred with the introduction of the multi-sensor technique in 1995 (Becker 1999). The earlier Varian V101- and Scintrex/Picodas CS2/MEP720 caesium magnetometer systems have been developed for one track gradio- or variometer configuration of the sensors, which ideally compensates the external geomagnetic variations. Every student in geophysics is still trained that the base for high sensitive magnetic prospecting is the complete reduction of the natural and technical temporal geomagnetic variations (micropulsations, diurnal variation, power lines, etc.) by measuring the difference between two sensors in vertical gradio- or variometer mode. It took the author almost two years realizing, that the two sensors of the gradiometer CS2/MEP720 could also be moved parallel in fieldwork covering two tracks for total field measurement at same height above ground. This was the first application of the so-called duo-sensor configuration (horizontally) with the CS2/MEP720 Picotesla-system in 1995, but this magnetometer gave still lots of problems in the data sampling in the field because of the long cables between the sensor-unit and the processor. Much more reliable for this purpose was the Scintrex Smartmag SM4G-special, which was used by the Munich group since 1996. A proto-type of the SM4G-special in the duo-sensor configuration was tested the first time at Monte da Ponte in Portugal in March 1996 with great success. In this duo-sensor configuration the two sensors of the vertical gradiometer are used horizontally with a spacing of 0.5 m, whereby the total geomagnetic field can be recorded in two tracks at one run. This simple "trick" doubles the sampling-speed. Every sensor added to the system multiplies the survey speed and opens a wide range for magnetic prospecting over large areas with limited time.

The key to this new technique is given by the magnetometer processor MEP720 (Picodas, Canada) with electronic bandpass filters selectable for 0.7, 1 and 2 Hz for cancellation of high frequency magnetic disturbances. Similar filters are used with Smartmag SM4G-Special (Scintrex). This offers also the opportunity for magnetic prospecting with Picotesla sensitivity directly underneath power lines or beside electric railways. Also the natural temporal high frequency geomagnetic variations (micropulsations) are cancelled by the same method of electronic bandpass filtering. Only the diurnal geomagnetic variation is reduced by the calculation and differentiation of the line means in a 20(40) m grid, which follow the main course of the geomagnetic field. At the moment the diurnal geomagnetic variation still shows an extremely smooth curvature because of minimal sunspot activity. For control one has to calculate also the square mean over a 20(40) m square because the line mean would cancel a magnetic alignment in line direction. The square mean reduction might be also important for the detection of deeply buried features. Only temporal variations with a wavelength compatible to the measuring time for a 20 m line (15–20 sec) cannot be cancelled by this method. But for the identification of archaeologically relevant anomalies there may be no problem, because these long wavelength disturbances will not show up in the next line and can be identified easily.

135

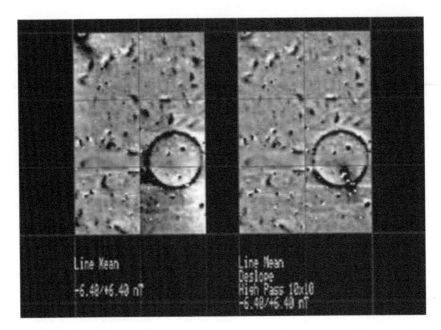

Figure 4. Wolfertschwenden (Bavaria).

The first example for a duo-sensor measurement with CS2/MEP720 system shows the magnetic prospecting in July 1995 for a Roman villa near Wolfertschwenden/Bavaria. The area containing a ring ditch possibly of a Roman burial or mausoleum is situated directly under a 500 kV power-line. The high frequency noise had been completely cancelled by electronic filtering with 1 Hz bandwidth, and the diurnal geomagnetic variation by numerical reduction on the line means in the 20 m grids. Only the strong static magnetic anomaly of a huge steel carrier in 25 m distance had been removed by high pass filtering (10×10 points) and desloping. Today this archaeological monument is partly covered by a cement paved road which can be identified in the magnetogram by its low noise signature.

In the meantime the duo-sensor configuration is applied as the standard method for magnetic prospecting carried out by the Munich team. The limits of this powerful method for large coverage in archaeological prospection are found on areas with nearby moving strong magnetic sources like trucks, caterpillars or tank lorries. But for "normal" applications in agricultural areas the duo-sensor configuration for caesium magnetometers with selectable band pass filters may be used for double speed or double spatial resolution.

3 ULTRA HIGH RESOLUTION CAESIUM MAGNETOMETRY AT MONTE DA PONTE, CONCELHO EVORA, PORTUGAL 1994–1996

During a prospection flight with O. Braasch in May 1989 the archaeologists team of the project Vale de Rodrigo (Ph. Kalb and M. Höck) realized that the site of Monte da Ponte must be something special. The place was used for centuries as canada for locking the sheep during night when driving them over long distances. But only for keeping sheep this building would be overconstructed consisting of several rings of high stone ramparts with a central tower and radial divisions. The site became a test area for the prototypes of the Picotesla-caesium magnetometer-system

CS2/MEP720 in 1994 and the duo-sensor configuration with SM4G-Special in 1996. A test excavation started in 1996.

For magnetic prospection the prototype of the ultra high sensitive CS2/MEP720 caesium magnetometer (Scintrex/Picodas, Canada) was used the first time. This instrument was for almost 10 years the most sensitive magnetometer used on the ground marks the step from Nanotesla- to Picotesla-systems (Becker 1995). The measurement was done in variometer mode (one sensor fixed as base station for cancelling the geomagnetic time variations). The instrument was switched to 10 measurements per second, which gave a spacial resolution of about 10 cm on the line. Traverse inverval was choosen with 0.5 m. Distance triggering was made manually every meter using a switch. The whole process was controled by the subnotebook computer Olivetti Quaderno, which was used for data logging too. A 12 V car battery was sufficient for running the system one day. Also a sun collector was added to the power supply, so there were no problems with energy in the field. However at this first test many problems mainly concerning the distance trigger and data logging had to be solved. The main problem under difficult surface conditions remained due to the separation of the sensor-unit and the (magnetometer, power supply, computer)—unit connected by a long cable which got stuck very often and had to be handled by a third person. The ideal magnetometer for rough surface conditions became the Scintrex SMARTMAG SM4G-Special, which can be operated by one person carrying the whole system, and which was used the first time in March 1996 at Monte da Ponte.

The site of Monte da Ponte shows a geometric construction of a huge oval fortification with 5 ring walls including the central tower, which measures 190 to 170 m. The plateau area between the second and the forth wall, which may have been the main habitation area, is divided into several sectors by radial walls with negative magnetization contrast, which indicates stone walls. The well preserved forth wall shows at their northern front a series of bastions, which are no more visible at the surface. The main gate may be identified on the east side in the fifth wall and the earth rampart extended in front of it, with the trace of the gateway leading to the interior plateau between tower and the second wall. In front of the fifth wall there is another curved structure, which could be an earth-work. Another 20 and 30 m outside of this structure there can be partly identified the trace of a palisade and an outmost ditch mainly on the northeast quarter of the fortification (Fig. XX). The third ring wall is only preserved in the northern part, but has vanished from the surface in the remaining area. The fifth wall cannot be seen above surface any more, but is clearly visible in the magnetogram.

Early in 1996, when the SMARTMAG magnetometer was to be tested, all stone ramparts (walls) were cleaned from their blackberry bushes, which resulted in an almost complete plan of the whole fortification (Fig. XX). With the use of the duo-sensor configuration the SMARTMAG magnetometer allowed also the prospection of huge areas in the surroundings, where the above mentioned palisade and ditch system was found. The idea of finding more external separate fortifications far outside the site was not confirmed by the prospection.

The combination of several prospection und survey methods like aerial photography, field walking, topographic surveying, digital terrain modeling and geophysical prospecting resulted in an idea and plan of the important archaeological monument of a Copper Age fortified settlement at Monte da Ponte. In addition to these nondestructive methods archaeological test excavation can be concentrated on specific areas for answering questions, which should give optimal additional information about this site.

The experiments with the duo-sensor configuration may have demonstrated, that modern caesium magnetometers like SMARTMAG offer the opportunity also for a quadro-sensor configuration simply by arranging the four sensors of two gradiometer systems horizontally. The whole set up of such a system consisting of the four sensors A, B, C, D with four magnetometer/sensor electronics, two consoles AB and CD and four batteries have been mounted on a non magnetic cart. The quadro-sensor system on wheels reach a total weight of 48 kg (non magnetic cart = 18 kg, batteries = 14 kg and 4 magnetometer systems = 16 kg) and can still been operated in the field by one person.

Figure 5. Monte da Ponte 1996. Caesium magnetometer SMARTMAG SM4G-Special (Scintrex) with duo-sensor configuration, sensitivity 10 pT (0.01 nT) at 0.1 sec cycle (10 measurements per second).

Figure 6. Monte da Ponte 1997. Aerial view of the Copper Age fortified settlement with signalled control points (20 m-grid) on the ground by Rupprecht and Michaela Steinman as base for a photogrametric evaluation for a plan of all stones.

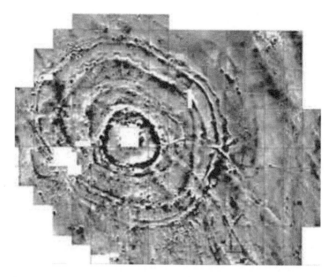

Figure 7. Monte da Ponte 1994–1997. Magnetogram as digital image. Caesium magnetometer CS2/MEP720 (technical details see above) and SMARTMAG SM4G-Special with duo-sensor configuration, sensitivity 10 pT, raster after resampling 0.5/0.25 m, dynamics –6.4/+6.4 nT to –3.2/+3.2 nT (outer area) in 256 greyscales (white/black), 20 m grid.

Figure 8. Quadro-sensor configuration on a "caretto" with 2 Scintrex Smartmag systems with the 4 sensors arranged horizontally at 0.5 m distance. Position on the line is triggered by the rotation of the wheel with 2 cm accuracy.

A first test of a quadro-sensor system was carried out in August 1996 at Ostia Antica, the ancient harbour of Rome. An test area of 15 ha was measured in the regio V of Ostia during 7 days of field-work. In the meantime under smooth surface conditions the prospection of 1 ha with 0.1/0.5 m spacial resolution may be done with the quadro-sensor chariot in 2 hours. The project in Ostia resulted in the discovery of the basilica of Constantinus I.

4   PROSPECTING IN OSTIA ANTICA (ITALY) AND THE DISCOVERY OF THE BASILICA OF CONSTANTINUS I. IN 1996

After the huge excavations in 1938 to 1942 in Ostia Antica, the ancient habour of Rome for the World Exhibition 1942 in Rome, there remained about 40 ha of the area of the ancient city untouched. This is about 50 to 60% of the original built up area. An area of about 15 ha, the biggest untouched area, in regio V in the southeast of the ancient city was selected for a first test for geophysical prospecting in August 1996. This was also the first test for a quadro-sensor caesium magnetometer system mounted on a non magnetic chariot—the so-called magneto-scanner. The interpretation of the archaeological structures in the magnetograms are in some parts very simple and clear, but in others rather problematic possibly caused by the multi-layer structure of many building phases of this important city over many centuries. The peculiar wide positive-negative anomalies (black-white stripes in the magnetogram) are geologically caused by the shore lines of the Tiber delta with a concentration of geological magnetite due to the wash of the waves. Their effect can be slightly improved by highpass filtering of the data. But there are also many archaeological structures to be seen in the magnetograms. Very dominant shows traces of the Via del Sabazeo (from north to south) possibly due to a channel made by backed bricks in the underground (cloaka maxima), but there are also some other streets. The Late Republican city wall is drawn only by a narrow line corresponding to the little width of the wall, which was made in opus quasi reticulatum technique. But if one looks

Figure 9a.   Ostia-Antica 1996. Part of the magnetogram of regio V with many structures of archaeological evidence. The newly discovery basilica of Constantinus I., the Great, is clearly visible in the corner at the gate of the Via del Sabazeo and the city wall to the south. Caesium magnetometry SM4G-Special in quadro-sensor configuration, sensitivity 10 pT (= 0.01 nT Nanotesla), Dynamics −50.0/+50.0 nT in 256 grayscales (white/black), raster after resampling 0.25/0.5 m, 1 Hz bandpass filter, reduction of the diurnal geomagnetic variation by line-mean value, 40 m grid.

Figure 9b.   Ostia-Antica 1996. Part of the magnetogram of regio V after highpass filtering 10 × 5 pixel, same technical data as Fig. 3a, but dynamics −10.0/+10.0 nT.

at a very oblique angle exactly in the direction of the wall (in Fig. 3a, 3b in 290° xx from the centre to the west = left) is to be defined as a very clear black line (positive magnetic anomaly caused by the building mode as opus quasi reticulatum made from volcanic tuff). Outside of this southeastern part of the city wall there was found a road leading from the porta secondaria directly to the Via del Sabazeo. To the west this road seems to be a bypassing route directly to the Via Laurentina. Adjacent on the outside there is a row of rather early burial monuments. In the interiour area of the city there a several buildings arranged in a insulae.

The most significant discovery in 1996 was a early christian basilica, which may be the basilica of Constantinus I., the Great, also mentioned in the liber potificalis in the vatican, which is clearly visible in the corner at the gate of the Via del Sabazeo and the city wall to the south. The overall dimension reaching nearly 90 m in length provides strong evidence having found the basilica of Constantinus I. indeed. This nearly eastwest oriented building consists of 3 arches with a apsis, but without the lateral hall. The part of the basilica adjacent to the Via del Sabazeo is not clearly visible, but west to the main building there is clearly visible the atrium. At the southern side of the atrium there may be a round building with 9.0 m diameter, which could be a baptisterium. Also clearly visible is another older building underneath the basilica which may be leveled for the foundations of the basilica. In the meantime a directed sondage excavation early in 1998 proved this interpretation of having discovered the basilica of Constantinus I.

5   RESAMPLING PROCEDURE AND DATA PROCESSING

Fast moving sensor systems need special procedures for sampling and data processing. The major advance for fast field measurements with high spatial resolution is the time mode sampling instead of the event triggered sampling at distinct sample intervals at 0.5 m. Modern magnetometers allow 10 measurements per second with pico Tesla sensitivity (MEP720/CS2, Picodas/Scintrex), 10 pT sensitivity (SMARTMAG SM4G, Scintrex) and 50 pT sensitivity (G-858G, Geometrics). The high frequency geomagnetic time variations are cancelled by band pass filtering 0.7, 1, 2 Hz for Picodas MEP720 or 1, 2, 8 Hz for Smartmag SM4G. As mentioned above the diurnal variation is reduced to the mean value of a 40 m line and also to the mean value of a 40 m square to be sure not cancelling anomalies directly in the line. The cycle of Picodas MEP720 and Scintrex SMART-MAG SM4G can be set to 0.1 sec (10 measurements per second) which means a spatial resolution of 10–15 cm at normal to fast walking speed. With rather fast sensor moving systems the problem of a data shift must be solved, this means in zigzag mode a displacement of the sensor's position even after exact distance triggering. The measuring time of the magnetometer should be known for exact distance triggering, which is also dependent to the walking speed. This shift correction must be calculated with a time constant, which is typical for specific magnetometer types (0.25 for MEP720/CS2 and 0.75 for SMARTMAG). Only a speed dependent shift correction results in a 'sharp' image for the magnetogram. In 1996 an area of about 80 ha, but in 1997 an area of 140 ha with 0.5/0.1 m spatial resolution (70 Million readings) had been measured with CS2/MEP720 and two SMARTMAG SM4G-Special systems. About 200 ha per year should be possible, using duo- and quadro-sensor configurations at good surface conditions on wheels. Automatic position-ing systems consisting of GPS for beginning and end of a line combined with wheel-triggers for exact distances on the line may speed up field procedure even more. The two MEP720 systems with four CS2-sensors and five SMARTMAG SM4G-Special caesium magnetometers with three consoles which can be operated as 2 complete compensated quadro-sensors systems offer now the power for covering at least 200 ha per year with magnetic prospecting, which will attribute a important part in archaeological research and archaeological monument conservation. Although the discussion so far has focussed primarily on the development of measuring techniques using caesium magnetometry and on the presentation of magnetograms, it must be also emphasized that, with as much as a ten-fold increase in the speed of measurement, the problems of magnetic pros-pecting are now to be found on a totally different level. The areas prospected with the multi-sensor techniques have sometimes covered more than 200 hectares per year, which puts the processes of visualization, analysis and interpretation to a new test.

Figure 10. XXX Roman castellum near Ruffenhofen, Middle Franconia. Defensive walls, towers, gates, ditches, and parts of the interior stone buildings shown as crop marks in a grainield. Aerial photograph from 5 July 2001, photographer K. Leidorf.

Figure 11. XXX Roman castellum near Ruffenhofen. Magnetogram of the fortress with the vicus and the baths. Smartmag SM4G-special as quadro-sensor. Sensitivity 0.01 nT (10 picotesla), raster 0.1/0.5 m interpolated to 0.25/0.25 m, dynamics −7.0/+7.0 nT in 256 grey scales (black/white), 40 m-grids, magnetic prospecting H. Becker, 2000.

Even some specific single sites like the Ramesses capital in the Nile delta were reaching an area over 2 square kilometer (200 hectares). A further increase in the speed of measurements was achieved by using the Smartmag-SM4G-special magnetometers either as three complete duo-sensor system, or one quadro-sensor system: up to four complete magnetometer systems with the two gradiometer processors, data storage and power supply for four hours of operation can be fitted on a newly constructed cart. The analysis process corresponds to that of a double duo-sensor configuration.

The rotation of the wheel automatically triggers the position on the line with a precision of better than 10 cm. Thus the speed of measurement in the field could be increased fourfold. Very strong geomagnetic variations during magnetic storms can be compensated through a fifth magnetometer system acting as the base in a variometer configuration. In rough or steep topographical conditions Smartmag is carried handheld on a wooden frame only as a duo-sensor. Since a long cable between the sensor-unit and the magnetometer processor is not necessary, this arrangement can be even used under very extreme topographical conditions by one person. The Smartmag SM4G-special in the duo-sensor configuration has also proved highly successful for many projects abroad, which have been carried out using this techniques since 1996 until today (2007).

This duo-sensor system is considered as the most successful development so far for magnetic prospecting. Not only the huge area in Qantir-Piramesse using up to three duo-sensor systems, but also the measurements at Uruk in Iraq 2001 and 2002 with two duo-sensors covering 4 hectares per day, and many other sites in the Mediterranean, Turkey, Syria, Siberia, China, etc. etc. were measured with the Smartmag duo-sensor. The magnetic prospecting on open fields mainly in

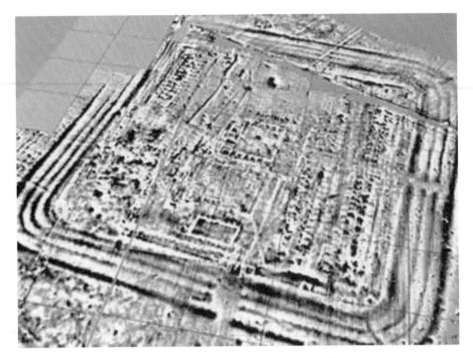

Figure 12a.   XXX Roman castellum near Ruffenhofen. Magnetogram of the fortress from a bird's eye view as the basis for reconstruction and computer animation.

Figure 12b.   XXX Roman castellum near Ruffenhofen. Still image from the computer animation with the inner structures of the castellum, computer animation H. Becker, A. Pohl, 2000.

Bavaria, but also abroad (Celone Valley Project in Italy, or Hindwell in Whales) were done using the quadro-sensor cart if possible. Without this device it would have been quite impossible to measure the huge site of the Roman castellum with vicus and necropolis near Ruffenhofen in Middle Franconia (c. one half square kilometre) in the short pauses available between the agricultural activity.

6 PIRAMESSES—THE LOST CAPITAL OF RAMESSES II. IN THE NILE DELTA (EGYPT)

During the prospection of Piramesse, the city of Ramesses in the Nile delta, an area of nearly two square kilometres was covered for the first time (100 million measurements in a raster 0.1/0.5 m, interpolated to 0.25/0.25 m). With Piramesse estimated to extend c. 30 square kilometres, the area measured so far cannot be considered representative, but nevertheless entire urbane quarters could already distinguished, with temples, palaces (probably including one of the main Ramesses palaces), villa districts, dense residential areas and a shore line more than one kilometre in length.

Up to four settlement strata are superimposed in some places. The data evaluation of the measurements from 1996 to 2004 were standardized so that the magnetograms of the prospected areas were presented as the overlay of the high pass filtered data on the original total field data with a transparency of 60%. The compilation of these compiled magnetograms onto the topographic map best reflects the "city map" of the Ramesses metropolis.

Figure 13. Qantir-Pi-ramesses. Caesium-Magnetometer Scintrex SM4G-special as handheld duo-sensor of a wooden frame over the feet of a colossal statue of Ramesses in Qantir-Piramesse (Egypt).

145

Figure 14. Qantir-Piramesse. Compilation of the magnetograms from 1996 to 2004 on the topographic map reaching an over all area of nearly 200 hectares. Smartmag SM4G-special as duo-sensor. Sensitivity 0.01 nT (10 picotesla), raster 0.1/0.5 m interpolated to 0.25/0.25 m, dynamics −7.0/+7.0 nT in 256 grey scales (black/white), overlay of the highpass filtered data with the original total field data with 60% transparency, 40 m-grids, magnetic prospecting H. Becker (1996–2003), J. Faßbinder (1996–2000), Chr. Schweitzer (1999, 2000), topography V. Fuchs, D. Kaltenbach and M. Burgmayer.

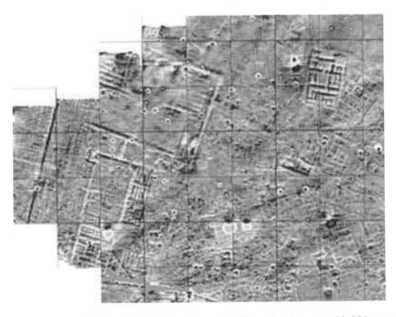

Figure 15. Qantir-Piramesses East 1998. Magnetogram (detail) as digital image with 256 greyscales of a district of the Ramesside capital with the main palace, Amarna type villas, streets, channels, houses, workshops and a temple building (?) in the centre. Raster 0.5/0.2 m., dynamics −12.8/+12.8 nT (black/white), square mean, edge matching and desloping, 40 m-grid.

146

Figure 16.   Qantir-Piramesses-Tell Abu Shafir 1998. Magnetogram (detail) of a temple building with several phases. Same technical details than Fig. 15.

7   CICAH—A SCYTIAN FORTIFIED SETTLEMENT WITH NECROPOLIS IN THE BARABA STEPPE 1999 AND 2000

The legendary Scythians, controlling in the first millenium b.c. the vast steppes of Central Asia, were first described by Herodotus (5th century b.c.) as mounted nomads and feared warriors. This view was only little altered through the times until today. Even modern archaeology tries to verify this picture from antique times. Archaeological research nowadays is still considering the Scythians as nomads and concentrates mainly on the investigation of their burial buildings—so called kurgans—and on their admirable craftmenship and art style especially for metal work. Although one would think that these capabilities, the organisation and management of numerous people for constructing the huge kurgans, and the highly developed art style in metal work are not likely for people living in the saddle. But the idea of searching for permanent habitations or settlements of the Scythians still would cause a mild smile by most scholars in the field of Central Eurasian archaeology.

In the course of a joint project the Russian colleagues offered the opportunity for investigating a small fortified settlement of the Scythians which was recently discovered in the Baraba steppe south of Barabinsk in Southern Siberia near Cicah. Trial trenches excavated by the Russian archaeologists unearthed a grubenhaus inside a rather small ditched enclosure at the steep shore of a lake. Dating by typological reasons of the ceramics indicates a narrow spectrum in the 8th and the 7th century b.c., which would be clearly Scythian period. It seems rather astonishing that there are still archaeological structures from the late Bronze Age or the Early Iron Age visible on the surface and well preserved, but the steppe seems to be almost resistant against erosion.

In preparation of the planned excavation of the site at a bigger scale in 2000 the Department for Archaeological Prospection and Aerial Archaeology of the Bavarian State Conservation Office was asked for a geophysical prospection measurement in 1999. Already in 1998 a successful prospection campaign for Scythian and older burial monuments in the Minusinsk area and Tuva using caesium-magnetometry mainly had been organized as such a joint project.

The Scythian site of Cicah, partly ploughed in the surrounding area, was also surveyed by field walking through our Russian archaeologist colleagues under Marina Chemyakina from the Siberian Academy, which resulted in a vast distribution of ceramics, stone tools and slags far beyond the ditched site visible on the surface. On the base of this distribution a 40 m grid over 400 × 120 m., latterly enlarged to 400 × 200 m. (8 hectare) covering the whole area was topographically surveyed and marked by wooden pegs.

The first campaign for magnetic prospection took place during three days in June 1999 using a Scintrex Smartmag SM4G-Special caesium magnetometer with 10 Picotesla sensitivity at a cycle of up to 0.1 sec. The magnetometer system was run the whole day from morning till evening by the authors covering the whole area of the visible ditched settlement and the surrounding area with the ceramic fragments at an extent of 7.5 hectare (about 1.5 million measurements) (Fig. 1). This was only possible by using a non compensated duo-sensor configuration covering two tracks at one run. The sensors were configured at 0.5 m horizontal distance, sampling rate was set to 0.2 sec, which gives at normal walking speed a spacial resolution of 0.2 × 0.5 m. The distance control was made manually by switching every 5 m over the 40 m line. The high frequency part of the diurnal variation (natural micro-pulsations and technical noise) was cancelled by setting a bandpass filter of 1 Hz in the hardware of the magnetometer processor. The slower magnetic changes of the daily variation of the geomagnetic field was reduced to the mean value of all measured data of a 40 m line and also to the mean value of all data of a 40 m-grid. All data were interpolated to 25 cm in each direction and on the line, dependent on the walking speed. All data were dumped and finally processed on a notebook computer in the main tent of the camp during night, which resulted in an almost complete visualization of the measurement in grey shading technique. The fit of adjacent grid sides were corrected by digital image techniques like edge matching and desloping, which resulted in a rather smooth image for the magnetogram even of the raw data (Fig. 2a). Highpass filtering resulted in an even clearer image showing some interior structure of the grubenhauser like post holes, fireplaces and walls.

The magnetic prospection 1999 and 2000 of the Scythian fortified settlement with the adjacent necropolis near Cicah in Siberian was considered as a sensation. After the major part of the Scythian urban site from the 8th to 7th centuries BC could be measured in just three days in 1999, in 200 the entire city consisting of more than hundred grubenhauser and its necropolis were to be discovered. The size of the grubenhauser with 8 × 10 m normally was found to be almost similar to the houses which were excavated previously in the trial trenches in the "citadel" still visible at the surface by the archaeologists from the Russian Academy of Science, Sibirian Branch, Novosibirsk. The whole settlement is clearly divided into several sectors by ditches and palisades, which also show some gates. The houses seem to be aligned along streets, but these were not visible in the magnetogram. Outside (north) of the main ditch of the settlementa series of smaller houses is aligned, which may be storage houses of workshops, because of their size. Considering the different signature of the grubenhauser in the area of the citadel, where the houses are still open and the ploughes area in the lower city, where the grubenhauser were cut in the Sibirian loess and are filled by top soil (Chernozem), the main magnetization might be dominated by the Le Borgne effect (Le Borgne 1965).

The overall set-up of this fortified settlement divided by ditches and palisades consists of the "citadel" with a main ditch still open in the southern and northern part and situated directly at the steep shore of a lake and the rather complex "lower city" which may have been developed in several steps over a longer period. Especially the northern extension of the city bordered by two palisades rather than a main ditch may have been built in the final period.

Directly outside to the external northern gate and oriented to this two burials appear, which lead to a large necropolis of different periods towards the area of two very big kurgans in several hundreds meters in the distance, which are still visible above ground. But there was a line of several other big kurgans not more visible at the surface and hundreds of smaller burials of different types.

Uruk—the biblical Erech—still remains one of the most famous sites in Mesopotamia. Even when Babylon became the capital of Akkad and Sumer, Uruk was always the main religious and cultural centre of ancient Mesopotamia. The dawn of civilisation is connected with the name of Uruk—the development of urbanism and the beginning of writing and literature in the 3rd millennium. Gilgamesh, hero, half man half god, was King of Uruk. In the epic of Gilgamesh a lot of information about the ancient city and ancient life in Mesopotamia can be found. In this early text we also find the remark of Gilgamesh, that he has used baked bricks for the city wall he has built. With this wall Gilgamesh made Uruk to the biggest and most important city of the world in these times, at the beginning of the 3rd millennium. The wall with a length of about eleven kilometres and with hundreds of bastions surrounded an area of 5,5 square kilometres. We assume, that the early city wall of Uruk had only a mantle of baked bricks, filled with the cheaper mud bricks (adobe) which would give an ideal base for magnetic prospecting, because of the high susceptibility and remanent magnetization of burnt clay. Therefore we tried to prospect the city wall also in an area where the wall is no more visible at the surface, and we planned to locate at least one of the city gates, which were still unknown.

Again from the Epic of Gilgamesh, as well as from many iconographic illustrations we have descriptions of rituals using a bark on a canal, e.g. from Uruk on a cylinder seal from the beginning of the 3rd millennium a ritual scene is shown with a boat on a canal carrying a shrine and an altar mounted on the back of a bull. Like in other cities of Mesopotamia, a sanctuary of a special type was situated in Uruk, outside the walls of the city, but belonging to it and called the New Years Chappel (bait akitu). Once a year (on New Year's day, when the King was allowed to enter the inner temple area) the statue of the principal deity of the city (in Uruk she was Ishtar) was carried to the sanctuary in a procession, accompanied by throngs of worshipers. In certain instances, a sacred road through a special gate linked the outer sanctuary to the main temple of the goddess (Oppenheim 1964). For Uruk we have a description of Gilgamesh leaving the city on a bark for a ritual course. That means, that we should expect a canal or even a system of canals in the city. This question led to controversial discussions among archaeologists, because nowadays Uruk is situated in completely dry dessert. But also Margarete van Ess succeeded to identify a series of canals in the RAF aerial photographs from 1935.

Based on these questions of how to understand some of the descriptions about the ancient topography of Uruk in the Epic of Gilgamesh from an archaeological point of view, and in preparation of the planned archaeological investigations for the future, the first Iraqi—German joint archaeological project after the Gulf War was started in February 2001, followed by a second campaign in 2002. Unfortunately this promising restart of scientific investigations at Uruk after a long pause came to a sudden end in 2003 with the 2nd Gulf War by the American—British Allies, when it became absolutely impossible to work at Uruk again on the ground. But these two campaigns 2001 and 2002 were concentrated on high sensitive magnetetic prospection, topographical survey and a drilling project for sedimentological questions in 2002. These tests for magnetic prospection by caesium magnetometry were mend to answer the question, if this method would be suitable even for a multi layered site consisting mainly of debris from sun-dried mud bricks—the almost unknown problem of prospecting archaeological structures made of mud in a surrounding of mud. The experiences from other oriental mud brick sites have shown, that high sensitive—high-resolution caesium-magnetometry could do the job.

Results of the magnetometry already in the two test areas were really fantastic. Already the first evening of the 2001 campaign we realized that we have traced the main canal as well as some house structures, including the very clear plan of an Old Babylonian house built of baked bricks. The following days we found, that the old city of Uruk had a complete canal system with a nearly 5 m wide main canal from north to the south and several secondary canals in the area to the west. Besides this canal system some streets existed at different levels, but the canal system seems to be much more important. Another interesting discovery mainly in this area I (Middle) were vast

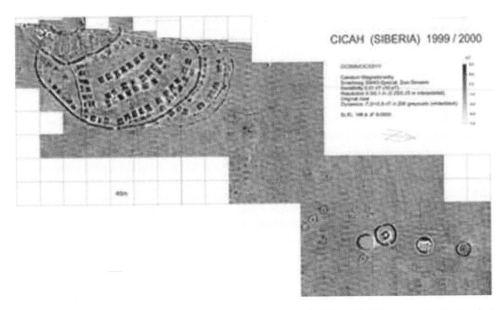

Figure 17. Cicah-Siberia 1999 & 2000. Magnetogram in grey-shading with 256 greyscale of a fortified settlement with a "citadel" and a "lower city" consisting of grubenhauser divided by ditches and palisades and a large necropolis of different periods with some big kurgans and hundreds of smaller burials. Caesium magnetometry Smartmag SM4G-special in duo-sensor configuration, sensitivity 20 Picotesla, raster 0.5 × 0.1 m interpolated to 0.25 × 0.25 m, dynamics −5.0/+5.0 Nanotesla (white to black), line-mean over 40 m, desloping and edge matching, 40 m grids.

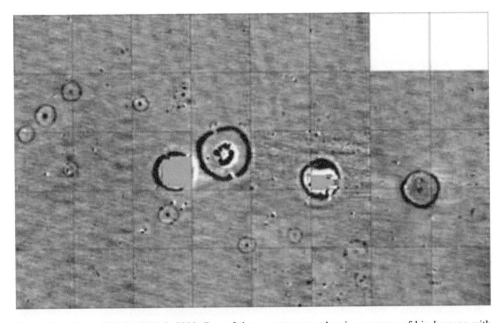

Figure 18. Cicah-Siberia 1999 & 2000. Part of the magnetogram showing a group of big kurgans with details of the entrance, surrounding ditch and burnt burial chambers. Technical details from Fig. 17.

ravages of the settlement patterns by floods mainly on the western side of the main canal. It seems that in Uruk it was always a severe problem to get out of the water. Not a single city gate for streets were found, but on the outer side of the city wall another canal parallel to the wall was found. Possibly all the transportation at Uruk was only on the water.

In the area II (South), but also in the 2002 extension of area I (Middle) the city wall gave a clear magnetic signal mainly caused by the mantle of burnt bricks on both sides as described in the Epic of Gilgamesh. The southern city gate shows a wide opening of nearly 16 m with towers on both sides and built completely by baked bricks. The whole structure can be interpreted as the main "water-gate" for the passage of the main canal through the city wall, which helps to imagine a splendid building like the Ishtar gate in Babylon. The coring in this water-gate in the 2002 campaign resulted in the depth of 4 m for the base of the main canal, and showing the existence of baked bricks down to this depth.

In the area outside of the city wall towards the so-called bait akitu building, a large cemetery was identified in the magnetogram. Ceramics on the surface from numerous robbery pits in this area indicates that this was part of a large necropolis of the Kassite epoch. The burials show very clear in the magnetogram, because of the used single or double pitoi for coffins.

Also the supposed New Year's Chapel (bait akitu) situated nearly 200 m south of the city gate, showed very clear magnetic anomalies due to the high magnetization of burnt bricks. The architectural details with three cellular structures on the western side of the building with a large court and a surrounding wall are quite different from the known plans of temples in Uruk. There existed a direct canal from the "Ishtar-gate" through the city wall to a smaller gate (of burnt bricks) beyond the New Year's Chapel and a smaller canal leading to its western side and forming an elongated basin for boats to anchor. Everything seems to correspond exactly to the illustrations and descriptions of the New Year's ritual, except the fact that we are dealing

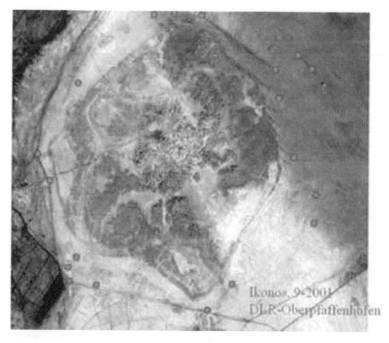

Figure 19. Ikonos imagery from September 2001 showing the ancient city of Uruk. Multi-channel image, sharpened by a panchromatic channel for higher resolution (less than 1.0 m). Courtesy by DLR-DZD (Oberpfaffenhofen, Germany).

Figure 20. Ikonos imagery from September 2001. Detail of Fig. 19, showing the southwestern part of the city, the city-wall, "Ishtar-gate" and some straight lines with 40 m distance, representing the footsteps of the magnetic prospecting on the ground about half year earlier.

with the first millennium instead of the third. But there may be an older building underneath Nebuchadnezzar's temple. Unfortunately in 2003 after the outbreak of the Gulf War II it became absolutely impossible to reach Uruk on the ground for further investigations, and this will be the situation for many years in the future. Again it was the idea of M. van Ess of the German Archaeological Institute to contact the German Aerospace Centre, German Remote Sensing Data Centre, (DLR-DFD), Oberpfaffenhofen near Munich for a joint project "Detection of Looting Activities at Archaeological Sites in Iraq using Ikonos Imagery" financed by the German Foreign Ministry since 2005. Uruk was chosen for the main test side for using this very complex combination of different prospecting techniques like archaeological and topographical ground and surface survey, geophysics (caesium-magnetometry) in 2001 and 2002, coring for sedimentological investigations in 2002, historical aerial photography by the Royal Air Force in 1935 and high resolution satellite IKONOS imagery from 2001 and 2004. Definons AG and European Space Imaging, both in Munich, made some of the data processing and pattern recognition from the satellite imageries. The specialists of DLR-DFD put all the information from these various sources together in GIS.

The IKONOS scene from September 2001 of Uruk was chosen as a multi-spectral picture with a resolution on the ground of several meters from 681 km height, but this was sharpened by the panchromatic channel with less than one meter resolution on the ground. Nobody of the participant scientists will forget the moment when a first zoom into the picture was made in the laboratories of DLR-DFD, recognizing the plan of the entire city of Uruk with its long city-wall, the central area with all the traces of the excavations since 1912, the canal systems, the "Ishtar-Gate" and even details of some buildings like an architectural plan. Even the posts of the surrounding fence were visible, which led to the geo-referencing of the satellite image by the edges of the fence, which were topographically surveyed on the ground some years before. In the area in the southwestern part almost outside of the city wall long stripes in north-south direction were detected with a measured distance of exactly 40 m.

Figure 21a.   Ikonos imagery from September 2001 with the superimposed magnetogram of the southwest area from 2001 and 2002, showing many details of the ancient city like the city wall with the bastions, "Ishtar gate" as a water gate for the main canal, leading to another gate outside the wall, tracing the ritual course of a bark to the New Years Chappel (bait akitu).

Figure 21b.   Figseal impression showing a ritual bark on its course (from Uruk, 3rd millennium).

153

Figure 22.   Detail of the Ikonos imagery from Fig. 19 and its interpretation showing details from the survey area North.

Figure 23a.   Part of this area of fig. 22 (Ikonos imagery), as base for superimposing the RAF photo, and the magnetograms of 2002.

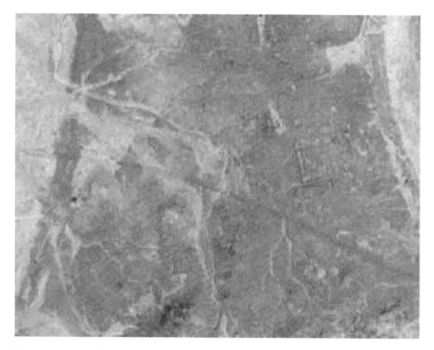

Figure 23b.   Same as fig. 23a, but with the superimposed RAF photo from 1935.

Figure 23c.   Same as fig. 23b, but with the composite of the magnetic prospection of 2002, showing many details of this area, the canal, the buildings (Ur III palaces), a street and a concentration of big ovens for the production of bricks and ceramics. Interpretation and drawing of the archaeological features by M. van Ess, German Archaeological Institute, Oriental Institute.

Figure 23d. Extraction of archaeological structures from the magnetogram 2002 by pattern recognition. Courtesy of Definions, Munich.

9 THE CELONE VALLEY PROJECT (2003–2006)—PARADISE FOR AERIAL AND GEOPHYSICAL PROSPECTION IN THE TAVOLIERE (APULIA)

Between 1943 and 1945 during the allied liberation of Italy a significant chapter for aerial archaeology was opened especially by John Bradford, when he served as an intelligence officer in the Royal Air Force. His great interest for archaeology led to a great lot of training flights and aerial photo interpretation projects specially flown by RAF even during the last years of World War II. After 10 years work John Bradford published his very important book "Ancient Landscapes" (1957) which was based on numerous articles mainly published in Antiquity. Bradford was the first to realize, that especially the Tavoliere in Apulia represents one of the most important archaeological landscapes in Europe. In that rather small area he could detect 150 to 200 ditched enclosures, and the greatest part of them belong to the Neolithic epoch. It is hard to believe, that this extremely rich archaeological area did not become one of the most important European areas for archaeological investigations. Almost 50 years after Bradford there were again the aerial archaeologists—Jim Pickering, Otto Braasch and Chris Musson—rediscovering the Tavaliere as the paradise for aerial archaeology and drawing a strong archaeological interest to this area. Again it was Otto Braasch suggesting an archaeo-geophysical project by sending an absolutely fantastic

Figure 24.   A Caesium—magnetometry in the Celone Valley Project with 2 Scintrex Smartmag SM4G-special mounted in quadro-sensor configuration on a non-magnetic carretto. The position on the line is triggered by the rotation of the wheel with 2 cm precission.

aerial photo from the Early Neolithic ditched enclosure Salveterre near Ascoli Satriano, which he called "bastione baroque".

Early in summer 2003 a fruitful cooperation started with 1. Department of Geology, Campus Universitario, Bari (Marcello Chiminale, Danilo Gallo), 2. Bavarian Monument Conservation Authority, Department for Archaeological Prospection and Aerial Archaeology, München (Helmut Becker) and 3. University of Foggia (Guilano Volpe, Valentino Romano). Until now (2006) an area of more than one square-km was investigated by high resolution-high sensitive caesium-magnetometry. The main part of this area (about 90%) was measured with two Scintrex Smartmag SM4G-special from Munich mounted in the quadro-sensor configuration on a non-magnetic "carretto". This instrument has a sensitivity of 10 Picotesla at a cycle of 10 Hz, which resulted in a sample interval on the line of 10 to 12 cm. The distance on the line (40 m) was triggered by the rotation of the wheel, and the line interval was set to 0.5 m. The data sets with the spatial resolution 0.5×0.1 m were interpolated to 0.25×0.25 m for further data processing (see chapter about the duo- and quadro-sensor configuration). We also applied the caesium-magnetometer of Bari (Geometrics G-858G) from the very beginning, but it became evident, that the original set-up of the sensors as a vertical gradiometer and the aluminium staff etc. was not suitable for this large areas. A new wooden frame was built following the Munich example of a duo-sensor frame. But the positioning on the line was only done by manual switching every 5 m. The sensitivity of Geometrics G-858G is only 50 Picotesla at the same 10 Hz sampling rate.

The data from the Munich Smartmag SM4G-special in quadro-sensor configuration were always processed during the night with the above described technique for a non-compensated total field measurement, so the next morning there existed already the almost final magnetogram for any purposes. This fast procedure is very important for the controlling of the quality of the primar data, but also for following linear archaeological structures like ditches or limitations. This is even more important under "normal" conditions in rescue archaeology, where only one chance is given to do the job. The "Bari group" used a specially developed software for despiking, destaggering and destriping. The destaggering was programmed by an auto-correlation of linear anomalies in lines side by side, which gave good results for data sets after bad switching of the position (5 m marks). But this method seems to be so time consuming, that the data processing of some sites could not be finished even after 2 years later. This is in severe contradiction with the speed of the 3 s demands, which were discussed in the beginning. Speed is very important—also the speed in data processing!

In the meantime (until 2006) about 8 sites of various epochs have been measured with high resolution/high sensitive caesium-magnetometry. In 2003 and 2004 the prospecting was focussed on 4 big sites (Montedoro, Montearatro, Monte San Vicenzo and Masseria Lo Re), followed by Torre Bianca, Masseria Anglisano and Salveterre (near Ascoli Satriano) in 2005. Because of technical and organizational problems at Munich in 2006 the prospecting was continued in Masseria Salentino only with two Geometrics G-858G systems, both in duo-sensor configuration. But this double duo-sensor configuration with four sensors on the site allowed the coverage of more than 20 hectares in about 6 days.

9.1   *Monte San Vicenzo 2003 and 2004*

The demand for high-speed field survey and fast data processing almost on the site became evident also in the Celone Valley Project even with the first measured site, Monte San Vicenzo. In 2003 the archaeological monument seemed to be almost undisturbed and not endangered by destruction. The site produced very clear crop marks showing the archaeological structures in great detail. The magnetometry made this picture still clearer and one could try to identify the first round

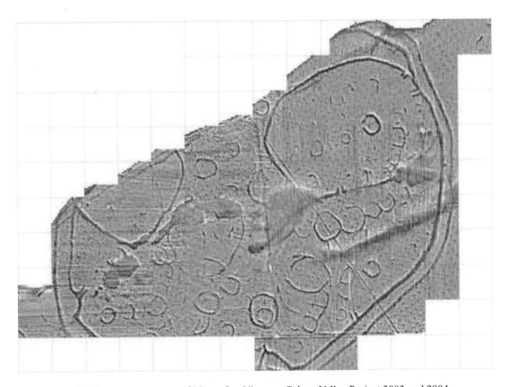

Figure 25.   Caesium-magnetometry of Monte San Vicenzo, Celone Valley Project 2003 and 2004.

Figure 26. Monte San Vicenzo. Destruction of the archaeological monument by building an "Aeolian Park".

houses from the Neolithic epoch in Apuglia. Some parts of the aerial photo in the centre were rather unclear, but in the magnetometry the Neolithic ditches and also the pits of a Roman olive garden showed up very well, but with inverse magnetization contrast. In general one can state, that in many examples the magnetometry shows the archaeological structures in greater detail. But already in 2004 a armada of caterpillars went through the site preparing the tracks for the heaving machinery for the construction of an "Aeolian Park"—which sounds quite nice, but let to an almost complete destruction of the site, especially by cutting deep trenches for the basements of the huge rotors—and marking the end of an archaeological monument of European importance.

9.2 Salveterre (near Ascoli Satriano) in 2005

The aerial photograph of this so-called "bastione baroque" taken by O. Braasch during the summer school in Foggia in 2003 (Plate 11), marks the beginning of the geophysical part on the ground in the Celone Valley Project. Surface finds from a field-walking project (V. Romano, Foggia) indicate 2 phases of an early Neolithic site. Although it seems impossible to beat this fantastic aerial photo by any geophysical method on the ground, the magnetic prospecting of the whole site was done in 2005 during 3 days only. This high speed prospection was made possible by the quadro-sensor caretto in the usual configuration. Like all the others sites in the tavoliere the caesium-magnetometry resulted in an even more detailed and clearer image. In Salvaterre also some of the Neolithic round houses became visible in the magnetogram. The so-called bastiones in the ditch contain a round house each. However the signature of the houses in the magnetogram is showing only extremely faint, so they may be only visible on a high resolution monitor with a strong luminance (not possible on flat screens!).

9.3 Masseria Anglisano in 2005

Also Masseria Anglisano was already well known by series of aerial photographs taken by O. Braasch and K. Leidorf from 2003 onward (Plate 12a). Again this site is also dominated by

Plate 11. Aerial photograph of Monte San Vicenzo at the Celone river. Cropmarks showing complex Neolithic ditched enclosures and Roman olive gardens. O. Braasch 2002. (See colour plate section)

Plate 12a. Masseria Anglisano 2005. Compilation of the oblique aerial photographs on the base of the orthophoto in GIS. Aerial photos by O. Braasch, GIS application and image processing by D. Gallo. (See colour plate section)

complex ditched enclosures of the Neolithic epoch, but there are also Bronze—Iron age burials, Roman wine yards and traces of medieval archaeology. Magnetic prospection with caesiummagnetometry took place in 2005, using mainly the "magneto scanner" with quadro-sensors Scintrex Smartmag SM4G.

In an olive garden the Smartmag was also applied in duo-sensor configuration on a handheld frame, which was the only way of working under the trees. A smaller area was also measured by the Geometrics G-858G caesium magnetometer in duo-sensor configuration on a handheld frame. But an area of about 1.5 hectare was also measured by both Scintrex and Geometrics in the same grid for comparison.

The rectification of the oblique aerial photos, image enhancement and the compilation of the aerial photos with the magnetometry on the base of an orthophoto was done by D. Gallo (University of Bari) in a GIS application (Plate 12b). Again it became evident that the caesium-magnetometry on a large scale application adds a lot of information to the aerial photos, but not only under the trees of the olive garden. Also very interesting was the direct comparison of the 2 caesium magnetometer systems Geometrics G-858 and Scintrex Smartmag SM4G measured in the same grid and under exactly the same conditions. At the first look the 2 magnetograms seem to be almost identical, but a deeper investigation shows differences in detail, contrast and visibility of some very faint traces of archaeological structures like the Neolithic round houses, which became visible with the Sintrex Smartmag magnetometry only. At the moment it is not quite clear, if these differences are due to the sensitivity (Smartmag 20 pT and Geometrics 50 pT at 10 Hz sampling rate), to a different height of the sensors above ground (Smartmag 0.30 m and Geometrics 0.35 m), or to the different methods of data processing. Also the direct comparison of the raw input data shows significant differences between the 2 data sets. There are also differences in the quality of the exact positioning of the sensors on the line. The distance of the handheld Geometrics system was triggered by a manual switch every 5 m on the line, whereas the Smartmag on the magneto scanner was triggered by the rotation of the wheel with about 0.1 m precision over 40 m. In the meantime the author also uses the Geometrics G-858G in duo-sensor configuration on a handheld frame. The

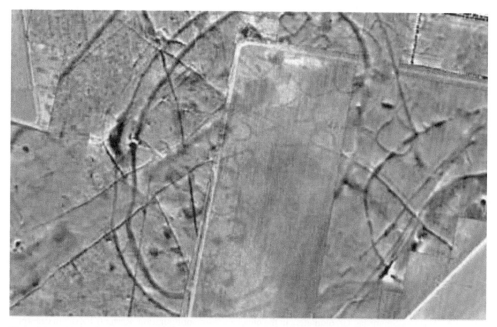

Plate 12b.   Masseria Anglisano 2005. Same as Fig. Xxa, but with the addition of the magnetogram. Caesium-magnetometry by H. Becker with Scintrex Smartmag in quadro-sensor configuration. (See colour plate section)

161

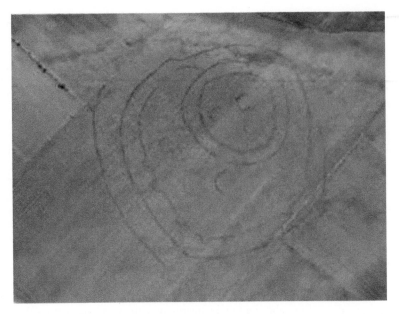

Figure 27. Salveterre (Ascoli Satriano) 2003. Aerial photograph taken by O. Braasch in 2003. Cropmarks of an early Neolithic complex ditched enclosure consisting of 2 building phases.

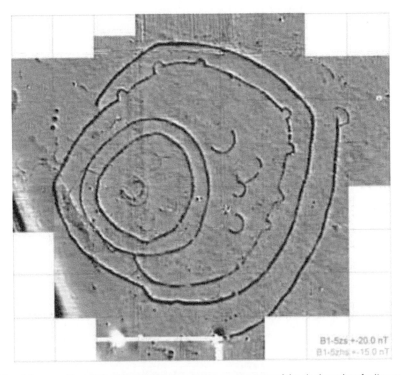

Figure 28. Salveterre (Ascoli Satriano) 2005. Caesiummagnetometry of the site by using the "magnetoscanner" with 2 Scintrex Smartmag Gradiometers in the quadro-sensor configuration. Raster 0.1 × 0.5 m (interpolated to 0.25 × 0.25 m), overlay of the total field measurement with the high pass filtering (60% transparency) for sharpening the picture, dynamics ±20.0 nT (total field) and ±15.0 nT (high pass filtered), 40 m-grids.

Figure 29a.   Masseria Anglisano 2005. Same as Fig. Xxa, but zooming in for more detail.

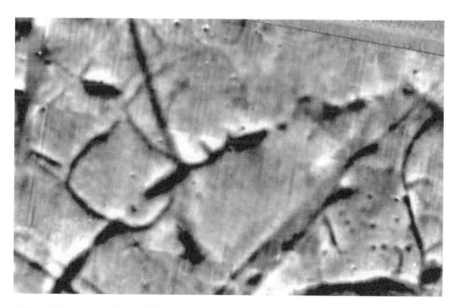

Figure 29b.   Masseria Anglisano 2005. Same as Fig. Xxb, but zooming in for more detail.

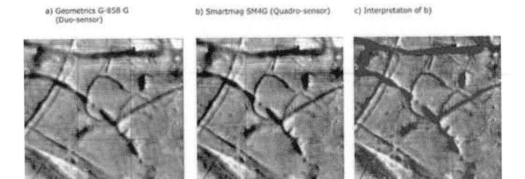

a) Geometrics G-858 G (Duo-sensor) b) Smartmag SM4G (Quadro-sensor) c) Interpretation of b)

Figure 30a–c.   Masseria Anglisano 2005. Direct comparison of the 2 caesiummagnetometer systems Geometrics G-858 and Scintrex Smartmag SM4G. c. Archaeological interpretation of the Smartmag magnetogram, showing some Neolithic round houses.

positioning is done by synchronizing the beeps of a metronome to the meter-marks on the line over 40 m and with a control beep every 5 m. Beside sensitivity, spatial resolution and speed, the precision of the positioning of the sensors especially in zig-zag mode is very important.

10 FUTURE ASPECTS FOR DATA PROCESSING, VISUALIZATION AND ARCHAEOLOGICAL INTERPRETATION

The huge amount of data derived from high speed magnetometry of large archaeological sites and for landscape archaeology every year arise new problems for data processing, visualization and archaeological interpretation. Even when the pre-processing of all data during the night after the measurement was done, which marks normally the end of a prospecting campaign with a ready to print magnetogram, there is a lot of work left for final visualization and interpretation. Unfortunately there is little hope that the interpretation could be done by automatic methods based on pattern recognition and artificial intelligence. Considering the data processing of the Uruk project by definions (Munich) and the German Space Centre (Oberpfaffenhofen) combining geophysics on the ground, low altitude aerial photography and high resolution satellite imaginary it became clear, that even the very specialists couldn't find a convincing way for an automatic interpretation of the archaeological relevant information in the different images of the various remote sensing methods. This was not possible either for the well known place of Uruk in Iraq based on an excellent quality of the remote sensing, nor could it be shown, that the achieved level of recognition could be transferred to other compatible sites in Mesopotamia or elsewhere. It seems, that the human eye (and the brain behind) may be the best sensor for archaeological pattern recognition. But the eye needs an optimised image for this job. Human experience since several hundred thousands of years handling images is only available for positive images. Therefore it is important to transform the computer images of a magnetogram to this format of optimisation for the human eye. But at this point many problems are coming up. The images should be in grey tone (like black and white photography), because the receptors for grey tone images of the human eye have a better resolution. Colour image processing is only needed for special application. The information of the grey tone of an image normally is only stored as 8 bit per pixel (256 grey scale from 0 = black to 256 = white). But the human eye can only distinguish between 60 grey tones at the best, which is far below under the 256 greyscale. The problem of the data processing for the visualization at this point means, how we can transform the relevant information about the archaeological structures into this rather narrow window—and which sort of technique would be the best for showing this image to the eye for further interpretation. But if we really have solved this problem, how we could draw the archaeologically relevant information? We know, that the homo sapiens was drawing with handheld equipment since Palaeolithic times – there is only little experience for drawing with a mouse on a computer screen, which cuts the direct line from the eye, via the brain to the hand with the pencil etc. Everybody knows about these problems, but is there any convincing method for drawing in computer images?

In the meantime the development of caesium-magnetometry is going to even higher sensitivity and speed (spatial resolution). Modern optical pumped magnetometers have reach a sensitivity of 0.1 Picotesla at a sampling rate of 100 Hz (100 measurements per second). But at the moment we can only handle a sensitivity of 0.1 Nanotesla for visualization for the eye. There are strong needs for a more sophisticated software for data processing, visualization and graphic drawing. This makes magnetic prospecting for archaeological prospection still interesting—we are not at all at the end.

Seeing the Unseen – Campana & Piro (eds)
© *2009 Taylor & Francis Group, London, ISBN 978-0-415-44721-8*

Why bother? Large scale geomagnetic survey and the quest for "Real Archaeology"

D. Powlesland

Landscape Research Centre, West Heslerton, UK

ABSTRACT: It is now just 30 years since the discovery of an AD 5th to AD 7th century cemetery during aggregate extraction at West Heslerton, North Yorkshire, triggered one of the longest running and most extensive projects in field archaeology in Britain, The Heslerton Parish Project managed by The Landscape Research Centre (LRC). The *ad hoc* discovery of this Early Anglo-Saxon or Anglian cemetery within what, at that time, appeared to be an archaeologically barren landscape, reflected the reactive nature both of the archaeological record and of the archaeological response in Britain during the 1960s and 70's. The research agenda developed during the late 1970's and still underpinning the projects developed by the Landscape Research Centre, was driven from an environmentally deterministic viewpoint concerned with the relationship between humanity, the environment, the landscape and its many resources. The ongoing research programme combines many different rescue excavations and independent research projects, within a research framework which relies heavily upon the ability to integrate a wide variety of data from multiple sources at varying scales within a single data management and presentation environment.

The distinctive contribution made by the work in and around West Heslerton, situated on the southern side of the Vale of Pickering owes most to the long term support of English Heritage, the scale of excavation and other research undertaken and the exceptional archaeological resource which has been the focus of the work. Incorporating nearly 30 Ha of open area excavation, intensive airborne remote sensing programmes covering more than 100 sq km, more than 1000 Ha of ground based geophysical survey and more than 400 Ha of subsurface deposit modelling undertaken to identify areas with well preserved stratigraphy beneath blown sands; the combined data-set is the most comprehensive of its kind in Britain.

1 INTRODUCTION

The landscape centred on the village of West Heslerton, on the southern side of the Vale of Pickering, has been the setting for one of the most ambitious projects in landscape archaeology undertaken in Britain. Large area excavations covering more than 22 Ha, 25 years of air photographic survey, two large area multi-spectral and vertical photographic surveys, a Laser Imaging Detection and Ranging (Lidar) survey, more than 1000 Ha of contiguous gradiometer survey and a 200 Ha subsurface mapping auger survey, have revealed the most comprehensive body of archaeological landscape evidence for any area of its size in Britain. The majority of the evidence already gathered covers an area of alkaline sands and gravels following the southern edge of the Vale of Pickering, and work is in progress focussing attention on the sands and gravels in the flood plain of the Vale of Pickering. In this area geophysical responses are different but when combined with other sources are providing methods of isolating areas of archaeological interest, including crop-marks, relict stream channels and fragments of surviving peat deposits (which offer the potential for the recovery of the environmental evidence so desperately lacking in the alkaline sand and gravel areas). The combined dataset derived from more than 25 years of rescue and research archaeology, mostly funded by English Heritage, and, in particular, the recent geophysical and sub-surface surveys, funded from the Aggregates Levy Sustainability Fund (ALSF) through English Heritage, offer the potential to assist in the development of approaches to the extraction of aggregates that secure the sustainability of the archaeological resource. In order to develop a pro-active approach to managing the archaeological landscape, simply knowing it is there is not enough, if intelligent decisions are to be made then the sheer quantity of known archaeology must be set against the chronological depth and in particular to the 'quality' of the resource. Nowhere is this more important than in buried landscapes whether covered by colluvium, alluvium or aeolian sands. This paper discusses the results of this massive project, the various methodologies applied in an attempt to discover the 'real' archaeology of the Heslerton research area, and issues regarding archaeological 'quality' with reference to the sustainability of the resource within an aggregate bearing landscape.

2   CONTEXT

The area around the villages of East and West Heslerton on the southern side of the Vale of Pickering, North Yorkshire, England, has been the focus of an ongoing and intensive programme of archaeological research following the accidental discovery of an early Anglo-Saxon cemetery during sand and gravel extraction at Cook's Quarry, West Heslerton in 1977 (Fig. 1). The discovery,

Figure 1.   Location of the Heslerton research area on the Southern side of the Vale of Pickering, North Yorkshire, England.

during removal of overburden comprising plough-soil and a sealing layer of aeolian sands, was one of many unexpected major sites discovered in Britain during aggregate extraction in the 1960's and 70's. The discovery prompted a rescue excavation, funded from the public purse, by the Department of the Environment (now English Heritage) and provided the setting for the development of a long term research project combining large scale rescue excavations with broader research into the landscape context of the excavated areas (Powlesland et al 1986).

The light soils so characteristic of the aggregate bearing subsoils found in the valleys and river terraces of lowland England were ideally suited for prehistoric and later settlement and agriculture and thus provide the setting for the highest density of past activity in Britain. Aggregate extraction is by its nature totally destructive of the archaeological resource, and poses particular problems on account of the large areas of ground affected and the relationship between the sands and gravels and the past patterns of land use. Although excavation and recording programmes funded by the quarry operators, through arrangements established under Planning Policy Guidance 16 (PPG16) published in 1980, have replaced the large rescue projects funded from the public purse, the discovery of unexpected and nationally or internationally important archaeological sites during aggregate extraction has continued to pose significant problems both for archaeology and the aggregate industry. The greatest challenge for planners, aggregate operators and archaeologists, is created by the lack of detailed knowledge as to exactly how much archaeology there is. It is important to realise that the contents of the nations Sites and Monuments Records, the primary basis upon which planning conditions relating to archaeology are set, are derived from ad-hoc evidence, almost all of which has also come to light through accidental discovery in the past. Air photographic records, which have revealed much in aggregate landscapes in particular, are likewise ad-hoc in nature relying upon air photographers being in the right place at the right time to be able to record cropmark evidence; evidence which appears in response to a complex combination of crop, soil, long term climatic and lighting conditions. It became very clear during the initial rescue excavations at Cook's Quarry, West Heslerton, that our understanding, comprehension and ability to interpret the excavated evidence was compromised by the lack of information that allowed us to place the 'site' in its landscape context. Moreover it was also appreciated that investment in this one excavation reflected a deliberate decision to examine this 'site' acknowledging that others would lost at the same time without record. The conscious decision to concentrate efforts in Heslerton reflected the realisation that the 'site' was significant not only on account of the multi-period and multi-faceted archaeological deposits, that would otherwise be lost without record, but also because, blown sands had buried the archaeological features over large areas of the 'site' in a manner not commonly found in Britain. These blown sands had for instance, preserved an upstanding Early Bronze Age barrow, which it appeared, had not been seen since the Roman period; it was not simply another quarry site with already ploughed out multi-period settlement and funerary activity, and made the 'site' exceptional. In 1980 the Heslerton Parish Project (HPP) was established to define a research strategy within which to frame the ongoing excavations (Powlesland 1980, 1981, 2001, 2003a). It was amongst the first major projects in what was then the emerging discipline of 'Landscape Archaeology'. The project was driven by the need to establish the landscape context of the large multi-hectare and multi-period excavations then in progress. Our ability to interpret the, in reality, very small sample excavations was compromised by a lack of comprehension of the contemporary landscape beyond the trenches. The Vale of Pickering, with the exception of the major Late Palaeolithic/Early Mesolithic sites at Star Carr and those being examined by Tim Schadla-Hall at Seamer Carr on the margins of the ancient Lake Flixton (Schadla-Hall 1987a, 1987b, 1988), the Late Bronze/Early Iron Age palisaded enclosures excavated by Tony Brewster at Staple Howe and Devil's Hill (Brewster 1963, 1981), the Roman centre at Malton, medieval manorial elements examined by Brewster in Sherburn and Potter Brompton (Brewster 1952) and some Romano-British and Early Anglo-Saxon settlement at Seamer, Cross Gates (Pye 1976, 1983), evidence was largely blank in maps reflecting the archaeology of almost any period. It was argued in the research design that prehistoric and later land-use and settlement patterns were to a large extent environmentally determined, using environment in its widest sense combining soils, climate, vegetation and ground water conditions and proximity to other ecosystems supporting hunter gathering and transhumance systems (Powlesland et al 1986, 1987a, 1988). The landscape of the southern Vale of Pickering and the northern edge of the Yorkshire Wolds can be readily assigned to a number of distinctive ecozones, which extend both to the east and west of the core research area covering the parishes of East Heslerton, West Heslerton and Sherburn.

3 ANALYSIS OF THE BLOWN SAND

A small excavation (DigIT project) comprising three very restricted trenches was conducted in 2001, this allowed us to examine the blown sand in detail, assess the degree of plough damage and test a range of different features identified in the survey whilst at the same time undertaking a detailed review of digital recording techniques (Powlesland & May, forthcoming). The DigIT project incorporated a detailed study of the blown sand over part of a Late Iron Age and Roman 'ladder settlement'; it is an amorphous deposit of red-ochre sand, sometimes with thin clay varves running through it, which at first glance may be considered not worthy of careful excavation. The blown sand deposit, which at this point was up to .3 m thick beneath a .3 m thick plough-soil, was carefully troweled away and all finds individually three- dimensionally plotted. Subsequent analysis reveals that although there is little visible structure in the deposit the distribution and fragmentation of the finds demonstrate a gradual build-up over the top of the debris from the deserted ladder settlement. Here the blown sand appears to have accumulated from the Roman until the Late Medieval period. A group of Iron Age sherds recovered from the upper few centimetres of the deposit may be derived from areas of plough damage occurring near by (Fig. 2).

At Cook's Quarry it was clear that the blown sand was a characteristic of the late prehistoric landscape, initially becoming mobile during or before the Late Mesolithic (probably very much earlier) and that it formed moving dunes which sealed old ground surface fragments from the Neolithic until the Late Iron Age. The evidence from the detailed study seems to indicate that during the Post-Roman period the ground levels must have risen very gradually, presumably whilst the area was being ploughed the organic components within the lower plough-soil leached out, reducing what must at times have been a plough-soil back to its principal blown sand component. Reworking of the soil during the period of build-up has removed any clearly defined stratigraphy within the deposit. Observation work associated with the laying of a new water main, to the east of Sherburn, revealed a number of locations where stratified deposits could be identified within the blown sand, confirming that the formation and de-formation processes involved are neither simple or uniform throughout the research area.

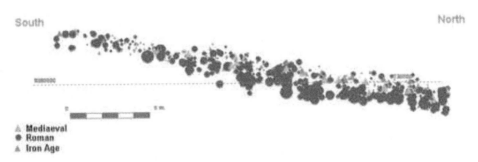

Figure 2. Sectional view of the ceramics distribution in the blown sand in DigIT area 028AC (dots sized by weight) (Z magnified × 10).

4   A HOLISTIC APPROACH TO AN ARCHAEOLOGICAL LANDSCAPE

In order to provide the landscape context of the excavations and build a spatial, chronological, social and economic model of landscape development we needed to be able to identify, quantify and understand the condition of the archaeological resource. A programme of remote sensing and other fieldwork was begun during the first excavation season at Cook's Quarry in 1978. From the outset it was argued that a holistic approach to investigating the landscape should be adopted employing whatever methods were available to build up a complimentary body of evidence from multiple sources. This view still lies at the core of the rescue and research philosophy of the Landscape Research Centre (LRC). In reality it has only been possible to build what we could term a holistic dataset through a large number of different projects funded from different sources over more than 25 years. Over this period the approach has constantly evolved in response both to discovery and also to changes in technology, particularly with reference to remote sensing, field recording, dating and GIS technologies (Powlesland 1986, 1987b, 1991). Whether it be improved instruments for multi-spectral imaging, the development of 'intelligent' total stations, better C14 dating or entirely new technologies such as high precision GPS, Lidar or 3D Scanners each technology has been assessed to see what it can add to the dataset, rather than simply to replicate something that could already be done using established technologies. The building and use of the Heslerton Landscape Dataset without modern up-to-date computers would have been entirely impossible, as it combines multi-sensor remote sensing data of different types and at different resolutions, multi-excavation databases, plans, interpretive drawings, dating and stratigraphic sequencing models, finds and environmental evidence within a single linked resource. Although a wide range of different software tools are utilised for different applications the projects have benefited from data integration provided through the application of GIS technologies for more than 20 years.

5   AIR PHOTOGRAPHY THE PRIMARY REMOTE SENSING RESOURCES

Air photography from a light aircraft with a hand-held single lens reflex camera is relatively cheap, can cover large areas, can produce splendid results and is fun to do. An ad-hoc programme of air photography was begun in 1977 and continued with varying frequency on an annual basis. Despite the often poor conditions for air photography in the Vale (which is liable to sea mists or frets which often reduce visibility at exactly the time when crop-marks would be showing at their best) by 1980 a large number of crop-mark 'sites' had been identified. Our own air photographic record was enhanced with new 'sites' being added on an annual basis following flights by other air photographers. Even as recently as 2005 major new crop-mark complexes have been added to the record despite saturation of the cropmark coverage having been effectively achieved over most of the area by the early 1990's. Flights in 2005 and 2006 revealed that the admittedly splendid returns from oblique air photography failed to show as much as high resolution vertical photography taken at the same time, during the early crop ripening phase when the majority of the crop-mark is to be found in the vegetation at the base of the crop. The blown sand deposit that makes the research area unique, is in places more than a metre deep and can be traced for more than 10 km, along the southern side of the Vale of Pickering. However by 1980 the role played by the blown sand in concealing buried archaeology from the visible landscape and its potential to restrict crop-mark formation was recognised and alternative approaches were sought to enhance the patchy nature of the crop-mark record.

6   GEOPHYSICAL SURVEY, THE EXCAVATION CONTEXT

Trial geophysical surveys using gradiometery at Cook's Quarry in 1980 and again using resistance methods near Sherburn in 1984 failed to produce good or even convincing results. Had the initial geophysical survey results been taken as indicative of the general potential, work on this front may have stopped at this point; however, following the immensely successful gradiometer survey carried out by English Heritage ahead of the West Heslerton: Anglian Settlement Excavation in 1989, we purchased our own gradiometer, the results from which have been outstanding. Initially our attention was focussed on the application of gradiometery within the context of the excavation. It was argued that high resolution gradiometery undertaken during the excavation process, following the removal of the disturbed plough-soil and at a ground resolution of .25 × .25 m, should give highly detailed results that could be used to assist in developing the excavation strategy (Fig. 3) (Lyall and Powlesland 1996). The survey showed that without the masking influence of the disturbed plough-soil, post-holes and minor features as well as structure and sequence in ditch complexes that could at first barely be seen on the surface, were clearly visible. The greatly increased returns from gradiometer surveys undertaken after removal of the plough soil, which acts as a filter, confirms the benefits of this approach as an active contributor to the management and process of large open area excavations.

7 MULTI-SPECTRAL IMAGING

Whilst attention was directed towards the large scale (13 Ha) excavation of the West Heslerton Anglian Settlement there was little opportunity to apply magnetometry to the wider landscape, air photography continued as and when the opportunity arose both to document the excavations from the air and to identify further crop-marks. Many crop-marks appeared repeatedly and the although the number of new discoveries reduced over time, exceptional new discoveries were occasionally made which, in one case, could be linked directly with a deeper ploughing event in the previous year, so that an apparently blank field produced a large complex of clear and extensive crop-marks. The patchwork nature of the air-photographic plots, with clear gaps in what appeared to be extensive and coherent crop-mark complexes, indicated that we were not seeing the full picture—some fields were permanent pasture, in others it was felt that the blown sand sealing the deposits was effectively reducing the potential for crop-mark formation. If we were to increase the

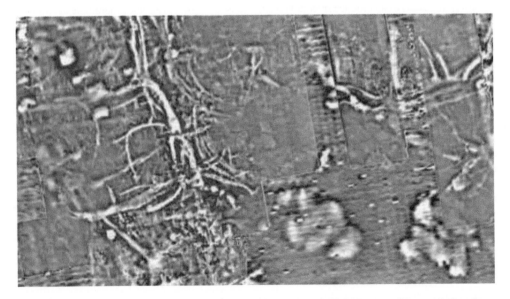

Figure 3. Gradiometer survey results gathered before removal of overburden at 1 × .25m resolution above and the same area below with areas gathered at .25 m × .25 m resolution after removal of the overburden prior to excavation, West Heslerton Anglian Settlement. Data collected by English Heritage, upper image, LRC lower image.

knowledge base new approaches were required which were either more sensitive to crop luminosity and reflectance than conventional 35 mm oblique photography, or employed wholly different techniques to map the buried resource. In 1992, in collaboration with Durham University, Department of Geography, the Landscape Research Centre was awarded a NERC remote sensing grant and the area was flown capturing high resolution 12 band multi-spectral data and high resolution large format vertical photography (Donoghue and Shennan 1988a, 1988b, Donoghue et al 1992, Powlesland et al 1997). The singular difference between the NERC survey and previous oblique air photography was the edge-to-edge coverage, the returns from the high resolution vertical photography were outstanding, far better than had been anticipated. It had originally been planned to undertake the survey flight early in the growing season to look for both soil marks and germination marks; in the event the flight was delayed until June when large areas of crop-marks showed clearly. The multi-spectral data, although of lower resolution than one might wish for, not only showed the features seen in the air photographs but also a considerable amount of new information or additional detail in the infra-red and thermal wavelengths.

8   GEOPHYSICAL SURVEY, BEYOND THE EXCAVATION

Following the completion of the large excavations in West Heslerton in 1996 and the initial post-excavation and analysis, there was time to develop new projects. A series of tests using the gradiometer revealed that the sands and chalk gravels which lie beneath the blown sands and the chalky areas at the foot of the Wolds were highly responsive to geomagnetic prospection techniques. Reduction of the magnetic response caused by the blown sands was far less evident than we had anticipated and it was clear that this method could greatly enhance the picture that had emerged from the airborne surveys.

In contrast to airborne remote sensing approaches including air-photography and multi-spectral scanning which can be considered reactive, requiring a complex combination of conditions to exist within the soils, crops, climate and critically the time of year and even day of observation, geomagnetic remote sensing is proactive and provided the soils and archaeological features exhibit sufficient magnetic contrast survey can be carried out at any time and give effectively

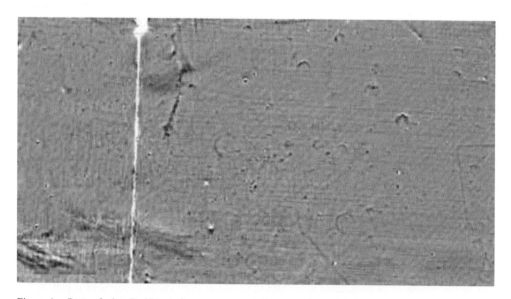

Figure 4.   Parts of a late Prehistoric barrow cemetery discovered in the initial large area geophysical survey comprising ring ditches ranging from 10–25 m in diameter.

the same results. Sufficient magnetic contrast between the natural soils and sub-soils and the anthropogenically modified fills of buried archaeological features is the single condition required. On the sands and gravels on the southern side of the Vale of Pickering the conditions seem perfect for geomagnetic survey either with full field magnetometers or gradiometers. Excavation has revealed that some features, particularly prehistoric features earlier than the Iron Age do not always show, perhaps because the magnetic contrast was originally low or because the contrast has faded over time; it is thus critical to appreciate that even with very effective methods we are not seeing a total picture of the buried landscape. Since 2000, four different English Heritage funded projects have contributed to a radical new understanding of the landscape. Initially a large area (c.350 Ha) geophysical survey covered the area from the lower slopes of the north face of the Yorkshire Wolds to the edge of the ancient wetland areas in the base of the valley between the villages of East Heslerton and Sherburn 3.2 km to the east. This project was designed to identify the distribution of settlement, field, and burial activity between the foot of the Wolds and the margins of the ancient wetland that occupied the lower valley, establish a basic chronological sequence and determine whether the combined air and geophysical survey evidence was comparable with the evidence excavated at various locations in West Heslerton (Fig. 4).

9 PROBLEMS OF TERMINOLOGY, WHAT IS A 'SITE'

As the geophysical survey developed it became clear that the term 'site' with its connotations of dots on maps and isolated features in the landscape reflected poorly on the reality in which the whole landscape was the site. A barrow or settlement complex for instance did not exist in isolation, either from events happening elsewhere in the landscape, chronologically before its existence, or from the space around it. If there was a need to undertake geophysics to establish the context and settings of the excavated areas this could only usefully be achieved through large area coverage. Geophysical survey and active fieldwork covering every field, garden and paddock within an area is only possible with the co-operation and support of the landowners and tenants. The vast amount of new evidence emerging from the geophysical surveys in particular and the lack of chronological reference to most of the mapped evidence meant that simple or traditional feature-based databases, whilst useful for quantification and mapping clearly classifiable features such as Anglo-Saxon Grubenhäuser, and suitable for internal use within the LRC, would mean little to those planning aggregate extraction or to landowners. In LRC projects the term 'Site' is synonymous with field, with each site covering a field as it is defined when survey or other fieldwork is initiated. In order to bring together the evidence from all archaeological research in any single field the results are compiled to form standardised Site Dossiers which include narrative descriptions of all work undertaken, the interpretation of the results, and the supporting evidence. The Site Dossiers now exist for every field in which evidence has been recorded. The Site Dossiers have been designed to support the long-term management objectives through distribution to the individual 'Landkeepers' (this may be the owner or a tenant and often both) and as supporting evidence to the county Sites and Monuments record. We need to clearly acknowledge the role of the Landowners and Landkeepers in securing the long-term future of the archaeological resource, as none of the work undertaken in and around Heslerton can be completed without their support and cooperation. By providing field-by-field datasets that can be incorporated into farm records we hope not only to nurture interest in the land and its past but also to support the case for preservation and management as and when it is necessary.

10 ASSESSMENT OF 'QUALITY'

The survey results when considered in relation to the small excavations undertaken as part of the DigIT project raised important issues regarding the 'quality' of the archaeology identified through remote sensing and the need to inform the long-term management process. Within the research area we will neither be able to preserve all of the archaeological resource or do justice to those areas that get destroyed through preservation by record (the costs of which, in an intensively farmed rural landscape, defy calculation). If we are to secure sustainability of the archaeological resource then we need to understand the nature and potential of the resource before we can determine which parts of the resource are most worthy of long term management. In the 1960's and 70's many outstanding crop-mark sites were afforded some legal protection through scheduling in the belief that we were preserving the best examples. However, we now understand that many of these crop-marks were outstanding on account of the amplification of the crop-mark formation process resulting from very high contrast between the buried ditches and other features and the natural into which they were cut, as a result of truncation through plough damage. Little thought was given to trying to find less damaged parts of these cropmark complexes in adjacent areas where the crop-mark returns may have been less visible but were better preserved. To assign 'quality' as an attribute of the archaeological resource may seem a hopelessly subjective approach. However, the distinctive characteristic of the archaeology of the research area, with the high levels of stratigraphic survival beneath the blown sands facilitating the preservation of whole monuments, from round barrows, to surviving sealed floor deposits in prehistoric and later settlement complexes, can be assessed on a quantative basis. The survival of floor deposits and old land surfaces in rural settlement sites is very much the exception and it is the degree of survival and therefore our ability to intelligently interpret the results of excavation, which drives our interpretation of 'quality'. In addition to the survival of the physical and visible stratified deposits other aspects of 'quality' relate to the survival of faunal evidence and environmental evidence either on site or in a nearby offsite location.

During the 1970's when work begun in Heslerton the 'sand-fields' on the alkaline sands and gravels with their relatively sterile capping of blown sand were considered of poor agricultural quality, the very poor soils produced unreliable and usually poor yields. A change in the focus of agriculture and the extensive use of irrigation made these fields suitable for increasingly mechanised production of root crops, with associated deep ploughing and furrowing caused by the machines used to lift the crops is having a devastating effect on the buried landscape. The DigIt project revealed tremendous variability in the extent of plough damage over a single field and that a single year of root cropping left broad plough scars cutting 15 cm into the archaeological features.

11 THE NEED FOR BETTER INFORMED SURVEY RESULTS

In 2002 environmental taxation applied to the aggregates extraction industry led to the development of the Aggregates Levy Sustainability Fund (ALSF), part of this fund was allocated to English Heritage to cover projects concerned with archaeological sustainability in aggregate extraction areas. It was realised that whilst our airphotographic, multi-spectral and geophysical survey plots demonstrated a completely unanticipated density and complexity of archaeological evidence it was simply a plot. The plot demonstrated density and spatial variation in the buried archaeological evidence but was without any information regarding the relative survival of the buried deposits or, beyond basic classification, a detailed chronological framework. If any attempt was to be made to build a dataset that could be used to assist in developing a more archaeologically sustainable approach to aggregate extraction it was critical to identify those areas that were best preserved and those that were already under serious threat or damaged by agriculture. If we were to develop an approach to archaeologically sustainable aggregate extraction we needed more complete evidence than we could provide for the areas already surveyed. It was agreed with English Heritage that the active survey project should be extended and combined with a more comprehensive survey applied to the area between East and West Heslerton centred on the active Cook's quarry where the work in Heslerton began. In addition to contiguous geophysical survey undertaken under the supervision of James Lyall, assisted by Maria Beck and David Stott, an auger survey, supervised by Guy Hopkinson assisted by Aidan Harte, was undertaken to map the depths of the buried deposits and thickness of the blown sands, as these had a bearing on the potential survival of the sub-surface evidence and on the response of the various remote sensing technologies applied.

Whilst the geophysical surveys revealed field upon field covered with features the auger survey was intended to map the extent and variability in depth of the blown sands and plough-soils. In addition experiments were undertaken to examine the nature of the evidence in the topsoil and a number of observation trenches were opened simply to observe the condition of the buried resource. This project was amongst the most ambitious so far undertaken in Heslerton comprising c550 Ha of .25 m × 1 m gradiometer survey, examining nearly 2,500 auger cores, nearly 300 Ha of high precision topographic survey data, gathered using a vehicle mounted Kinematic GPS system, and the stripping of more than 1 km of observation trenches. The exceptionally dry conditions that prevailed throughout the eighteen month duration of the project meant that it was possible for a single team to undertake the geophysical survey within the times when the land was available for survey. However the dry conditions made the auger survey extraordinarily difficult, and the observation trenches, which could only be excavated during the short period between the harvest and replanting, were less successful than we had hoped as the ground was so dry and the sand so well bonded that the exposed areas were like concrete to work (Fig. 5). The auger survey had a second objective, in the area to the north of the lower spring line, which emerges at the boundary of the alkaline sands and gravels which slope gently down from the foot of the Yorkshire Wolds to the flood-plain, a completely different deposit of glacial sands and gravels, clays extend to the River Derwent. This area had once been covered with extensive deposits of peat some of which survives either as infill in relict stream channels or small relict lakes and occasionally in archaeological features. Augering in this area allowed us to map the northern limits of the blown sands and also to identify areas of peat that would be compromised by the localised drop in the water-table associated with aggregate extraction. The importance of these peat resources should not be underestimated as they have the potential for the recovery of environmental evidence, particularly pollen and plant remains that can inform the interpretation of evidence in the broader landscape.

Figure 5.   Thickness of wind blown sand deposits in relation to geophysical results.

Environmental evidence of this type is conspicuous by its absence in the excavated areas and it is likely, most of the areas covered with dense complexes of mapped features. Analysis of the offsite pollen offers the potential to add environmental context to the enclosures, fields and cemeteries emerging through the remote sensing programme. A recently completed pilot project undertaken to sample and test the potential of the palaeochannels and archaeological features identified in the flood-plain from the air for the recovery of good and dateable environmental evidence has produced promising results. The dating of some of the features indicates that there was more activity in the floodplain from the later prehistoric period onwards than we had thought. The present topography of the floodplain, which has been documented using both high precision Kinematic GPS and more recently through a far more extensive Lidar survey, is problematic and it seems that the present land-surface may have dropped over large areas as the deeper peats have dried out.

12   EXTENDING THE GEOPHYSICAL SURVEY

The extension to the geophysical survey made possible with grant funding from the ALSF has enabled the collection of more than 1200 Ha of gradiometer data including a single strip covering much of a 9.5×2 km area on the alkaline sands and gravels and a second area on the acidic sands and gravels in the lower valley. The extended project also allowed a complete review of the air-photographic and multi-spectral data and further analysis incorporating Lidar data, secured as part of a second NERC funded multi-spectral campaign undertaken during 2005. The 2005 NERC data confirmed that whilst multi-spectral survey incorporating thermal imaging could add details not visible within the visible spectrum the method is fundamentally reliant on the same factors as conventional air-photography, and the survey having been conducted at a time when crop-mark visibility was minimal by comparison with the 1992 survey the returns were very limited. The extended survey area now forms the largest survey of its kind, but its importance is not simply related to the scale of the survey or the sheer quantity of features identified. If our objectives are to establish an evidence base that facilitates long term planning and management of the archaeological resource then we need to be able to determine to what degree the evidence is repeated within the larger landscape and identify the scale of the components or land blocks that define the structure of the archaeological landscape. In the case of Heslerton it appears that the landscape was divided up into large tracts bounded by major boundaries and by water courses emerging from the foot of the Wolds and draining into the Vale at roughly 3 km. intervals during the Neolithic. The overall structure which appears to include gaps or open space associated with the limits where defined by streams was maintained right through into the medieval period with some boundary features still in active use today. Amongst the key research objectives within the combined surveys is the need to identify the archaeological capacity of the landscape and identify the scale and nature of the land blocks. It was argued that if consistent land blocks could be identified these would have a long term impact on the development of future sampling strategies both within the area of the survey and beyond it; the extended survey covers less than a quarter of the blown sand zone that fills the gap between the foot of the Wolds and the edge of the wetlands on the southern side of the Vale of Pickering. By attempting to secure a saturated archaeological dataset derived from multiple remote sensing sources and missions we have been able to identify a population of archaeological evidence that can be reliably sampled.

There is insufficient space here to discuss in detail the results of the large area surveys and it would also be premature given the limited nature of the ground truthing exercises so far undertaken. However, with approximately 20,000 features identified over an area of about 5,000 Ha, 16,000 of which are derived from the geophysical surveys covering nearly 1000 Ha, the results have been spectacular. A single example showing the distribution of Anglo-Saxon Grubenhäuser, features that have recognisable and distinctive geophysical characteristics and are amongst the few anomalies that can be classified without hesitation in the geophysical survey data, demonstrates the nationally important nature of the results plate 13. Nowhere else in England has such a dense cluster of Early Anglo-Saxon settlement evidence been identified, even if we exclude those

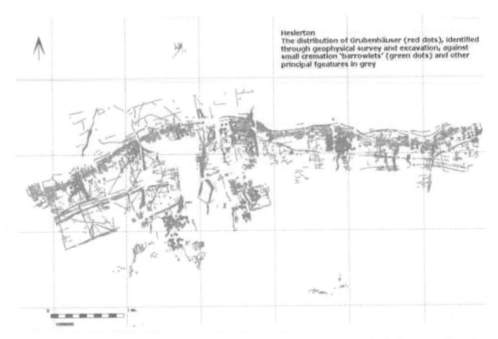

Heslerton
The distribution of Grubenhäuser (red dots), identified through geophysical survey and excavation, against small cremation 'barrowlets' (green dots) and other principal features in grey

Plate 13.   The distribution of Anglo-Saxon Grubenhäuser identified across all geophysical surveys. (See colour plate section)

clusters that may be Middle Saxon, the evidence indicates that the huge excavated settlement at West Heslerton is one of a number similar sites reflecting a density matching that of the present villages, moreover these large settlements have to viewed against an even greater frequency of smaller sites comprising 20–50 buildings which follow the Late Iron Age and Roman 'ladder set-tlement' at roughly 800 m intervals (see below) (Powlesland 1991, 1998a, 1998b, 2000, Haughton and Powlesland 1999). The complexity and density of the evidence as a whole can bee seen in Plate 14, which shows the geophysical evidence combined with the digitised interpretation as managed within the LRC GIS.

13   CONCLUSIONS

Thirty years ago the known archaeology of East and West Heslerton comprised the East Heslerton long barrow, discussed by Canon Greenwell and re-examined by the Vatchers in 1966, a number of round barrows rather poorly described, also by Greenwell, one of which gave its name to a Neolithic pottery type, Heslerton Ware, and the Late Bronze Age palisaded enclosure of Devil's Hill, excavated by TCM Brewster; the rather more famous Staple Howe also excavated by Brewster is situated in the next parish (Greenwell and Rolleston 1887, Mortimer 1905). A fine Late Neolithic/Early Bronze Age jet necklace discovered during aggregate extraction at Cook's Quarry, West Heslerton, in the early 1960's that lay in fragments in Scarborough museum was the only indication of the important archaeology that ran through the area on the edge of the flood-plain of the River Derwent. Following the discovery of a multi-period settlement and cemetery complex during removal of overburden at the same quarry in 1977, a programme of very large scale rescue excavations and landscape survey was begun which have revealed the most detailed picture of an archaeological landscape for its scale in Britain.

During the first few years of excavation it became clear that much of the archaeology of Heslerton was important not simply because it is there, but because large areas of settlement and cemetery

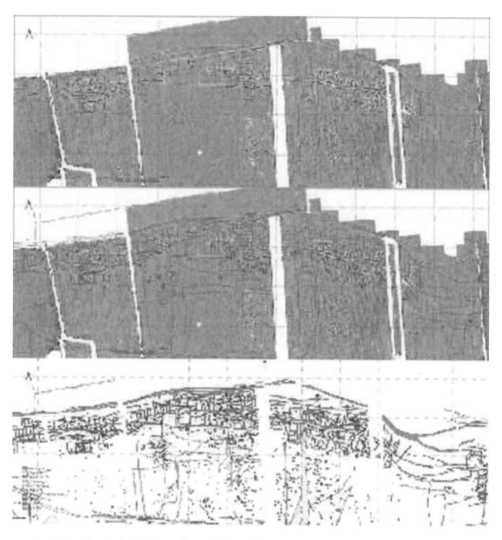

Plate 14. A section of the 'ladder settlement' shown through the geophysical results (top), overlain with the digitised polygon plot (centre), and classified according to basic phase (bottom). (See colour plate section)

evidence for the Neolithic to Early Medieval periods lay sealed beneath deposits of blown sand which preserved evidence such as intact floor deposits not commonly found on rural sites in Britain. A programme of remote sensing initiated during the first season of excavation on an ad-hoc basis but later supported by research grants from NERC and English Heritage and most recently through the English Heritage managed ALSF programme have revealed flaws in our appreciation of the archaeological landscape based on isolated 'sites', mostly identified through accidental discovery. A problem particularly associated with buried landscapes is that accidental discovery is the norm, especially in an area where the discovery of plough scars cut into an apparently featureless reddish sand, could be mistaken for undisturbed ground. Landscapes sealed by aeolian or blown sands are particularly difficult to assess since the present land surface gives few clues about what lies beneath. The work undertaken in Heslerton has been driven by a desire to identify the archaeological resource as a whole, to provide context for the known and excavated areas and gain insight into the evolution of the landscape unconstrained by chronological boundaries. Attention has, until recently, been primarily focussed on the examination of a 1.5 km wide strip of land on the sand and chert/chalk gravels extending from the foot of the Yorkshire Wolds to the lower spring line on the edge of the flood-plain of the Vale of Pickering. Work is currently in progress assessing the archaeology of the flood-plain, itself another source of aggregates. We need to be very aware that with the exception of the hostile locations (by virtue of the high water table or steep slopes of the Wolds), where other potentially important evidence may still lie buried in relict lakes and pools beneath the ploughsoil, the blank areas in our remote sensing plots do not represent areas without a past, this was has been strikingly demonstrated at Cook's Quarry in the last four years, where the quarry owner now funds excavation ahead of extraction. A 30 m diameter hengiform enclosure of probable Neolithic date with a number of associated cremations was undetected by magnetometry, but had been seen as a poor cropmark in the 1970's, the multitude of other features including barrows and field systems had not been detected by any method (Fig. 6).

To those charged with securing the future of the past, Heslerton must seem like a nightmare, the tip of an archaeological iceberg, with totally unanticipated density of activity covering over 1000 Ha, to others it is a dream. In a landscape which appears from recent pollen studies to have remained cleared and open from the Late Mesolithic period onwards, only those areas too hostile for use by virtue of the high water table or steep slopes of the Wolds show no evidence of past activity. Once large area contiguous geophysical survey was in progress it rapidly became clear that any attempt to develop a management strategy for the resource required much more than an air-photographic and geophysical survey plot. The buried landscape is under threat from large scale aggregate extraction, road schemes and the gradual, and effectively invisible, erosion cased by modern agricultural practices. Given that only a tiny percentage of the mapped remote sensing data has been directly examined, adding chronological depth and the ability to accurately interpret the mapped features, we need not only to identify the distribution of the resource but also to determine its likely condition and potential to add new knowledge. A singular weakness in the work at Heslerton it is the lack of comparable landscape datasets from other parts of the country, as without them we may end up confusing what is the norm with something, which at present looks unique.

The Landscape dataset for Heslerton still largely amounts to a series of maps which require ground truthing; however the combined evidence does offer us the opportunity of identifying

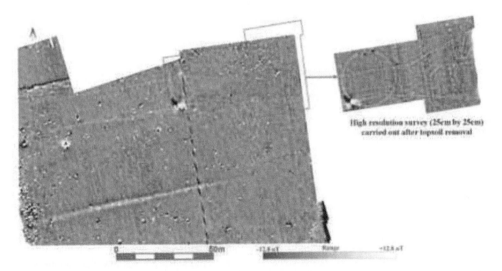

High resolution survey (25cm by 25cm)
carried out after topsoil removal

Figure 6.   An area of gradiometer survey at Cook's Quarry prior to excavation; of the three linear anomalies only the larger feature is ancient, a prehistoric pit alignment. The high-resolution survey (inset) carried out after topsoil stripping, revealed the full extent of a hengiform enclosure and pit alignment.

those areas where aggregate extraction would be uneconomic on the grounds of excavation cost, those areas where the resource is likely to be best preserved and those areas where a degree of archaeological risk can be calculated based on the evidence to hand, including some where the risk is likely to be minimal. During the life of the project we have seen great changes in the nature of archaeological funding particularly for projects that are development led, this seems sadly to have been seen as an opportunity to greatly reduce the level of funding for archaeology from central government. Today there is a focus on the management of landscapes and securing sustainability of the archaeological resource. However, you cannot manage what you do not know about, and you cannot sustain a resource without, at lest some level of understanding of the resource, and probably some degree of monitoring and therefore documenting it. A consequence of the work in Heslerton is that now that we gathered all this evidence we are responsible for its future. If it is to be sustainable then it serves no purpose without interpretation and, with the exception of certain classes of monument, this can only be secured through further work. A second consequence has been the absolute requirement to work in a highly computerised and integrated environment and develop new layers of narrative documentation both to engage the 'landkeepers' whose land holds this magnificent resource but also to underpin the conventional databases, maps and image banks.

With such a large project covering so much land that is in use it is vital that the 'landkeepers' are kept informed, in modern parlance they are the stakeholders, but to us they are so much more: none of them chose to have this archaeology on their land, but they are as interested as the archaeological professional in the fact that it is there.

The project, of course, has only just begun. The challenge now is to work with the aggregates industry, the farmers, landowners and planners to secure a future for a past. A future in which we hope to see the resource sustained through management, examination and interpretation as part of a perpetual landscape park, an archaeological and environmental living laboratory for everyone to enjoy, from 5 year olds to 95 year olds, and from the academic to the simply interested.

REFERENCES

Brewster, T.C.M. 1952. Two Mediaeval Habitation Sites in the Vale of Pickering, York: Yorkshire Museum.
Brewster, T.C.M. 1963. The Excavation of Staple Howe, E. Riding Arch. Res. Committee.
Brewster, T.C.M. 1981. 'The Devil's Hill', Current Archaeol., 76 (1981), 140–41.
Donoghue, D.N.M., Powlesland, D.J. and Pryor, C. 1992. Integration of Remotely Sensed and Ground Based Geophysical Data for Archaeological Prospecting using a Geographical Information System, in A.P. Cracknell and R.A.Vaughan (eds), Proceedings of the 18th Annual Conference of the Remote Sensing Society, University of Dundee 1992, 197–207.
Donoghue, D.N.M. and Shennan, I. 1988a. 'The Application of Remote Sensing to Wetland Archaeology'. Int.J.Geoarchaeology, 3, 275–285.
Donoghue, D.N.M. and Shennan, I. 1988b. The Application of Multispectral Remote Sensing Techniques to Wetland Archaeology. Oxford: BAR.
Greenwell, W. and Rolleston, G. 1877. British Barrows, Oxford: University Press 18.
Haughton, C.A. and Powlesland, D.J. 1999. West Heslerton—The Anglian Cemetery, Landscape Research Centre Monograph 1, 2 vols. Yedingham.
Hinchliffe, J. and Schadla-Hall, R. 1980. The past under the plough, Department of the Environment Occasional Paper No.3.
Lyall, J. and Powlesland, D.J. 1996. 'The application of high resolution fluxgate gradiometry as an aid to excavation planning and strategy formulation'. Internet Archaeology 1 (http://intarch.ac.uk/journal/issue1/index.html).
Mortimer, J.R. 1905. Forty Years Researches in British and Saxon Burial Mounds in EastYorkshire, London: A. Brown and Sons Powlesland, D.J. 1980. 'West Heslerton—the focus for a landscape project' Rescue News 21 12.
Powlesland, D.J. 1981. The Heslerton Parish Project: 1982–92 Strategy document circulated Mss.
Powlesland, D.J. 1986. "Random access and data compression with reference to remote data collection: 1 and 1 = 1", in M.A. Cooper, and J.D. Richards (eds), Current Issues in Archaeological Computing, 17–22, Oxford.

Powlesland, D.J. 1987a. 'Staple Howe in its setting', in Manby (ed) Archaeology Eastern Yorkshire, essays in honour of T.C.M. Brewster. 101–107, Sheffield.

Powlesland, D.J. 1987b. 'On-site computing: in the field with the silicon chip', in Richards, J.D. (ed) Computer usage in British Archaeology, 39–43, Birmingham.

Powlesland, D.J. 1988. Approaches to the excavation and interpretation of the Romano-British landscape in the Vale of Pickering, in J. Price, and P.R. Wilson (eds), resent Research in Roman Yorkshire: studies in honour of Mary Kitson Clarke, 139–151, Oxford.

Powlesland, D.J. 1998a. 'Early Anglo-Saxon Settlements, Structures form and layout, Towards an Ethnography of the Anglo-Saxons', San Marino I.S.S Seminar 1994.

Powlesland, D.J. 1998b. West Heslerton—The Anglian Settlement: Assessment of Potential for Analysis and Updated Project Design, Internet Archaeology 5 (http://intarch.ac.uk/journal/issue5/pld/index.html).

Powlesland, D.J. 1991. 'From the trench to the bookshelf: computer use at the Heslerton Parish Project', in Ross, S. et al (eds) Computing for archaeologists pp. 155–170.

Powlesland, D.J. 2000. West Heslerton: Aspects of Settlement Mobility, Early Deira: Archaeological studies of the east Riding in the 4th to 9th centuries AD, Oxbow, Oxford.

Powlesland, D.J. 2001. 'The Heslerton Parish Project: An integrated multi-sensor approach to the archaeological study of Eastern Yorkshire, England' Remote Sensing in Archaeology Forte & Campagna (eds), University of Siena, Firenze 2001, 233–235 19.

Powlesland, D.J. 2003a. 'The Heslerton Parish Project: 20 years of archaeological research in the Vale of Pickering', in The Archaeology of Yorkshire An assessment at the beginning of the 21st century T.G. Manby, S. Moorhouse & P. Ottaway (eds), 275–292, Yorkshire Archaeological Society, Leeds 2003.

Powlesland, D.J. 2003b. 25 years research on the sands and gravels of the Vale of Pickering. The Landscape Research Centre, Yedingham.

Powlesland, D.J., Haughton, C.A. and Hanson, J.H. 1986. 'Excavations at Heslerton, North Yorkshire 1978–82', Archaeol. J. 143, 53–173.

Powlesland, D.J., Lyall, J. and Donoghue, D. 1997. 'Enhancing the record through remote sensing: the application and integration of multi-sensor, non-invasive remote sensing techniques for the enhancement of the Sites and Monuments Record. Heslerton Parish Project, N. Yorkshire, England' Internet Archaeology 2 (http://intarch.ac.uk/journal/issue2/pld/index.html).

Powlesland, D.J. & May, K. DigIT, English Heritage Project 3065, Internet Archaeology, forthcoming.

Pye, G. 1976. Excavations at Crossgates near Scarborough. Trans Scarb Arch Hist Soc Vol. 3, no. 19, 1–22.

Pye, G. 1983. Further at Crossgates near Scarborough 1966–1981. Trans Scarb Arch Hist Soc no. 25.

Rackham, J. and Powlesland, D.J. 2006. Pilot Project: Environmental Assessment Project for the central Vale of Pickering, English Heritage Project 3038, unpub Mss.

Shadla-Hall, T. 1987a. Early man in the eastern Vale of Pickering. In S. Ellis (ed) East Yorkshire Field Guide, Cambridge: Quaternary Research Assoc.

Shadla-Hall, T. 1987b. Recent investigations of the Mesolithic landscape and settlement in the Vale of Pickering, CBA Forum 1987, CBA Group 4 Newsletter, 22–3.

Shadla-Hall, T. 1988. The early post-glacial in Eastern Yorkshire in T.G. Manby (ed) Archaeology in Eastern Yorkshire, University of Sheffield, 25–23.

Tipper, J. 2004. The Grubenhaus in Anglo-Saxon England. Landscape Research Centre Monograph Series Number 2: Volume 1.

Seeing the Unseen – Campana & Piro (eds)
© *2009 Taylor & Francis Group, London, ISBN 978-0-415-44721-8*

The complementary nature of geophysical survey methods

M. Watters

HP Visual and Spatial Technology Centre, Institute of Archaeology and Antiquity,
University of Birmingham, UK

ABSTRACT: Geophysical surveys are being used increasingly for large scale landscape coverage. Current developments in technology enable fast coverage of large areas with different geophysical survey methods. Each survey method responds to different earth properties with resulting maps revealing unique information on remaining archaeological features. Used in combination, the results from different geophysical methods complement each other to provide a rich resource to help answer archaeological questions. Effective survey project planning is a fundamental step toward integrated geophysical surveys. Once a fundamental comprehension of different geophysical survey methods is understood an effective methodology can be established for successful geophysical surveys. An example from the Catholme Ceremonial Complex located in Staffordshire, UK demonstrates the necessity of multiple geophysical survey methods for mapping the buried archaeological resource and how detailed project planning is essential for an organized, effective and responsible landscape survey.

1 INTRODUCTION

A great variety of geophysical survey methods are employed in archaeological prospection. The most commonly used methods include magnetometry, magnetic susceptibility, electromagnetic induction (or conductivity), resistivity and ground penetrating radar (GPR). Each geophysical survey method measures different properties of the earth's subsurface. When used in combination, these survey methods can provide detailed information on the nature of the remaining buried archaeological record.

This paper presents a brief introduction to geophysical survey methods, discusses details of geophysical survey project planning and presents a case study from the Catholme Ceremonial Complex located in the UK. The geophysical survey methods presented include magnetometry, resistivity and GPR. As an important part of geophysical surveys, project planning is discussed in detail and aimed to outline a comprehensive framework from which any type of geophysical survey can be designed. The case study presents a brief overview of a large scale geophysical survey undertaken with the aims to identify the buried archaeological features and to record the subsurface response to each geophysical technique used to provide insight to the nature of the archaeology, the local environment and the impact of contemporary land use on the remaining archaeological resource.

2 GEOPHYSICAL TECHNIQUES

2.1 *Magnetometry*

Magnetic survey measures the variation of the magnetic field of the Earth and the effects of near-surface features which might be superimposed upon it. Archaeological surveys map the contrasting values of buried or invisible anthropogenic activities generally characterized through magnetic susceptibility, thermoremanent magnetism and ferrous materials.

The magnetic gradient is generally collected in archaeological applications. The magnetic gradient acts as a filter to reduce the effects of background geological magnetic fields and diurnal effects caused by the interaction between the magnetic fields of the Earth and its atmosphere (Weymouth 1986; Clark 1996). Further information on the principles of magnetic surveys can be found in Breiner 1973, Kearey *et al.* 2002 and Mussett & Khan 2000 and for archaeological applications Clark 1996 and Gaffney & Gater 2003.

Various magnetometers are used in archaeological surveys which include the G858 caesium vapour magnetometer and FM256 fluxgate gradiometer. The caesium vapour magnetometer has single sensors that collect a total vertical magnetic field. When two sensors are used the gradient can be calculated thus reducing the background geological and diurnal effects. The gradient array is typically used with the caesium vapour magnetometer in archaeological applications (Weymouth 1986).

The caesium vapour magnetometer can be used with a variety of setups that include the 'Weymouth' technique with the instrument operated by two people, one holding the sensor array and the other following and a fixed distance with the computer and battery pack (Fig. 1) (Watters 1999).

The caesium vapour magnetometer can also be used by a single person with the sensors mounted vertically (or horizontally) on a pole as displayed in Figure 2.

The fluxgate gradiometer equipment contains two magnetic sensors and computer in a single casing (Fig. 3).

2.2 *Resistivity*

Resistivity surveys were first applied in archaeological prospecting in the UK by Atkinson in 1946 and by De Terra in Mexico in 1949 (Atkinson 1952; De Terra *et al.* 1949). Resistivity in archaeology developed primarily in the UK and Italy with several manufacturers producing relatively fast and affordable alternating current resistivity meters. Since then, the development of automated hardware with faster measurement speed and efficient mobile electrode arrays has increased the utility of resistivity surveys for large area coverage at high sampling rates (Schollar *et al.* 1990; Panissod *et al.* 1998; Dabas *et al.* 2000; Dabas 2006; Walker & Linford 2006).

A resistivity survey is designed to measure the electrical resistivity of the earth in order to provide information on the subsurface structure. The electrical properties of the earth are recorded as a function of depth and/or horizontal distance. An electrical current is introduced into the earth through electrodes and the resulting potential distribution is sampled at the ground surface. The measured apparent resistivity provides information on the magnitude and distribution of the electrical resistivities in the volume of the sampled subsurface (Griffiths & King 1981).

A number of resistivity arrays are used in archaeological surveys. The most common is the twin–electrode array. This array is used in horizontal traversing and also electrical imaging on archaeological sites. In horizontal resistivity surveys the spacing between electrodes is maintained but these are then moved along a line with fixed spacing intervals. The results can be plotted as a vari-

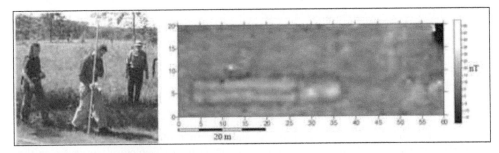

Figure 1. The caesium vapour gradient array with the 'Weymouth' setup and a magnetic data sample from a linear effigy mound from the Effigy National Park in Iowa, USA.

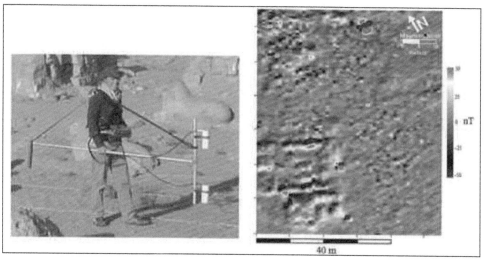

Figure 2.   The single person setup for the caesium vapour gradient array (A). The magnetic map (B) shows two structures from Gebal Barkal in Sudan.

Figure 3.   The fluxgate gradiometer (A) with a magnetic gradient map of the Roman villa at Forum Novum, Vescovio, Italy (B).

ation of apparent resistivity with distance along a line of survey. Exploration depth is determined by the electrode spacing. The data produced from this type of survey can very effectively map buried subsurface features at an approximate single depth across a site (Fig. 4).

A more accurate model of the subsurface is a two dimensional model where resistivity changes can be mapped in both vertical and horizontal directions (Barker 1993; Loke 2004). This is obtained from electrical imaging, or tomography, and is applied to archaeological applications in order to determine the depth and geometry of subsurface features in a vertical section. Data collection involves a series of constant separation traverses along a single line with electrode spacing increasing with each successive traverse (Fig. 5). As the increased electrode spacing penetrates to

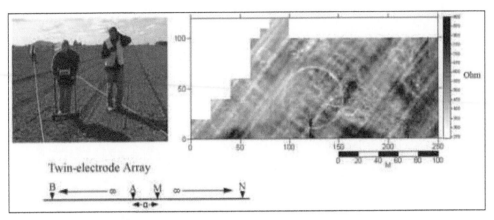

Figure 4.   The RM15 resistivity metre (A) with the twin-electrode array (B) and a Bronze Age ring ditch from the Catholme Ceremonial Complex, Staffordshire, UK (C).

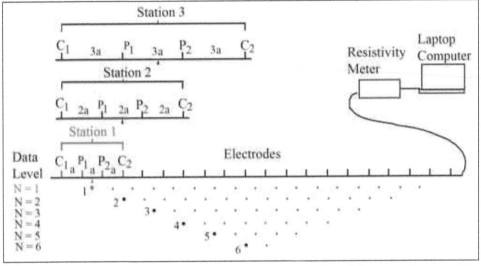

Figure 5.   Electrical imaging measurement sequence and equipment for building a pseudosection. After Loke 2004.

a greater depth, the measured apparent resistivities provide information on the lateral and vertical variation in resistivity along the section.

Early approximate interpretation techniques (Hallof 1957) involved construction of a "pseudo-section", a plot of apparent resistivity as a function of the position and separation distance of the electrode array. However, since the mid 1990s, inversion techniques have been introduced, which convert the pseudosection to an "image" of true formation resistivity against true depth (Loke & Barker 1996a, 1996b; Loke & Dahlin 2002; Tsourlos 1995; Tsourlos & Ogilvy 1999) as can be seen in Plate 15.

2.3   GPR

GPR can provide high resolution records of boundaries between subsurface features with contrast-ing dielectric properties. A standard method for detecting buried archaeological features, GPR is

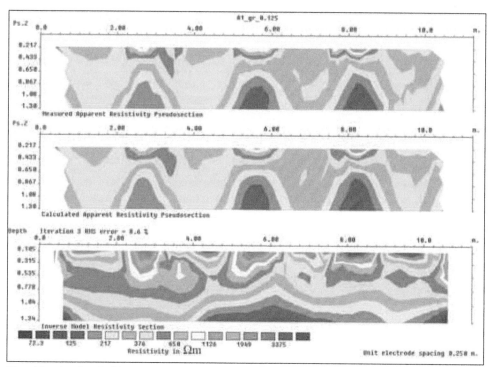

Plate 15. An electrical image over a Neolithic pit alignment in Staffordshire, UK. (A) is observed data plotted as a pseudosection, (B) is the pseudosection computed from a model and (C) is an image or model showing the true depth and the true feature resistivity. (See colour plate section)

able to collect large amounts of data, covering large areas, over a short period of time. GPR is a geophysical technique that can produce a three dimensional image of the subsurface and provide accurate depth estimates and information concerning the nature of buried features. The basic principles and theory of GPR operation have evolved through electrical engineering and seismic/geophysical exploration (Daniels 2000).

GPR hardware and survey methods were first established in the 1960s where they were utilized for widespread applications such as ice sounding, mapping subsoil properties and the water table (Annan & Davis 1976; Campbell & Orange 1974; Cook 1960; Barringer 1965). The first GPR systems for commercial sales were manufactured in 1972 (Morey 1974). Once systems were readily available, the use and applications of GPR increased dramatically (Olhoeft 2000). During the past few decades research and applications utilizing GPR include work in mining, permafrost mapping, cave and tunnel detection, utility and infrastructure mapping, concrete mapping, highway and bridge deck mapping, environmental monitoring and site characterization, agriculture, land mine detection, groundwater mapping, archaeological and forensic investigation and many other applications.

GPR was first applied in an archaeological context by Vickers and Dolphin (1975; Vickers *et al.* 1976) at Chaco Canyon, New Mexico, in 1974. Since its inception, GPR has been used in a number of archaeological investigations (e.g. Bevan & Kenyon 1975; Kenyon 1977; Conyers & Goodman 1997; Goodman & Nishimura 1992b, 1993; Leckebusch & Peikert 2001; Leckebusch 2003; Neubauer *et al.* 2002).

GPR maps the form of contrasting electrical properties (dielectric permittivity and conductivity) of the subsurface and records information on the amplitude, phase and time of electromagnetic energy reflected from subsurface features. The results are presented as 2D vertical profiles in

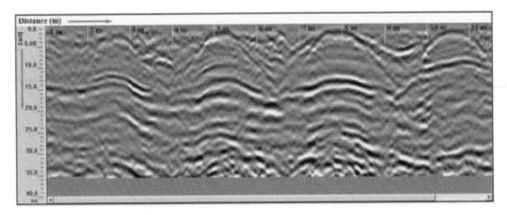

Figure 6.   GPR profile with distance along the horizontal axis and twtt, along the vertical axis.

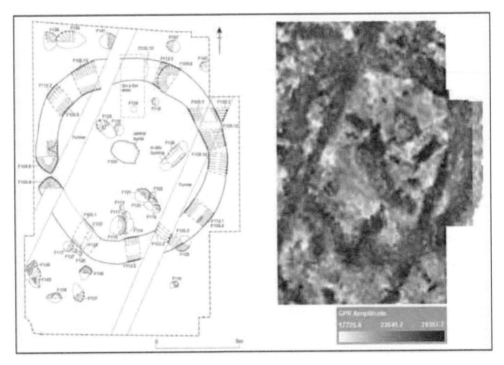

Figure 7.   Archaeological plan (A) and GPR map (B) of a Neolithic ring ditch with central burial from the Catholme Ceremonial Complex, Staffordshire, UK.

the earth. The stronger the contrast between the electrical properties of two materials, the stronger the reflected signal in the GPR profile will be.

The GPR method measures the time elapsed between the transmission of an electromagnetic wave from a surface antenna into the earth and its reception back at the surface after reflection off a buried discontinuity; this is two-way travel time (twtt). The electromagnetic wave moves at a velocity determined by the electrical property of the material it is travelling through. The wave

188

spreads and travels downward until it hits an object with different electrical properties where part of the energy is reflected back to the surface while the rest of the energy continues to pass downward. Changes in mineral composition, water content and density are among the soil discontinuities that may cause reflections.

When collected at regular intervals across a landscape, GPR profiles can be combined and a data cube used to create plan maps (figure 6). Plan maps of GPR data can provide a more comprehensive view of the buried archaeological landscape and spatial relationships between features than is possible through vertical profiles.

Figure 7 shows an archaeological plan (A) with the corresponding GPR map (B) from the subsurface survey of a Neolithic ring ditch with central burial from the Catholme Ceremonial Complex in Staffordshire, UK.

Used singly, these different geophysical survey methods can map details of the subsurface specific to the method sensitivity (i.e. magnetic susceptibility, ground saturation or contrasting electrical properties). When used in combination, not only is there a greater potential for identifying finer archaeological details but preliminary feature clues can be identified based on geophysical responses (i.e. a buried terracotta pipe may have a strong magnetic response and high resistivity response; if the clay pipe was not fired, the magnetic response would be weak).

3 SURVEY PROJECT PLANNING

One of the most important aspects of a good geophysical survey, regardless of the techniques utilized, is project planning. A number of steps are useful when considering surveying a site that include the:

1. determination of project goals
2. consideration of the survey area
3. selection of geophysical methods, instrumentation and data collection parameters
4. consideration of data processing, interpretation, reporting and archiving methods.

3.1 *Determination of project goals*

Many times, geophysical surveys are contracted for archaeological research. In these instances the archaeologists have specific research questions they are attempting to answer. It is the responsibility of the geophysical contractor to learn what these questions are and to design an effective survey methodology that will provide the most effective information to help answer them. In some circumstances the archaeologist may not have a full comprehension of all of the various aspects and finesse of geophysical surveys; therefore they may not know what to ask for. The geophysical surveyor must take the lead in providing information and survey methodologies that will best suit the required project.

It is also very important not to *oversell* the potential results of geophysical surveys. When speaking with a client and developing a survey methodology one must fully comprehend the fundamental geophysical principles for each method and realize that these are what will determine the end results of the survey. Having stated this, theory is an essential guideline to geophysical surveys, but it is always good to use a selection of different geophysical methods on a site. Most archaeo-geophysicists have a collection of stories where they thought a method may or may not produce good maps based upon the fundamental geophysical principles, only to be surprised by final results[1].

---

[1] The first two weeks of the magnetic survey at Gebel Barkal in Sudan (Fig. 2B) did not map any distinctive structural features although they were clearly visible on the ground surface. In the final day of the project the author was told to survey wherever she wanted as previous work (guided by the archaeologist) had not been effective. Two new temples were mapped in a morning's survey.

Useful questions to ask include[2]:

- what are the final project goals?
- how will the data be used? (i.e. simple interpreted maps, a detailed report, interpretations to integrate into a project GIS[3] or ArcIMS[4] project).

3.2   Consideration of the survey area

Once the project goals are understood the physical character of the archaeological site and potential remaining archaeological features need to be described. Different geophysical survey methods cannot be used in some environments such as resistivity survey in very arid conditions, magnetometry in some urban settings or GPR survey in an area with dense ground cover or trees. If the geophysical surveyor is not able to visit the site prior to data collection very detailed information must be acquired in order to assess the use of different survey methods and techniques in order to obtain good results. Keep in mind that unless photographs of the site are provided the survey area is described based upon a single person's comprehension, detailed questions must be asked.

Useful questions include:

- what are the expected archaeological features, how deeply are they buried (to their top and bottom), what material are they made of?
- what is the local geology, hydrogeology, soil type?
- what is the nature of the survey area?
  - ground cover, topography
  - caves, cliff faces, standing structures
  - what are local interference factors (i.e. radio broadcasting, traffic, livestock, tourists, etc.)

3.3   Selection of effective geophysical methods, instrumentation and data collection parameters

With information on the survey goals, the nature of the survey area and potential remaining archaeological features the geophysical survey can be planned. Decisions must be made based upon factors such as time, available personnel and financial resources for effective data collection, processing and interpretation. Some of the factors that are determined during the project design include:

- selection of appropriate geophysical survey methods and equipment
- site recording
- survey grid location and size; establishing survey grids and recording positions
- data collection sampling rates and technique (zig-zag, uni-directional, GPS-recorded).

3.4   Consideration of data processing, interpretation, reporting and archiving methods

It is important to establish how the geophysical survey data will be handled prior to data collection. All geophysical data that are collected these days are digital. Therefore all data processing is done on a computer. Computer power and appropriate data processing, imaging and interpretation software are essential in order to obtain reasonable interpretations. Consideration of the use of data, such as integration into a project GIS, helps to determine necessary output data formats.

Equally important is data archiving. Once data are collected, interpreted and reports sent to the client the data and results must be preserved for potential use in the future. Archival practices should be clearly established making retrieval and review of original data a simple process. In the UK the Archaeology Data Service (ADS) is an excellent resource that provides archival space

---

[2] A good example of a detailed survey question list can be found on the Archaeo-physics web site at Web 1.
[3] Geographic Information System.
[4] Basic GIS information available over the internet. For an ArcIMS example see Forum Novum at Web 2.

for data and promotes good practice in the use of digital data through technical advice and use of digital technologies (more information can be found at the ADS web site, Web 2).

A number of useful publications contain guidelines for effective geophysical survey and provide results from a wide variety of geophysical surveys. Some of these resources include:

- Geophysical Data in Archaeology: A Guide to Good Practice (Web 3)
- Geophysical Survey in Archaeological Field Evaluation, Research and Professional Services Guideline No. 1 (David 1995)
- the Geophysical Survey Database—English Heritage (Web 4)
- the North American Database of Archaeological Geophysics (Web 5)
- the Archaeo-Imaging Lab (Web 6).

4   A STUDY OF THE CATHOLME CEREMONIAL COMPLEX GEOPHYSICAL SURVEYS

The Catholme Ceremonial Complex is a collection of ritual monuments (Fig. 8) listed as Scheduled Ancient Monuments (SAM), that include a sunburst (A), a timber circle (B) and a series of pit alignments that cross the landscape (C) and is located in Staffordshire, UK. This particular area was targeted for investigation as part of the Where Rivers Meet (WRM) project due to the impact of an aggregate quarry adjacent to the ceremonial complex. The aggregates extraction process is not only destroying the archaeological record through excavation but it is also seriously impacting the integrity of the remaining SAMs through the effects of de-watering. Preliminary investigation of the project area through analysis of aerial photography and LiDAR images enabled a smaller focus area for ground-based geophysical investigations to be targeted. The goal of the geophysical component of this project was to better understand the relationship between geophysical survey

Figure 8.   The Catholme Ceremonial Complex sunburst (A), timber circle (B) and pit alignment (C) monuments.

data, below-ground archaeology and site soil properties as a means to enhance the interpretation of geophysical survey results and as a guide for future geophysical survey practice.

The geophysical surveys were conducted in two phases. The first phase covered approximately 16 ha using a combination of resistivity, magnetic gradient and GPR surveys. The surveys covered broad areas centred on the ceremonial monuments in order to obtain information on the context of the monument complex within the broader landscape in addition to details of the individual monuments themselves (Watters 2004).

The second phase of geophysical surveys targeted specific areas within the ceremonial monuments based on cropmark information and results from the first phase of geophysical surveys (Watters 2005). In this phase of investigation geophysical surveys were first conducted on the ground surface and then conducted for a second time in the same area on the gravel/sand subsurface after removal of the topsoil. Excavation (Hewson *et al.* 2005) and soil analyses (Houndslow & Karloukovski 2005; Jordan 2005) in combination with the intense geophysical surveying in this case, seek to gain insight into the response of different geophysical methods to subsurface properties.

The geophysical data have been interpreted with an emphasis not only on the identification of archaeological features, but of equal importance, on the nature of the subsurface response to each geophysical technique. Future investigations and analyses of the project data hope to provide a unique insight into the nature of the archaeology and environment and the impact of contemporary land use on it.

The geophysical surveys at the Catholme Ceremonial Complex were only one of the numerous techniques used for mapping and assessment of the region. Careful consideration of the scope of the project and all of its component parts was essential in the planning and success of the geophysical surveys.

4.1   *Catholme project goals*

The geophysical surveys at the Catholme Ceremonial Complex were conducted for a number of reasons. The project goals included the:

- assessment of geophysical survey methods in alluvial environments such as that of the Trent and Tame River plains
- modification and recommendation of more effective geophysical survey methods for mapping the buried archaeology in an alluvial environment
- mapping and identification of the remaining archaeological resource
- assessment of the impact of aggregate extraction (and subsequent de-watering) upon the remaining archaeological resource
- development of effective data collection methods for three dimensional data visualization.

The end products for the geophysical surveys included:

- official reports for English Heritage
- integration of geophysical survey maps and interpreted features into a larger project GIS and ArcIMS site
- data archiving with ADS.

The well defined goals and desired end products for the Catholme geophysical surveys were central to planning the data collection and management methodology for the project.

4.2   *Consideration of the Catholme survey area*

Once the project goals were established a review of regional archaeology and past geophysical surveys helped establish a detailed picture of the environmental, geological and archaeological nature of the targeted survey area. Remaining archaeological features were estimated to be buried at approximately 0.4 to 0.75 m depth and excavated into a sandy/gravel alluvial subsurface

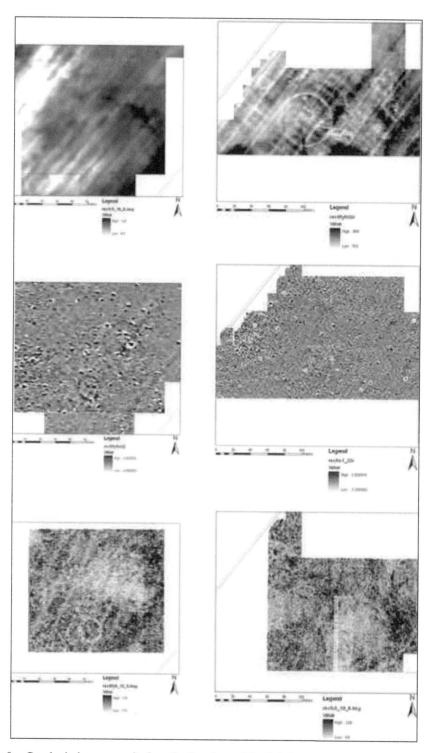

Figure 9. Geophysical survey results from the first phase of the Catholme survey. Where resistivity did not map the Sunburst ring ditch (A) but did map the large ring ditch (B), magnetometry mapped the Sunburst ring ditch (C) but not the large ring ditch (D) and GPR mapped both the ring ditch features (E and F).

material. Based upon similar features in the region and crop mark evidence features were thought to have a humic fill material.

Past geophysical surveys conducted in this region did not produce consistently good results over archaeological features. This failure to map archaeological features could be attributed to a number of reasons such as ineffective sampling rates, environmental conditions and survey grid locations.

4.3 *Selection of geophysical methods and data collection parameters*

Based upon the project goals and consideration of the Catholme survey area the first phase of geophysical surveys included resistance, magnetometry and GPR. Basic commercial survey data sampling rates were established with the intent to cover as much area as possible and to establish a base performance for each survey method.

Once the first phase of surveys was complete and analysed the sampling rates were re-considered and increased for the second phase of geophysical surveys. Survey methods included electrical imaging, magnetometry and GPR. The survey methodology for this phase of investigation was influenced by:

- irregular mapping of the ring ditch features by resistivity and magnetic surveys during the first phase surveys (Fig. 9 where resistivity did not map the Sunburst ring ditch (A) but did map the large ring ditch (B), magnetometry mapped the Sunburst ring ditch (C) but not the large ring ditch (D) and GPR mapped both ring ditch features (E and F))
- complete failure during the first phase of surveys to map the timber circle monument

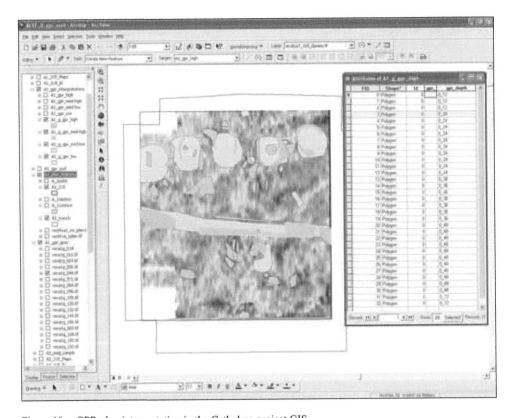

Figure 10.   GPR plan interpretation in the Catholme project GIS.

194

- integration of geophysical data with soil sampling and ground truthing through excavation (Houndslow & Karloukovski 2005; Jordan 2005 and Hewson *et al.* 2005)
- demands for three dimensional visualization research (Watters 2007).

4.4 *Catholme data processing, interpretation, reporting and archiving methods*

The data treatment protocol was established prior to any data collection. As this project had two phases, multiple geophysical surveys and additional research components (LiDAR, hydrogeology, excavation and soil sampling) data were required to integrate into a project GIS and be accessible through an ArcIMS site.

Data were processed in dedicated geophysical processing software and rectified into the project GIS (more details on this process can be found in Watters 2004, 2005 and 2007) and interpreted (Fig. 10). Reports contained detailed information on all aspects of the survey from review of past geophysical surveys, data collection and treatment parameters, interpretations to recommendations for future work. Once completed, the Catholme project geophysical data was sent to the ADS for archiving and is also archived at the University of Birmingham in digital form, on DVD and archival tape.

5 RESULTS OF THE CATHOLME GEOPHYSICAL SURVEYS

The full scope of the WRM project to study the surviving archaeological resource at the confluence of the Trent and Tame rivers in south eastern Staffordshire continues to be pursued through various research channels (Buteux *et al. forthcoming*). The combined geophysical surveys, excavation and

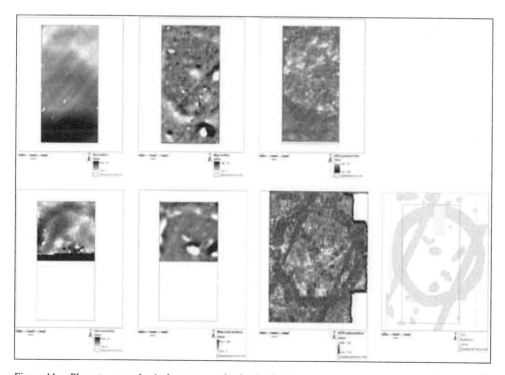

Figure 11.   Phase two geophysical survey results for the Sunburst monument topsoil surveys (A resistivity, B magnetometry and C GPR) and subsoil surveys (D resistivity, E magnetometry and F GPR) with a plan of the archaeology G.

Plate 16.   A three dimensional visualization of the Sunburst GPR survey results. (See colour plate section)

soil sampling of the first and second phases of this project have provided a robust and complex database for various methods for interpretation and analysis of the archaeological monuments of the Catholme Ceremonial Complex that should be used in the future to obtain a more fundamental understanding of the soil properties, site processes and geophysical recording.

The first phase of geophysical surveys identified a number of the Catholme archaeological features and helped assess the effectiveness of the geophysical methods and techniques employed. In the course of this work not only did all of the geophysical methods succeed in mapping at least one or two of the crop mark features, GPR provided additional information that led to the conclusion that modern (and past) land use is directly impacting the integrity of the Neolithic archaeological features, protected through English Heritage as SAMs. The broad area coverage employed in this phase of surveys helped identify the crop mark features as well as additional possible archaeological anomalies that contribute to the understanding of the larger dynamic landscape.

The second phase of geophysical surveys aimed to obtain insight to the effectiveness of the various geophysical methods employed for mapping the archaeological features and underlying geology of the survey areas. Resulting geophysical data and maps provided a base guideline of how the archaeological anomalies appeared in order to provide background knowledge for a re-examination of the results of the topsoil surface survey. In addition to a background for surface survey consideration, the second survey conducted on the natural subsoil interface provided detailed information of not only the geophysical properties of the archaeological features, but also additional information contributing to the form and content of the archaeological features themselves (Fig. 11).

It is clear that this suite of tools mapped the archaeological features of the Catholme Ceremonial Complex. In both phases of investigation GPR was the most effective and efficient method for archaeological mapping. GPR was able to identify different features but also provided important three-dimensional information related to the structure, form and depth of the buried archaeological features and the impact of contemporary land use on the archaeological resource (Plate16). Due to contemporary land use as a working agricultural farm, the magnetic survey was the least effective for mapping the mixture of archaeological features of the Catholme Ceremonial Complex through the topsoil due to the strong magnetic noise from ferrous debris and ploughing effects. However, once the topsoil was removed magnetometry surveys mapped all but the pit alignment pits because of a strong (overshadowing) magnetic signature attributed to a geological feature adjacent to the pits. The resistivity surveys conducted during the second phase of investigations used electrical imaging. This technique for resistivity survey did not map any of the archaeological features through the topsoil surveys but effectively mapped them once the topsoil was removed. Used in combination, these three geophysical methods provided detailed information on the subsurface archaeological features.

In combination with the overarching WRM GIS (including hydrogeological, geoarchaeological and geophysical survey visualization results) the archaeological excavation and soil sampling have provided the material for a re-interpretation of the Catholme Ceremonial Complex in terms of scientific dating, palaeoenvironmental reconstruction and its relation to the broader WRM landscape (Buteux *et al. forthcoming*; Evans 2006). Continuing research includes work on isolating factors that are affecting the integrity of the archaeology in order to assist in future land use guidelines for the least amount of impact upon the remaining archaeological features.

6 CONCLUSIONS

The Catholme Ceremonial Complex geophysical surveys were fundamental to the larger Where Rivers Meet project. The geophysical surveys conducted during the two phases of work covered approximately 16 ha and generated hundreds of data files. This project succeeded in part due to the detailed and thoughtful project planning that was one of the first and most important steps toward mapping the archaeological landscape.

The results of the Catholme geophysical surveys should contribute significantly to project planning and design of future surveys in a similar alluvial river plain environment. The Catholme surveys provide solid evidence that GPR was the most effective tool for mapping the archaeology in this area. A distinct lesson was learned in regard to the resistivity and magnetometry surveys that despite assumed performances based on geophysical principles, both of theses geophysical methods did not map significant archaeological monuments. Continuing research into the comparison of soil properties to the geophysical survey results hopes to explain why this phenomenon and provide additional information to be considered for even more effective future landscape mapping.

REFERENCES

Annan, P. & J. Davis. 1976. Impulse Radar soundings in permafrost. *Radio Science* 11:383–394.
Atkinson, R. 1952. Méthodes eléctriques de prospection en archéologie, in A. Laming (ed.). *La Découverte du Passé,* Picard, Paris.
Barker, R. 1993. A simple algorithm for electrical imaging of the subsurface. *First Break* 10(2):53–62.
Barringer, A. 1965. Research directed to the determination of subsurface terrain properties and ice thickness by pulsed VHF propagation methods. *Air Force Cambridge Research Laboratories contribution AF19* (628) 2998.

Bevan, B. & J. Kenyon. 1975. Ground probing radar for historical archaeology. *MASCA Newsletter* 11:2–7.

Breiner, S. 1973. *Applications Manual for Portable Magnetometers*. Sunnyvale: Geometrics.

Buteaux, S., H. Chapman & M. Hewson. *Forthcoming. Where Rivers Meet. The Catholme ceremonial complex and the archaeology of the Trent-Tame confluence, Staffordshire.*

Campbell, K. & A. Orange. 1974. A continuous profile of sea ice and freshwater ice thickness by impulse radar. *Polar Record* 17(106):31–41.

Clark, A. 1996. *Seeing Beneath the Soil, Revised Edition*. London: B.T. Batsford Ltd.

Conyers, L. & D. Goodman. 1997. *Ground-Penetrating Radar: An Introduction for Archaeologists*. Walnut Creek, Calif.: AltaMira Press.

Cook, J. 1960. Proposed monocycle-pulse VHF radar for airborne ice and snow measurements, in *AIEE Trans. Commun. and Electron.* 79(2):588–594.

Dabas, M., A. Hesse & J. Tabbagh. 2000. Experimental resistivity survey at Wroxeter archaeological site with a fast and light recording device. *Archaeological Prospection* 7:107–118.

Dabas, M. 2006. Theory and practice of the new fast electrical imaging system ARP©, presented at the *XV International Summer School in Archaeology, Geophysics for Landscape Archaeology, Grosseto, 10–18 July.*

Daniels, J. 2000. Ground Penetrating Radar Fundamentals, prepared as an appendix to a report to the U.S. EPA, Region V, Nov. 25, 2000.

David, A. 1995. *Geophysical survey in archaeological field evaluation, Research and Professional Services Guideline No. 1*. London: English Heritage.

De Terra, H., J. Romero & T. Stewart. 1949. The Tepexpan Man. *Viking Fund Publications in Archaeology* 11. New York: Werner-Grenn Foundation.

Evans, T. 2006. "Ritual Monuments and Natural places: An Assessment of the Early Neolithic and Early Bronze Age landscape architecture at the confluence of the Rivers Trent, Tame and Mease, Staffordshire." *Unpublished MA Dissertation*. University of Birmingham, UK.

Gaffney, C. & J. Gater. 2003. *Revealing the Buried Past, Geophysics for Archaeologists*. Stroud: Tempus Publishing Ltd.

Goodman, D. & Y. Nishimura. 1992b. Radar archaeometry and the use of synthetic radargrams to investigate burial grounds in Japan in *Proceedings of the 28th International Symposium on Archaeometry, Los Angeles*: 167.

Goodman, D. & Y. Nishimura. 1993. A ground-radar view of Japanese burial mounds. *Antiquity* 67:349–354.

Griffiths, D. & R. King. 1981. *Applied Geophysics for Geologists & Engineers*. London: Pergamon Press.

Hallof, R. 1957. "On the Interpretation of Resistivity and Induced Polarization Measurements." *PhD thesis*. Massachusetts Institute of Technology.

Hewson, M., E. Hancox & K. Bain. 2005. *Catholme Ritual Landscape Ground Truthing Project 2004: Post-excavation assessment. Aggregates Levy Sustainability Fund (ALSF) Where Rivers Meet Phase II Report 1*. London: English Heritage.

Houndslow, M. & V. Karloukovski. 2005. *Where Rivers Meet: Landscape, Ritual and Settlement and the Archaeology of River Gravels, Report for Magnetic Properties, Aggregates Levy Sustainability Fund (ALSF) Where Rivers Meet Phase II Report 4*. London: English Heritage.

Jordan, D. 2005. *Where Rivers Meet: Landscape, Ritual and Settlement and the Archaeology of River Gravels. The Geoarchaeology of Deposits at Catholme. Aggregates Levy Sustainability Fund (ALSF) Where Rivers Meet Phase II Report 3*. London: English Heritage.

Keary, P., M. Brooks & I. Hill. 2002. *An Introduction to Geophysical Exploration, Third Edition*. Oxford: Blackwell.

Kenyon, J. 1977. Ground-Penetrating Radar and Its Application to a Historical Archaeological Site. *Historical Archaeology* 11:48–55.

Leckebusch, J. 2003. Ground Penetrating Radar: A Modern Three-dimensional Prospection Method. *Archaeological Prospection* 10:213–240.

Leckebusch, J. & R. Peikert. 2001. Investigating the True Resolution and Three-dimensional Capabilities of Ground-penetrating Radar Data in Archaeological Surveys: Measurements in a Sand Box. *Archaeological Prospection* 8:29–40.

Loke, M. 2004. *Tutorial: 2-D and 3-D electrical imaging surveys*. Geotomo.

Loke, M. & R. Barker. 1996a. Rapid least squares inversion of apparent resistivity pseudosections by a quasi-Newton method. *Geophysical Prospecting* 48:181–152.

Loke, M. & R. Barker. 1996b. Practical techniques for 3D resistivity surveys and data inversion. *Geophysical Prospecting* 44:499–523.

Loke, M. & T. Dahlin. 2002. A comparison of the Gauss-Newton and quasi-Newton methods in resistivity imaging inversion. *Journal of Applied Geophysics* 49(3):149–162.

Morey, R. 1974. Continuous subsurface profiling by impulse radar, in *Proceedings of the Engineering Foundation Conference on Subsurface Exploration for Underground Excavation and Heavy Construction*. American Society of Civil Engineers: 213–232.

Mussett, A. & M. Khan. 2000. *Looking Into the Earth, an Introduction to Geological Geophysics*. Cambridge: Cambridge University Press.

Neubauer, W., A. Eder-Hinterleitner, S. Seren & P. Melichar. 2002. Georadar in the Roman Civil Town Carnuntum, Austria: An Approach for Archaeological Interpretation of GPR Data. *Archaeological Prospection* 9:135–156.

Olhoeft, G. 2000. Maximizing the information return from ground penetrating radar. *Journal of Applied Geophysics* 43:175–187.

Panissod, C., M. Dabas, N. Florsch, A. Hesse, A. Jolivet, A. Tabbagh & J. Tabbagh. 1998. Archaeological Prospecting using Electric and Electrostatic Mobile Arrays. *Archaeological Prospection* 5:239–251.

Scollar I., A. Tabbagh, A. Hesse & I. Herzog. 1990. *Archaeological Prospecting and Remote Sensing: Topics in Remote Sensing, Vol. 2*. Cambridge: Cambridge University Press.

Tsourlos, P. 1995. "Modelling interpretation and inversion of multi-electrode resistivity survey data." *Ph.D. Thesis*. University of York.

Tsourlos, P. & R. Ogilvy. 1999. An algorithm for the 3-D Inversion of Tomographic Resistivity and Induced Polarization data: Preliminary Results. *Journal of the Balkan Geophysical Society* 2(2):30–45.

Vickers, R. & L. Dolphin 1975. A Communication on an Archaeological Radar Experiment at Chaco Canyon, New Mexico. *MASCA Newsletter* 11(1):6–8.

Vickers, R., L. Dolphin & D. Johnson. 1976. Archaeological Investigations at Chaco Canyon Using a Subsurface Radar, in T. Lyons (ed.). *Remote Sensing Experiments in Cultural Resource Studies at Chaco Canyon*. Albuquerque, New Mexico: USDI-NPS and the University of New Mexico: 81–101.

Walker, R. & P. Linford. 2006. Resistance and Magnetic surveying with the MSP40 Mobile Sensor Platform at Kelmarsh Hall. *ISAP News* 9:3–5.

Watters, M. 1999. Geophysical Work Conducted with the Geometrics G858 Cesium Vapor Magnetometer at Effigy Mounds National Monument, IA. As part of the U.S. National Park Service Workshop: "Recent Advances in Archeological Prospection Techniques," May 10–14, 1999. Submitted to the Midwest Archeological Center, U.S. National Park Service, Lincoln, Nebraska.

Watters, M. 2004. *Where Rivers Meet: Geophysical Survey At Catholme. Aggregates Levy Sustainability Fund (ALSF) Where Rivers Meet Phase I Report 5*. London: English Heritage.

Watters, M. 2005. *Where Rivers Meet: Geophysical Survey At Catholme. Aggregates Levy Sustainability Fund (ALSF) Where Rivers Meet Phase II Report 2*. London: English Heritage.

Watters, M. 2007. "New Methods for archaeo-geophysical data visualization." *Ph.D. Thesis*. University of Birmingham, UK.

Weymouth, J. 1986. Geophysical Methods of Site Surveying, in M. Schiffer (ed.). *Advances in Archaeological Method and Theory, Vol. 9*. London: Academic Press.

WEB RESOURCES

Web 1: Archaeo-physics
   http://www.archaeophysics.com/methods/index.html
Web 2: Forum Novum ArcIMS
   http://ccc286.bham.ac.uk/website/fnv/fnvframe1.htm
Web 3: ADS Geophysical Data in Archaeology: A Guide to Good Practice
   http://ads.ahds.ac.uk/project/goodguides/geophys/
Web 4: English Heritage Geophysical Survey Database
   http://sdb2.eng-h.gov.uk/
Web 5: North American Database of Archaeological Geophysics
   http://www.cast.uark.edu/nadag/
Web 6: Archaeo-Imaging Lab
   http://www.cast.uark.edu/~kkvamme/ArcheoImage/archeo-image.htm

Seeing the Unseen – Campana & Piro (eds)
© *2009 Taylor & Francis Group, London, ISBN 978-0-415-44721-8*

The use of geophysical techniques in landscape studies: Experience from the commercial sector

C. Gaffney

Remote Vision Research, Thornton, Bradford, UK

ABSTRACT: Commercial geophysical work for archaeology has really only been workable for the last 20 years. In general the value of geophysical techniques involves a number of factors such as: speed, cost, technical accuracy, success rates. Although nowadays technology may allow whole landscapes geophysical survey, at present such options are rarely used in the commercial field. More commonly strategies are devised that are a mix of Level 1 (prospection) and Level 2 (evaluation). The assumption is that a landscape cannot be surveyed in detail, but areas of potential must be identified (Level 1) and tested by detailed survey (Level 2). Two methods have become commonplace for rapidly locating areas of archaeological potential; coarse grid magnetic susceptibility survey and 'scanning' along wide spaced transects using a magnetometer. In either case the 'hot spots' are then surveyed in detail using a magnetometer. In any geophysical survey the most important aspect of the work is the data interpretation. The methodology can be rigorous, but if the surveyor does not understand the data then the exercise is futile.

1 INTRODUCTION

Commercial geophysical work for archaeological purposes has really only been viable for the last 20 years, although significant area surveys were accomplished earlier. A publication that hints at the ensuing glut of geophysical work for commercial purposes is the Southern Feeder publication (Catherall et al, 1984). In that volume large scale magnetometer survey was undertaken at selected 'sites' across the landscape that was traversed by a proposed pipeline. The areas for survey were chosen via traditional archaeological prospecting methods such as aerial photographic analysis or field walking. Important technological changes had occurred prior to that work that allowed geophysical surveyors to help change a belief that was commonly held in British archaeology; it was thought that geophysical survey was too slow and the interpretation too imprecise to be of any great value. Since the early 1990s geophysical work has become meshed into many aspects of non-invasive archaeological investigation and in many parts of the world.

An unexpected by product of the numerous, but small-scale, geophysical surveys is the accumulative effect at the landscape level. The classic example is the value of geophysical survey in the area of the World Heritage Site of Stonehenge. It has been documented by David and Payne (1997) that many hundreds of hectares have been surveyed, the majority detailed magnetometer survey for planning purposes, along proposed road corridors. It is evident that by themselves the individual projects have a value in the planning system, but taken together the archaeological product is more compelling.

While a cursory look through any text book on geophysics will suggest that about 10 ground based geophysical techniques are on offer for near surface work, only a few are regularly used in the commercial field and even fewer are used to investigate landscapes. Most commonly the techniques of earth resistance, magnetometry, magnetic susceptibility sampling and GPR are used, although low frequency EM systems are also valuable in the commercial arena. Magnetometry, in its various forms, is the most widely used geophysical technique for detecting buried archaeology.

2 GENERAL RATIONALE FOR THE USE OF GEOPHYSICAL TECHNIQUES IN COMMERCIAL ARCHAEOLOGY

In general the value of geophysical techniques involves a number of factors such as:

- Speed
- Cost
- Technical Accuracy
- Success rates (with respect to finding unknown archaeology)

The first two factors are vitally important in determining which techniques (geophysical or otherwise) are used in landscape studies. Fortunately, geophysical techniques are widely believed to be efficient at assessing blocks of land and that means that they are often at least considered in the assessment of large areas of land. The perceived efficiency of geophysics has allowed an irresistible rise in both the number of surveys and the number of groups offering a commercial service in the last decade (see, for example, Gaffney and Gater 2003). The technical accuracy of surveys for archaeological purposes ie the quality of the data, is also generally good.

However, increasingly the success rate of geophysical techniques in determining archaeology is under discussion. The so-called 'Planarch' document (Hey and Lacey 2001), which was undertaken to establish the validity of prospecting methods on subsequently excavated sites, illustrates this point. Within that volume a detailed analysis of six sites in south east England by Linford and David (2001) indicates the scrutiny that the interpretation of geophysical data is now under. In broad terms the success rate, defined by 'true positives', in these six surveys ranged from 58.9% to 83.3%. In the overview of the project, analysed by Hey and Lacey, geophysical investigation was not highly commended. While it is true that most geophysicists would disagree with the interpretation of the findings (as would many of their clients) and possibly even query the structure of the analysis, there is no denying some of the summary points raised by Linford and David (ibid, p.87):

'...it is nonetheless clear that geophysical data can and often does overlook very important remains.'

'Geophysical data alone cannot be used to support the concept of negative evidence.'

In the particular case of landscape analysis there is a requirement (for surveyor, client and curator) to appreciate the potential pit-falls of any strategy that is implemented. A simplified sequence for undertaking any geophysical survey is shown in Figure 1. The crucial stage, if geophysical data is to be used effectively, relates to 'Specification: Experience: Guidelines.' The middle term relates not just to an individual's experience, but the shared pool from which curator's in particular can draw upon ... that includes the critical retrospective analysis of Planarch. A specification is usually agreed with a curator prior to a piece of work and sometimes refer to a community wide set of guidelines. In England, but also used elsewhere, the 'guidelines' refer to those produced by English Heritage (David, 1995). While still in common use they are in need of update and at the time of writing a draft of a new set of guidelines has been circulated. This is a very welcome development as agreed guidelines indicate a level of maturity in the discipline and give those with only occasional use of commercial organisations greater confidence. With these in place, and due diligence in the conduct of the work, it should be possible to devise a strategy that will satisfy the needs of all concerned.

3 RAPID STRATEGIES FOR LANDSCAPE STUDIES

Evidently, landscape investigation is different from site specific work due to the increased size of the area of interest and the often unknown level of buried archaeology. Although other contributions to this volume (for example Dabas 2007) or the GPS driven Foerster magnetometer cart demonstrated at the Summer School, provide technology that may allow whole landscapes to be analysed in detail, at present such options are rarely used in the commercial field. More commonly strategies are devised that are a mix of Level 1 (prospection) and Level 2 (evaluation) as defined

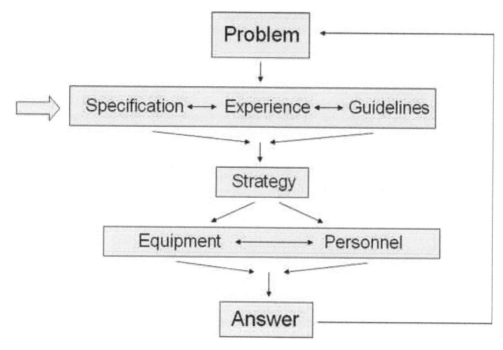

Figure 1.   Scheme for the use of geophysical techniques in commercial archaeology.

by Gaffney and Gater (2003). The assumption is that a landscape cannot be surveyed in detail, but areas of potential must be identified (Level 1) and tested by detailed survey (Level 2). Sampling strategies can be used over landscapes, but they are not regarded as an optimal strategy except in areas where the property contrast is low between target and surrounding medium. Two methods have become commonplace for rapidly locating areas of archaeological potential; coarse grid magnetic susceptibility survey and 'scanning' along wide spaced transects using a magnetometer. In either case the 'hot spots' are then surveyed in detail using a magnetometer. Both methods have their advocates and an attempt to understand the value of each method is illustrated in Gaffney and Gater (2003, pp. 96–101). However, for some 'purists', landscapes should not be evaluated in this way (eg Neubauer 2006). The argument against rapid assessment is that it favours the detection of long-lived settlement sites and this is evidently true if one considers low contrast sites. In such cases the scanning 'threshold', for example, may mean that those sites or features that produce only weaker signals may be missed in a rapid assessment. However, this view must be balanced against the both the speed / cost factors associated with the needs of commercial archaeology and the inescapable fact articulated by Linford and David outlined above; even detailed survey will miss significant buried archaeology. Of course, one additional fact that can be taken from the Planarch document is that all prospecting techniques miss significant archaeology. Archaeologists deal with this by structuring the analysis in a way that embeds many detection options within the overall strategy. That is, while some may favour one technique over another they will always assume that some archaeology will be missed and maintain a multi-technique approach. Those who do not will assess or evaluate landscapes in an inadequate fashion.

However, it must be acknowledged that the commercial factors that are at work do not allow surveyors to wait until conditions within a landscape are perfect for survey. Planning deadlines dictate both the timing and the speed of the analysis. If geophysicists are not prepared to devise rapid strategies then it is likely that geophysical involvement in landscape analysis will decline. There are additional factors that should be mentioned when discussing rapid analysis; clients often like such a strategy as problem areas, modern dumping for example, can be easily located and the speed of the survey allows extra time for mitigation strategies to be developed if archaeology is found. In the end these factors are minor to the fact that archaeology can be located using a rapid scheme; knowledge of where archaeology is located is important, even if the buried details are not all mapped.

In any geophysical survey, and the Planarch document highlights this, the most important aspect of the work is the data interpretation. The methodology can be rigorous, but if the surveyor does not understand the data then the exercise is futile. In order to achieve a suitable endpoint the archaeological and geological 'background' must be understood. Similarly, the delivery of the interpretation must be in a format that allows easy access for the client, otherwise the significance of the results may be downgraded. This last point is particularly true when analysing landscapes as the data sets can be huge (see Gaffney and Gaffney 2006 for a discussion of scale and visualisation). The all encompassing nature of digital data makes display, analysis and delivery easier as time progresses; the interpretation remains difficult as automatic interpretation algorithms have rarely proved successful.

4 CRITICISM AND THE FUTURE OF LANDSCAPE ANALYSIS

It seems evident that we need to report and discuss failures as well as successes in the methodologies that are used. If we do not then collective knowledge will not accrue, and this is vital if the subject is to advance. Additionally, if as a discipline we do not confront problems of interpretation, general consumers of geophysical data may extrapolate from individual surveys and find the techniques and methodologies lacking. A mature and balanced discipline such as ours should confront problems of interpretation and scale and develop improved strategies. To this end we need to establish better feedback from those who ground truth geophysical interpretations. It is generally true that feedback only occurs when something goes wrong; positive feedback would increase the confidence levels and hence allow greater certainty in future interpretations. In these circumstances it is clear that to embed geophysical techniques into landscape studies we cannot divorce geophysics from archaeology or archaeologists.

ACKNOWLEDGEMENTS

I would like to thank my former colleagues at GSB Prospection, especially John Gater, for many discussions on the subject of 'commercial' geophysics.

REFERENCES

Catherall, P.D., Barnett, M. and McClean, H. 1984 The Southern Gas Feeder: the archaeology of a gas pipeline. British Gas Corporation. London.
David, A. 1995 Geophysical survey in Archaeological Field Evaluation, Research and Professional Services Guideline 1, English Heritage, London.
David, A. and Payne, A. 1997 'Geophysical Surveys within the Stonehenge Landscape: A Review of Past Endeavour and Future Potential' Proceedings of the British Academy 92 pp 73–113.
Gaffney, C. and Gaffney, V. 2006 No further territorial demands: on the importance of scale and visualisation within archaeological remote sensing. In From Artefacts to Anomalies: Papers inspired by the contribution of Arnold Aspinall University of Bradford 1–2 December 2006. Published on line: http://www.brad.ac.uk/archsci/conferences/aspinall/presentations/Gaffney&Gaffney.pdf
Gaffney, C. and Gater, J. 2003 Revealing the Buried Past: geophysics for archaeologists. Tempus, Stroud.
Hey, G. and Lacey, M. (eds) 2001 Evaluation of Archaeological Decision-making Processes and Sampling Strategies: European Regional Development - Planarch Project. Oxford Archaeological Unit.
Linford, N. and David, A. 2001 Study of Geophysical Surveys, in Hey and Lacey (eds) pp 76–89.
Neubauer, W. 2006 Book Review: Revealing the Buried Past. Archaeological Prospection 13, 73–74.

Electromagnetic methods

Seeing the Unseen – Campana & Piro (eds)
© *2009 Taylor & Francis Group, London, ISBN 978-0-415-44721-8*

Short history, strategies and practical aspects of electromagnetic detection based on the description of field

A. Hesse

UMR Sisyphe, Université Pierre et Marie Curie—Paris VI, France

ABSTRACT: Electrical and magnetic methods for archaeological surveying have reached nowadays such a high level of efficiency in terms of speed and legibility of maps that other methods often appear to be of less interest. This is the case for electromagnetic (E.M.) methods despite their unique ability for detection of specific targets or under special circumstances. These methods can be disregarded for several reasons among which the complexity of theoretical aspects, the large variety of available instruments with not well-known specific abilities, a rather slow rate of data collection, etc. must be recalled. Several previous papers (Taggagh, 1990) have already dealt with the theoretical aspects which can also be found, by the same author, in this issue. I have myself presented almost at the same time (Hesse, 1991) a kind of classification of the instruments according to their shape, size, frequency or emitted signal. This will not be given again here in order to emphasise some historical aspects and case histories of significant interest.

1 SHORT HISTORY

Our first attempts to use E.M. instruments at the Centre de Recherches Géophysiques of C.N.R.S. (Garchy), dates back to 1969 when A. Tabbagh joined me and started, following my suggestion, the preparation of his PhD (Tabbagh, 1974 and 1977).

Thanks to his investigations which can still be considered as a reference on the subject, it became clear that the in phase component of the response of the so called "Slingram" instruments was directly correlated with the magnetic susceptibility of the surrounding soil and of underground features.

Our favourite instrument at that time was the EM 15 (Geonics) which provided in phase measurements only (Fig. 1).

However very good survey results were obtained with this single parameter on features with either a positive or a negative contrast of magnetic susceptibility with the surrounding subsoil (Fig. 2 and Plate 17).

Sometimes later it was clearly proved that electrical conductivity could be measured with the quadrature component and A. Tabbagh developed the SH3 at Garchy and a series of subsequent devices until the last ones, which are still under development at the Paris 6 university laboratory. At the same time the EM38 was designed by Geonics with similar abilities but proved to be less appropriate for archaeological surveying.

However it was not the first time that a physical response was obtained on magnetic archaeological materials with an E.M. instrument. At first, Robert du Mesnil du Buisson had a very early intuition (1934) of what could be later possible with geophysical methods (Hesse, 2000). Just after the second World War (1947), he did some experiments with a mine detector (i.e. a metal detector), on burnt clay in the excavations of two monuments at Senlis (France) and anticipated our nowadays results. Someone at that time reported on "a new very important ability of electromagnetic detecting instruments. The indication given by one of them having led to the unearthing of a piece of tegula, ... several verifications were undertaken ... They allowed the operators to establish that... ancient tiles and bricks can produce an effect similar to metals ... Thus arises a

Figure 1. EM15 (Geonics ltd) slingram apparatus used for apparent magnetic susceptibility measurements.

remarkable and unexpected consequence of the method recommended by Monsieur du Mesnil du Buisson: with the help of electromagnetic instruments, one could draw the shape of buried brick walls and the general map of antique rural buildings, ... which, ... as traces of their presence, did not leave more than a thick layer of tegula, collapsed roofing lying a few decimetres under the surface of the ground" (translation from Laming, 1952).

This conversion to earth features of instruments previously designed for metal detection is not unique and several ones such as the SCM (see Tabbagh, same issue, Fig. 1) or the Decco (Littlemore Scientific Engineering) gave very interesting results. This last one was based mainly on the induction of Foucault currents in the metallic conductors which were aiming to be detected but it also gave significant responses according to the magnetic viscosity of soil materials. Several previous researches by Emile Thellier (Thellier, 1966) and Eugène Le Borgne (Le Borgne, 1965) had shown the importance of these properties for archaeological prospecting and consequently, of the possibility of measuring them in the field.

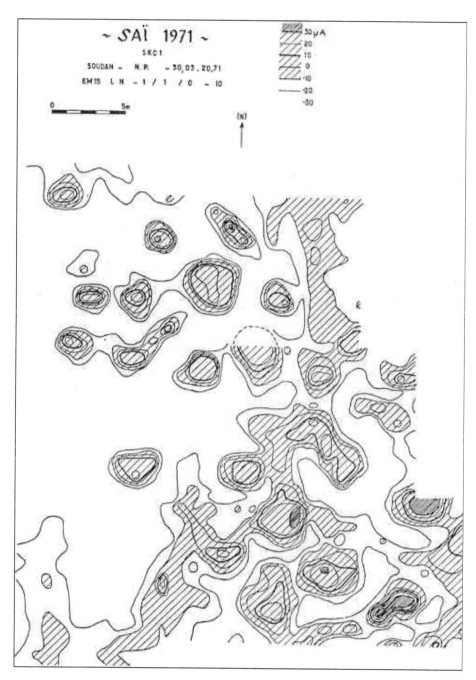

Figure 2. Saï Island (Sudan), EM15 apparent magnetic susceptibility variations (arbitrary units).

Plate 17. Saï Island (Sudan), excavation of the detected features. (See colour plate section)

2   STRATEGIES AND CASE HISTORIES

EM instruments make measurements at a rather slow rate and their automation revealed difficult to achieve in comparison of what was achieved with electrical and magnetic methods. One must also think that their depth of investigation is relatively small. For these reasons they cannot be recommended when other methods can be used.

In compensation, it must be emphasised that the Slingram instruments can provide a double information with the in phase signal (magnetic susceptibility) (see Alain Tabbagh, same issue, the example of Marchezieux, Fig. 5) and the quadrature signal (electrical conductivity) (see Alain Tabbagh, same issue, the example of Marchezieux, Fig. 6) (Tabbagh and Verron, 1983). The latter is particularly interesting for surveying damp areas where resistivity is low and consequently, resistivity meters with electrical currents driven into the ground by means of electrodes, are less sensitive.

Both physical properties correspond to a wide range of applications: conductivity of soils is subject to variations in function of their moisture content; on the contrary, magnetic susceptibility can be considered as an intrinsic property of soils and this consideration leads to the use of wide mesh

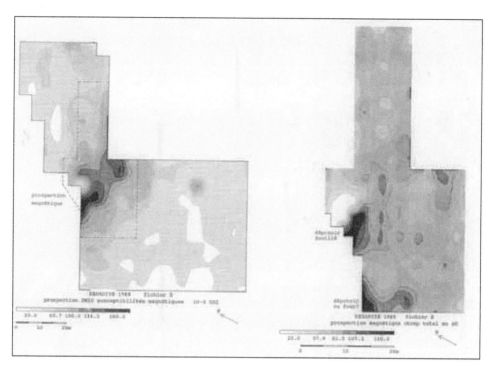

Figure 3. Apparent magnetic susceptibility map (EM15 apparatus) on iron smelting sites showing slag dumps (wide mesh grid survey).

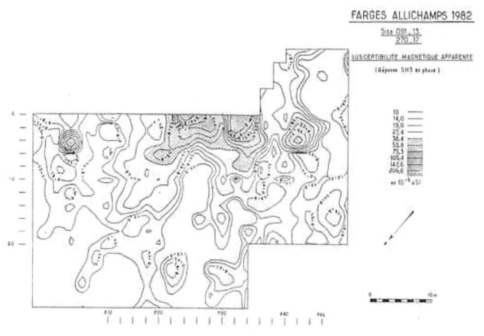

Figure 4. Farges Allichamps (Cher, France), apparent magnetic susceptibility map over an iron-smelting site (SH3 apparatus, wide mesh grid survey).

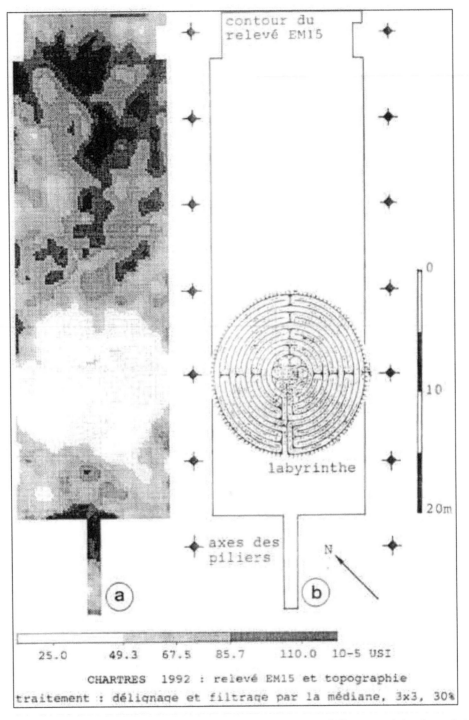

Figure 5. Chartres (Eure et Loir, France), apparent magnetic susceptibility map of the floor of the cathedral.

grids for preliminary investigations of certain types of sites such as pottery or metallic workshops (Hesse, Bossuet and Choquier, 1986) (Figs. 3 and 4). The enhancement of magnetic susceptibility over and around the structures is such that a clear delimitation of sites can be obtained in a very short time, allowing a more careful investigation for smaller structures to be conducted afterwards on a selected and limited area. A complete assessment of the archaeological significance of magnetic susceptibility on a pottery workshop was undertaken on the site of Jaulges-Villiers-Vineux (France) in 1982 and later published in detail (Hesse, 1999).

More usual surveys can be achieved with different types of E.M. instruments. This is the case for filled-in pits and ditches as quoted by Tabbagh in this issue (see the example of South Cadbury Castle, Fig. 1) or elsewhere (Tabbagh, Bossuet and Becker, 1988). Our example of Saï Kerma cemetery (Fig. 1) or the more complex survey made at Hisarönu in Turkey (Hesse and Doger, 1993) show that E.M. methods can also operate on reverse contrasts that is to say on features of low magnetic susceptibility embedded in a highly magnetic material.

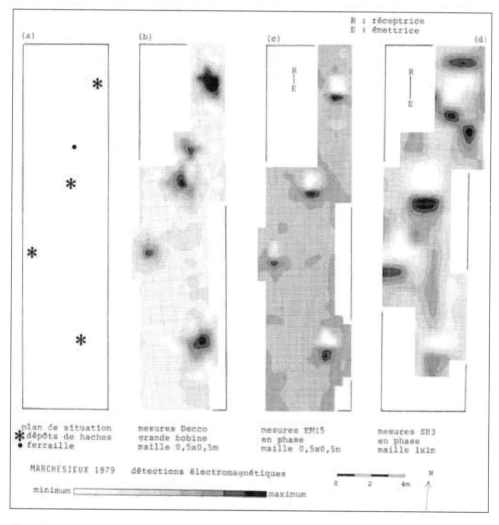

Figure 6.   Marchesieux (Manche, France), detail of the Late Bronze Age hoards responses to the different EM apparatus.

Another important advantage of E.M. instruments consists in the possibility of using them inside cities: they are often less sensitive than magnetometers to the E.M. disturbances generously provided by urban environments; unlike resistivity measurements they do not need electrodes to be driven into the ground which could be difficult or even forbidden on pavements or other hard

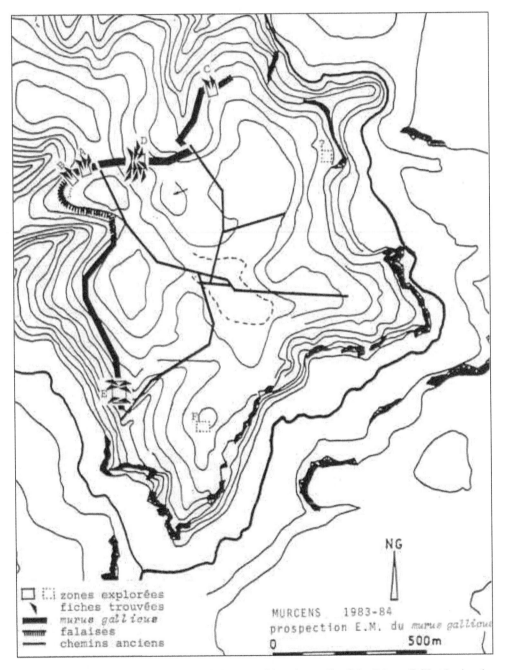

Figure 7. Murcens Oppidum (Lot, France), location of long iron nails of the 'Murus Gallicus' using the DECCO (Littlemore ltd) TDEM metal detector.

surfaces of the ground. We obtained fascinating results when we experimented the EM31 (Geonics) in Alexandria during our search for the Heptastadion (Hesse et al., 2002) or when exploring the floor of several churches (Tabbagh, 1974 and Dabas et al., 1993), (Fig. 5).

Last but not least, one must consider the E.M. devices for their specific ability to detect metallic remains. Strongly prohibited when applied to illicit researches by treasure hunters, they suffer from an undeserved bad reputation amongst many archaeologists. However, on occasion, they can help solve a number of stirring questions within the framework of an archaeological research (Hesse, 2007). Of course, detection of iron objects can often be performed with the magnetic method, but this solution is not valid as soon as non ferrous targets such as copper alloys (and why not silver or gold?!) are concerned. In that case, E.M. instruments must be preferred especially if they are designed for discrimination between ferrous and non-ferrous objects. The appropriate instrument must be chosen according to the shape of the target, its volume and depth (Fig. 6). Speed and spatial resolution of the survey must also be considered as is the case for all geophysical methods. Of course the aim of the survey should never be a rough search for precious objects even for museum collection, but rather consist in an attempt to solve an archaeological problem even if dealing with coarse objects of little value (Fig. 7).

REFERENCES

Dabas M., Stegeman C., Hesse A., Jolivet A., Mounir A., Casas A., 1993, "Prospection géophysique dans la cathédrale de Chartres." Bulletin de la Société Archéologique d'Eure et Loir no 36, 1er trim., 5–25.

Hesse A., 1991, "Les méthodes de prospection électromagnétique appliquées aux sites archéologiques. Geofisica par l'Archeologia.", Séminaire CNR 1988, Porano, Quaderni dell' I.T.A.B.C. 1, 41–52.

Hesse A., 1999, "Multi-parametric survey for archaeology: how and why, or how and why not?", Journal of Applied Geophysics, Proceedings of the 1st International workshop "Electric, magnetic and electromagnetic methods applied to cultural heritage", EMEMACH 97, sept. 29–oct. 1, Ostuni, Italie, no 2–3, march, 157–168.

Hesse A., 2000, "Count Robert du Mesnil du Buisson (1895–1986), a French precursor in geophysical survey for archaeology", Archaeological Prospection, vol. 7–1, 43–49.

Hesse, 2007, "De la détection des objets métalliques des Ages des Métaux", Volume en hommage à Henri Delporte, Editions du C.T.H.S., Paris, in press.

Hesse A., Andrieux P., Atya M., Benech C., Camerlynck C., Dabas M., Féchant C., Jolivet A., Kuntz C., Mechler P., Panissod C., Pastor L., Tabbagh A. et J., 2002, "L'Heptastade d'Alexandrie (Egypte)", Etudes alexandrines 6, dir. J.Y Empereur, IFAO Le Caire, "Alexandrina no 2", 191–273.

Hesse A., Bossuet G., Choquier A., 1986, "Reconnaissance électrique et électromagnétique de sites et de structures de métallurgie ancienne.", Propezioni archeologiche 10, 71–78.

Hesse A., Doger E., 1993, "Atelier d'amphores rhodiennes et constructions en pierre à Hisarönü (Turquie): un cas original de prospection électromagnétique." Revue d'Archéométrie 17, 5–10.

Laming A. (editor), 1952, La découverte du passé, A. et J. Picard, Paris, 363 p.

Le Borgne, 1965, "Les propriétés magnétiques du sol. Application à la prospection des sites archéologiques", Archaeo-Physica, 1, 1–20.

Tabbagh A., 1974, "Méthodes de prospection électromagnétique applicables aux problèmes archéologiques", Archaeo-Physika, 5, 351–437.

Tabbagh A., 1977, "Deux nouvelles méthodes géophysiques de prospection archéologique", Thesis, Université de Paris 6.

Tabbagh A., 1990, "Electromagnetic Prospecting", in I. Scollar, "Archeological Prospecting and Remote Sensing", Cambridge University Press, 520–590.

Tabbagh A., Bossuet G., Becker H., 1988, "A comparison between magnetic and electromagnetic prospection of a Neolithic ring ditch in Bavaria", Archaeometry, 30-1, 132–144.

Tabbagh A., Verron G., 1983, "Etude par prospection électromagnétique de trois sites à dépôts de l'Age du Bronze", Bulletin de la Société préhistorique française, 80, 375–389.

Thellier E., 1966, "Le champ magnétique terrestre fossile, Nucléus, t. 7, n. 1,2,3, 35 p.

Electromagnetic methods (low frequency)

A. Tabbagh
Université Pierre et Marie Curie, Paris, France

ABSTRACT: E.M. low frequency slingram instruments can measure both the electrical conductivity and the magnetic susceptibility of the ground. They also detect metallic objects. Moreover the coupling between magnetic and E.M. data opens interesting paths for a better characterisation of feature geometries and magnetisations.

1   INTRODUCTION

Question: Why using EM (LF) methods?
Answer: (at the beginning of the sixties) to replace electrical d.c. prospection which needs a good galvanic contact with the ground.
Question: Did it succeeded in archaeology?
Answer: No.
Question: Why?
Answer: Because of the magnetic signal.
This is illustrated in figure 1 where the results of a proton magnetometer survey and of an EM instrument (unwisely called SCM: 'Soil Conductivity Meter') survey are compared to the excavation evidence. The response of the EM instrument is clearly correlated to the soil magnetic properties.
To explain this, one needs to go back to a minimum of theory.

2   THEORETICAL BASES

In Maxwell's equations three different properties (complex properties and sometimes tensor) are considered:

- σ, the electrical conductivity, the inverse of the resistivity,
- ε, the dielectric permittivity,
- μ, the magnetic permeability, $\mu = \mu_0(1 + \kappa)$, κ being the volumetric magnetic susceptibility.

If one wants to measure σ, one needs to reduce as far as possible the influence of the two others.
The part of ε is governed by the frequency, f: σ and ε are in competition in the Maxwell-Ampere equation. In the frequency domain, for sine variation of the transmitted signal, one must have:

- $\sigma \gg 2\pi f \varepsilon$,   thus, $f < 300$ kHz

Which is the definition of the low frequency case where ε is neglected.
At a given low frequency, the respective magnitudes of the response due to magnetic properties and of the conductivity response are governed the geometry of the instrument. For Slingram instruments (Fig. 2), which have a coil as transmitter and an other coil as receiver, the distance between the two coils, L, is of the same order of magnitude as the required depth of investigation, which, in archaeological prospection, corresponds to one meter order of magnitude.
For coincident, or two loops, instruments (Fig. 2), the main geometric parameter, L, is the diameter of the (greatest) loop, usually less than one metre. Consequently one always stays in the "Low Induction Number" approximation $L2 2\pi f \mu \sigma \ll 1$.

Proton magnetometer and SCM surveys at South Cadbury Castle
with excavated features.

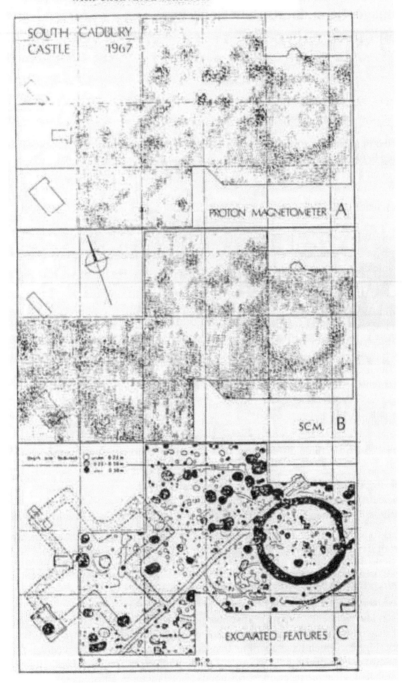

Figure 1. South Cadbury Castle (UK) survey, A) Proton magnetometer, B) SCM results C) excavated features.

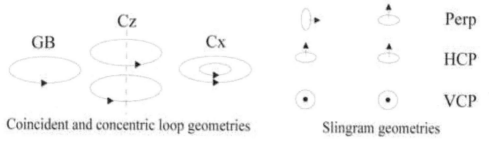

Coincident and concentric loop geometries Slingram geometries

Figure 2. Loop and Slingram geometries.

Figure 3. SH3 Slingram EM apparatus, 1.5 coil separation, 8.04 kHz frequency.

The conductivity response, which is directly proportional to this number, cannot be far greater than the magnetic one. This explains the results obtained using the SCM instrument.

However, archaeologists are lucky. In the LIN approximation the conductivity response is in quadrature out of phase while the magnetic susceptibility response is in-phase, these two responses are quite easy to separate using a proper electronics.

It is possible to build and to use Slingram EM LF instruments that measure the conductivity, the magnetic susceptibility and detect metallic objects (or groups of metallic objects). In the end, this class of instruments is the most flexible tool in archaeological prospection.

In LIN approximation the depth of investigation is different between conductivity and suscepti-bility measurements, it depends on L only and not of the frequency and is greater for conductivity than for susceptibility.

In figure 3 is presented the design of one of these instruments, the SH3, where the coil orienta-tion allows annulling the primary field effect on the receiver.

Question: Does E.M. magnetic susceptibility measurements open new possibilities that does not offer magnetic prospecting.

Answer: Yes.

The principle of magnetic prospecting is to compare the magnetic field intensity (either the magnitude or the gradient of one of its components) between adjacent points. One magnetic field measurement alone has no meaning in term of surrounding soil susceptibility and only lateral changes in the total soil magnetisation can be detected. On the contrary, E.M. in struments directly measure the magnitude of the surrounding soil susceptibility. This has two major consequences: (1) the presence of high susceptibility layers or lens like features can be evidenced (2) one measurement alone can be meaningful and wide mesh grid can be use for a rapid site identification over large areas.

Question: Is the depth of investigation of E.M. instruments sufficient for archaeological targets.

Answer: Yes but the geometry must be carefully considered.

One must repeat that as said above, the investigation depth depends on instrument geometry and not on frequency. Slingram geometries are far better than either coincident or two loop instruments. In practice the perpendicular configuration with a coil distance greater than 1 m is the most relevant for archaeological applications. If a perpendicular coil instrument is not available VCP configuration must be preferred to HCP because of its ability to detect 3D targets.

3   FIELD EXAMPLES

The first example corresponds to a wide mesh grid survey on the pathway of the future A66 motorway in the south of France. Both the magnetic susceptibility and the electrical conductivity were mapped using the EM38 VCP configuration Slingram instrument (1 m separation and 14600 Hz frequency).

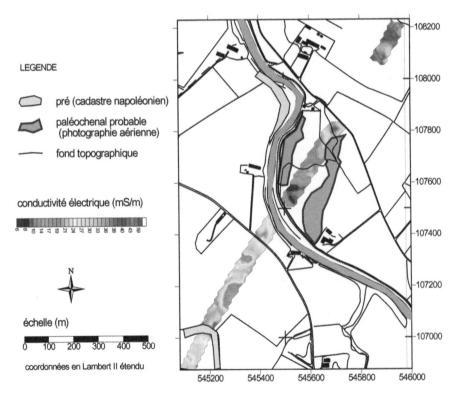

Plate 18.   A66 motorway route (Hautes Pyrénées, France) apparent magnetic susceptibility map, EM38 apparatus in VCP configuration. (See colour plate section)

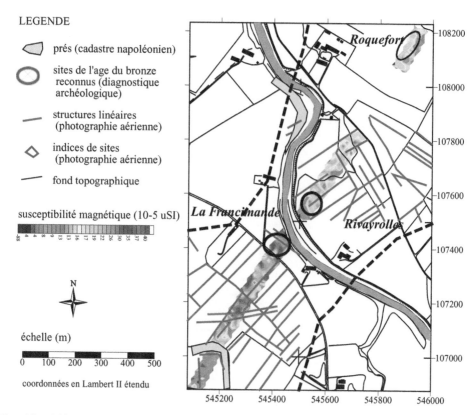

LEGENDE

prés (cadastre napoléonien)

sites de l'age du bronze
reconnus (diagnostique
archéologique)

structures linéaires
(photographie aérienne)

indices de sites
(photographie aérienne)

fond topographique

susceptibilité magnétique (10-5 uSI)

échelle (m)

0   100   200   300   400   500

coordonnées en Lambert II étendu

Plate 19.   A66 motorway route (Hautes Pyrénées, France) apparent conductivity map, EM38 apparatus in VCP configuration. (See colour plate section)

The magnetic susceptibility map, Plate 18, shows bronze age settlements on the west bank of the river, while conductivity, Plate 19, allows an assessment of the nature of the sediment, well drained on the west bank, clayed on the east bank.

The second example corresponds to the Bronze Age site of Marchésieux (Manche, France) where, before the survey, two axe hoards have been discovered in a marsh during the cleaning of drainage ditches. The EM Prospection permits, on the one hand, to locate six unknown hoards, figure 4, and, on the other hand, figure 5 to precisely determine their environmental context: the hoards were put into the bog along what can look as a path at several tenths of metre of the bog limit which correspond to the 29 Ω.m apparent resistivity line. The electrical soundings allowed estimating the thickness of the peat layer.

In the third example is presented the magnetic survey, figure 6b, and the EM susceptibility survey, figure 6a, of the 17th century pottery workshop of Dampierre sous Bouhy (Nièvre, France). Using the EM susceptibility results it is possible to calculate the magnetic anomaly corresponding to the induced magnetization, figure 7a, and to subtract it to the magnetic survey measurement to obtain the anomaly generated by the thermoremanent magnetization of the kiln, figure 7b.

4   SUMMARY OF ADVANTAGES AND DRAWBACKS OF EM LOW FREQUENCY SLINGRAM INSTRUMENTS

4.1   *Drawbacks*

- They are not as efficient as the DC electrical method in the characterisation of resistive features (high resistivity areas, walls…).

Hoards discovered at Marchesieux (Marais de St Clair) using the
EM15 and the SH3.

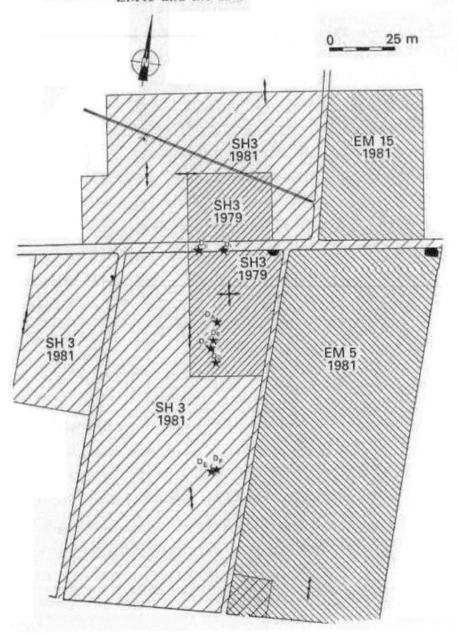

Figure 4.   Marchezieux (Manche, France) location of the late Bronze Age hoards discovered by EM survey.

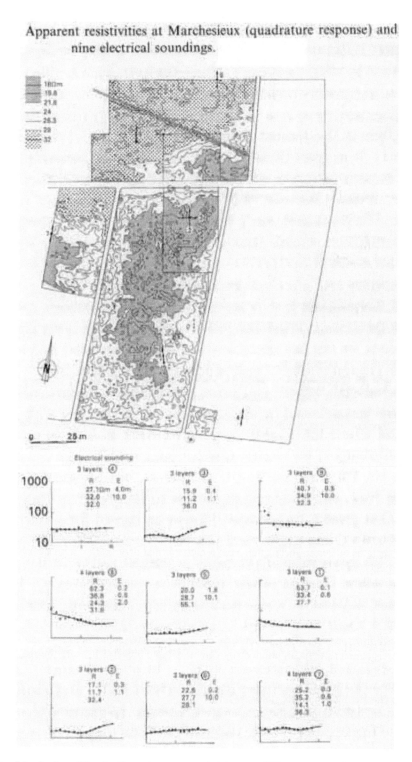

Figure 5.   Marchezieux (Manche, France) apparent resistivity map, SH3 apparatus and electrical soundings.

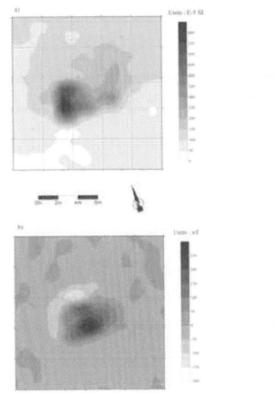

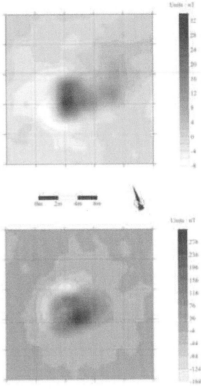

Figure 6. Dampierre sous Bouhy (Nièvre, France) kiln, a) apparent magnetic suscepti-bility map, SH3 apparatus, b) Total field magnetic anomaly map (G858 Geometrics).

Figure 7. Dampierre sous Bouhy (Nièvre, France) kiln, a) magnetic anomaly map of the induced magnetisation calculated from of the apparent susceptibility data, b) resulting magnetic anomaly map corresponding to remanent magnetisation.

- To correctly measure the magnetic susceptibility, they need to be very rigid and their mechanical building remains difficult. There often exist a high thermal drift on the in-phase susceptibility measurements (which can be but must be corrected).
- Investigation depth is more limited than the one of the magnetic prospection method.

4.2 *Advantages*

- Their capabilities are very wide, wider than any other geophysical techniques used in archaeological prospection as they simultaneously detect metallic objects, measure the magnetic susceptibility, measure the electrical conductivity.
- They allow wide grid meshing survey to locate site over large areas.
- They are cheaper than magnetometers.
- Combination of EM susceptibility measurements and magnetic measurement opens very wide perspective for a better characterisation of the magnetic features.

REFERENCES

All the publications anterior to 1988 are summarized in the reference book published in 1990:
Scollar I., Tabbagh A., Hesse A. & Herzog I., 1990, "Archaeological Prospecting and Remote Sensing", Cambridge University Press.

In English the most recent book is:
Gaffney C. & Gater J. Revealing the buried past: geophysics for archaeologists, Stroud, Tempus, 2003.

In French one can profitably consults:
Dabas M., Delétang H., Ferdière A. et W. Haio Zimmermann, La Prospection, édition ERRANCE, Collection « Archéologiques », 2006, 248 p.
Dossiers Archéologie et sciences des origines, La prospection Géophysique, n°308, novembre 2005, éditions FATON.

For more detail:
Benech C. & Marmet E., 1999, Depth of investigation and conductivity response rejection for susceptibility measurement using electromagnetic devices. Archaeological Prospection 6-1, 31–45.
Benech C., Tabbagh A. & Desvignes G., 2002, Joint interpretation of E. M. and magnetic data for near-surface studies. Geophysics, 67–6, 1729–1739.
Dabas M., 1999, Contribution de la prospection géophysique à large maille et de la géostatistique. Application aux ferriers de la Bussière sur l'A77. Revue d'Archéométrie 5: 17–32.
Dabas M., Herbich T., Hesse A., Misiewicz K. & Tabbagh A., 1993, "Electromagnetic prospecting at tow Polish sites (Slonowice and Milanowek) with the SH3 Slingram device". Archaeologia Polona, 31, 51–70.
Desvignes G. & Tabbagh A., 1995, Simultaneous interpretation of magnetic and electromagnetic prospecting for magnetic feature characterisation. Archaeological Prospection, 2, 129–139.
Tabbagh A., Bossuet G. & Becker H., 1988, "A comparison between magnetic and electromagnetic prospection of a Neolithic ring ditch in Bavaria". Archaeometry, 30–1, 132–144.
Annan, A.P. & J.L. Davis, 1992, *Design and development of a digital ground penetrating radar system*. In J.A. Pilon (editor), *Ground penetrating radar*, Geological Survey of Canada, Paper 90–4: 49–55.

Ground penetrating radar

GPR methods for archaeology

D. Goodman

Geophysical Archaeometry Laboratory, Woodland Hills, CA, USA

ABSTRACT: An introduction to Ground Penetrating Radar using simulation and imaging software is presented. Simulating GPR radargrams is used as a tool to show the importance of forward modeling of radar waves. The complisated radar patterns that can result from even the most simplest structures buried in the ground is shown using the typical GPR equipment that have broad transmitted beams. Simulation software is used to demonstrate the effects of GPR transmission, including refraction, reflection, attenuation, and to show the effects of multiply reflected energy paths in the ground on recorded radargrams. Simulation training of GPR is important in creating caution for the interpreters of GPR and helps as a guide post in avoiding interpretation pitfalls that can happen when only raw radargram information is read. The utility of visualizing GPR data using time slice analysis and other 3D imaging displays, indicates that large structural features can be accurately mapped with radar. GPR imaging of Roman, Japanese, and early Spanish structures are presented. GPR-GPS imaging of an Native American Indian cemetery is also provided using isosurface rendering of the site to indicate marked and unmarked burials. Overlay analysis in which comprehensive 3D reflection information is collected into single 2D images, is shown to be a very useful form of data display, particularly for sites in which continuous reflectors are not at a constant depth from the ground surface.

1 INTRODUCTION

The method of Ground Penetrating Radar (GPR) for archaeology is introduced using examples from simulation software and subsurface imaging software. The theoretical requirements to build a simulator for GPR which can predict the radargrams over known structures is presented. With a brief theoretical description introduced, examples of simulations of GPR from simple features and thus resultant radar patterns that occur from these subsurface structures are shown. Examples of what are referred to as reflection multiples, velocity pullups, radar shadow zones and several other geometric effects from recording with broad beam single channel GPR equipment is used as guide in avoiding interpretation pitfalls. The results from simulations will indicate to the students of GPR to be careful in interpreting raw unprocessed radargrams as the radar patterns recorded can be extremely different than the buried features causing such recorded patterns. Basic signal and image processing for GPR are also introduced along with examples of successful GPR imaging at Roman, Japanese and Native American Indian sites are presented. GPR volume imaging using only GPS navigation is introduced, along with advanced imaging methods for synthesizing useful 2D images from complicated 3D datasets.

2 BASIC GPR THEORY THROUGH THE EYES OF BUILDING A REAL SIMULATOR

If one were going to build a simulator for GPR the following things need to be consider in order to make a synthetic radargrams:

1. Subsurface Model: A subsurface structural model would need to be built that contains different materials. The different materials are identified by soil, clay or stones for examples, and each

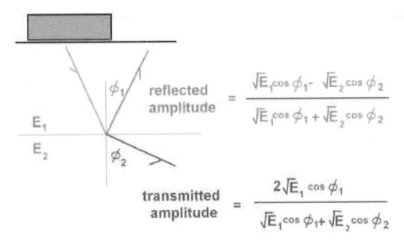

Figure 1. GPR microwave reflection off the surface in the drawn features in the simple model.

are identified by their physical electromagnetic properties called the relative dielectric (E) and the conductivity of the material. The velocity of a material is given by $v = c/E1/2$ where c is the speed of light.

2. Reflection: Once the subsurface model is drawn, we will need to understand how GPR micro-waves reflect off the surfaces of the drawn features in our model (see Fig. 1). The amount of reflected energy of microwaves is given by a simple equation that measures the change in the contrast between the 2 materials: Reflected Energy = $E11/2-E21/2/E11/2 + E21/2$ which is the equation for normal incidence of the microwave on an interface.

3. Transmission: In addition to reflection, we must know how much of the incident wave ampli-tude will be transmitted across the surfaces of subsurface model. This is given by the equation: Transmitted amplitude = $2 E2^{1/2}/E11/2 + E2^{1/2}$ (for normal incidence). It is worthwhile to note that if you add up the Reflected Energy + the Transmitted energy that the total energy is 1. These simple equations indicate how the energy is partitioned when a microwave pulse encounters a structure in our drawn model. The equations are also simplified and just consider a wave that is traveling and encounters one of our subsurface structures at a 90 degree angle. They are also simplified for zero conductivity.

4. Refraction: Well, there are other things that happen to the microwaves as they transmit from a material into a different material in our model—they can refract—e.g. change their angle of propagation. The angle that the wave will change direction is also a function of the ratio of the relative dielectric between the two adjacent materials and is $\sin(O1) = \sin(O2) E12/E22$.

5. Attenuation: The conductivity of a material controls how much the wave microwaves will dis-sipate—attenuate—as they travel along their travel paths. The equation which describes this is the loss tangent equation and rather to simply present this equation here, it is beneficial to simply state that the higher the conductivity—the higher the attenuation is.

6. Beam Radiation: All GPR antennas transmit microwave energy over a complicated radiation pattern (e.g. Fig. 2). The radiation is not a spherical wave front, but has different energy com-ponents in different directions. To accurately predict and make our GPR simulator, the beam response of antenna is required.

7. Impulse Response: The beam that is transmitted also has a pulse shape which is unique for every antenna. Although the antenna frequencies are stated as individual frequencies, e.g. 200 MHz or 400 MHz, these numbers refer to the central frequency of the antenna. GPR antennas have a broad spectrum of frequencies generated and real antennas are not single mono-frequency (e.g. Fig. 3—bottom-left diagram).

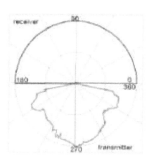

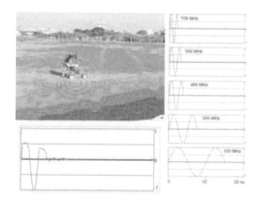

Figure 2. Schematization of the radiation pattern of the GPR antenna.

Figure 3. Examples of different wavelengths of GPR signal for different impulse frequencies.

8. Raypaths: The last ingredient to predict synthetic radargrams across a model is to know how the energy bounces around when energy impinges on the surfaces and whether the reflected waves eventually will return to the receiving antenna. Various raypaths can be inserted into a simulation to add up the energy that bounces once or even several times before getting recorded by the antenna (e.g. Fig. 4).

Once these 8 ingredients are developed, it is possible to run a GPR simulation to estimate the recorded radar patterns across subsurface structures. Using GPRSIM v3.0 Ground Penetrating Radar Simulation Software (Goodman, 1994), several example subsurface synthetic models and their corresponding synthetic radargrams are shown Fig. 5a–e:

Fig. 5a: A GPRSIM simulation of a model in which 2 round objects are buried at the same depth but in different materials (different microwave velocities) is shown. Each round object generates a hyperbolic reflection on the synthetic radargram. The shapes of the hyperbola can be used to determine the microwave velocity of the surrounding material. A faster material has a wider hyperbola; a slower material has a narrower hyperbola. Using software a whole progression of hyperbolas can be matched to the observed hyperbola which will then give us the velocity. Having the velocity allows us to assign depths to reflection targets. The depth d to a target is $d = v * t/2$ where t is the two-way travel time and v is the microwave velocity. Of course in the real world situation, we often do not have small round targets that allow us to easily measure the hyperbolic shape and thus give us the ground velocity.

Fig. 5b: The effects of refraction are exemplified in this figure where 2 models are drawn. In the top model, a layered structure with velocity increasing downward is shown. If the velocity increases with depth, then the microwaves will refract away from the downward direction. The effective beam of the GPR antenna is broadened in this instance. In contrast, a model in which the velocity decreases with depth (bottom diagram), causes the waves to refract downward and create a more focused GPR beam.

Fig. 5c: Simply having a buried object in the ground does not necessarily mean it will be detected. The most important feature of a buried object is its shape and orientation to the receiving antenna. In this example, a simple triangle simulation is shown. The reflected energy which is recorded from one side of the triangle is not recorded directly below the triangle, but at some distance to the left of it. One can imagine a more vertically oriented triangle would possible reflect no energy back to the receiving antenna. Other interesting things to note with regard to this model, if by some chance the sharp edges of the triangle were not sharp but were slightly rounded, this would cause some reflected energy to get recorded back at the antenna. This simulation indicates why some stealth fighter jets, such as the B1 Bomber, is made with no rounded edges and only flat and sharply connected plane surfaces. In this instance, the possibility of reflected energy returning to a detecting antenna is slim, and the bomber will remain "invisible".

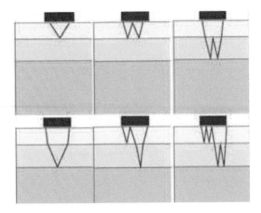

Figure 4.   Examples of various raypaths inserted into a simulation to add up energy that bounces once or even several times before getting recorded by the antenna.

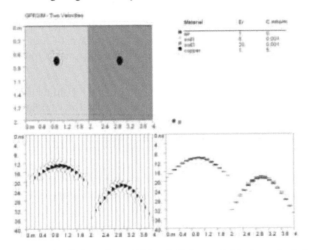

Figure 5a.   A GPRSIM simulation of a model in which 2 round objects are buried at the same depth but in different materials (different microwave velocities) is shown.

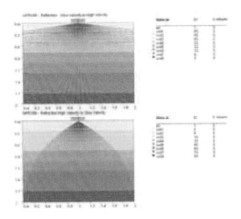

Figure 5b.   The effects of refraction are exemplified in this figure where 2 models are drawn. In the top model, a layered structure with velocity increasing downward is shown and in the bottom a layered structure with velocity decreasing downward.

Fig. 5d: Another subsurface structure which does exactly look like its synthetic radargram pattern is a simple V-trench. In this example, multiple reflections from within the V-trench, designated by the RR wave (in the raypath travel time plot) cause a rounded reflection pattern which has a butterfly appearance. In fact, the direct reflections recorded on the left side of the trench (designated by the R reflector) actually come from the right side of the trench and vice-versa. One can see that simple structures do not often look similar on the synthetic radargrams. For this example, changing either the depth or the narrowness of the trench will drastically change the recorded synthetic radargram (Goodman, 1994).

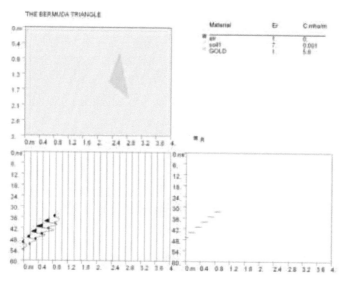

Figure 5c.   Example of a simple triangle simulation is shown. The reflected energy which is recorded from one side of the triangle is not recorded directly below the triangle, but at some distance to the left of it.

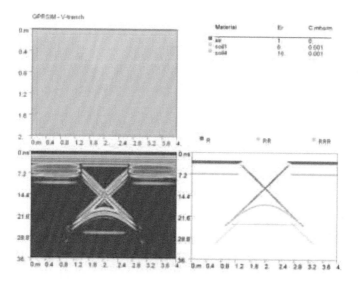

Figure 5d.   In this example, multiple reflections from within the V-trench, designated by the RR wave (in the raypath travel time plot) cause a rounded reflection pattern which has a butterfly appearance.

Fig. 5e: One of the effects which can drastically lead to interpretation mistakes of reading raw radargrams is the "velocity pull-up" effect. In this example, a subsurface layer which is flat is recorded as a warped reflection feature on a radargram. This is caused by a middle layer which has a very different velocity and variable thickness. The two-way travel time of a pulse that travels through this layer and reflects off the bottom flat layers, will have a variable travel time and give an effect that the subsurface layer is not a flat layer. The variable velocity and thickness causes the velocity pull-up effect. One must be careful in interpreting reflections as resulting from dipping or undulating structures, as these structures may be in fact be flat in contrast to what the radargram will show.

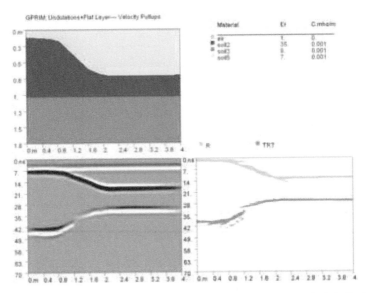

Figure 5e. In this example of a "velocity pullup" effect, a subsurface layer which is flat is recorded as a warped reflection feature on a radargram because of variable velocity/thicknesses of overlying layers.

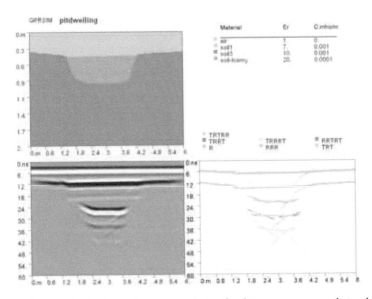

Figure 5f. A buried half circular trench facing upward can often have an appearance that makes the trench appear as though it has a downward parabolic reflection.

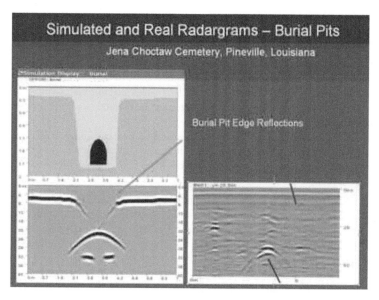

Figure 6.   Example of a comparison of a real radargram with a synthetic radargram is generated for a grave site at the Jena Choctaw Whiterock cemetery in the Kisatchi National Forest, Louisiana.

Fig. 5f: There are of course a variety of other structures that appear quite different on synthetic radargrams. A buried half circular trench can often have an appearance that makes the trench appear as though the trench is upside-down, yielding a parabolic reflection. There are examples of what are referred to as "shadow zones" where no microwave energy will travel into an area because of refraction effects. There are though some features which can often look very similar to their real world structures. An example of a pit dwelling simulation shows the synthetic radargram pattern looks very similar to the pit dwelling structure itself.

How can simulation software benefit one in understanding a site other than warning us of the possible interpretation pitfalls? Well, one of the primary uses of simulation software is to perform what is called forward modeling in Geophysics. In this process we use the simulation software in an iterative approach by guessing a model, running the simulation, and then comparing the simulation with a real recorded radargram over a real site. We keep adjusting the model till we get a good match between real and synthetic radargrams. Once we have accomplished this we can then say with some confidence, that the real structure responsible for our recorded reflections are given by the candidate model in the simulation program. However, several different candidate models may indicate the same radar pattern and the candidate model may not be a unique one.

Nonetheless, an example of a comparison of a real radargram with a synthetic radargram is generated for a grave site at the Jena Choctaw Whiterock cemetery in the Kisatchi National Forest, Louisiana is given in Figure 6. At this site there were many unmarked burials. GPR was used to not only locate these burials but to also give an estimate of the depth and size of the burial pit. Examination of the synthetic radargram and the real radargram shows that there is a good correspondence in the two and that the likely burial structure is similar to the model structure detailed from the GPRSIM v3.0 Simulator. One useful indicator for this particular survey is that the burial pit edge show up as faint half hyperbola reflection legs. Identification of these edge reflections can help one to identify to detect a burial pit, particularly in the case when the burial remains or casket have been destroyed over time and do not reflect back any GPR microwaves.

One must remember that what we call a 3D survey over an archaeological site, is really just a collection of reflected pulses recorded at a finite number of x and y locations at the site, with the reflection time along the individual pulses representing the 3rd dimension in z. To make useful images of these pulses several basic processes are first necessary to treat or filter these raw radar pulses. Once the pulses are processed then an images can be created. The first process we want to discuss is what is referred to as signal processing.

Signal processing are a set of mathematical operations that we can apply to the raw radar pulses we recorded to filter out noise as well as to help better map the real locations of the pulses which are collected from the broad beam antenna. The basic signal processes which are often applied to GPR data consist usually consist of the following:

1. Post Processing Gain: Raw radargrams often need to be re-gained after they are recorded since many GPR equipment record 16 bit ungained data. (16 bit refers to refers to the digitized pulses represented by the numbers from $-32768 + 32767$ which can also be written as -215 to $+(215 -1)$. The later arriving reflections to the GPR, the echoes that travel farther into the ground, are much weaker than the earlier recorded reflections. In order to see them on the computer screen, exponential gain curves must be applied to the later arriving raw radar pulses.

2. DC Drift Removal: Raw regained radargrams often contain a noise which is caused by shifting of the entire pulse from the 0 line. This can be caused by a variety of factors, most of which is the inherent manufacturing electronics which deal with microwave pulses. To remove this several filters either applied in the time domain, or in the frequency domain by cutting out certain frequency bands in the data will shift the radargram pulses back to the 0 line. An example of a radargram with postprocessing gain applied and with DC drift being removed is shown Fig. 7a.

3. Background Removal: The next signal process often used to process raw radar signal is the background removal. In this simple process the average radar scan across the profile is computed and then subtracted from every scan. This will remove the horizontal banding which represent parts of the pulse which are never change across the profile and are consider the background pulse or noises contained with the GPR control unit.

4. Migration: One of the processes which can help to better "locate" the radar scans in their proper position in the ground is called migration. As we discussed in the first section, the broad directional beam of the GPR antenna means that reflections from off the side of the antenna can also be recorded. The buried circular features in Figure 5a cause hyperbolic reflections. The process of migration works by simply fitting a hyperbola to recorded feature in the raw radargram, and then adding up all the energy along the hyperbola and placing this mathematical addition right at the apex of the hyperbola. This is done over every point recorded in the radargram. The net effect is to collapse the hyperbolas to smaller point source reflections. The process works since only those

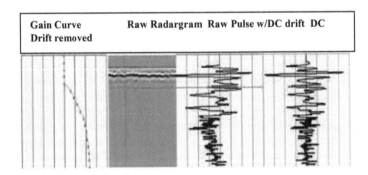

Figure 7a.   An example of a radargram with postprocessing gain applied and with DC drift being removed is shown.

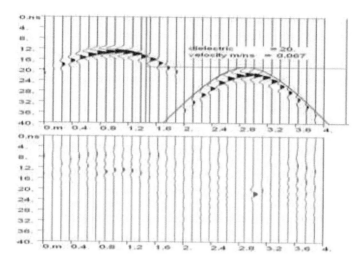

Figure 7b.   In this example the reflections from the hyperbola on the right are collapsed into a single point source reflection.

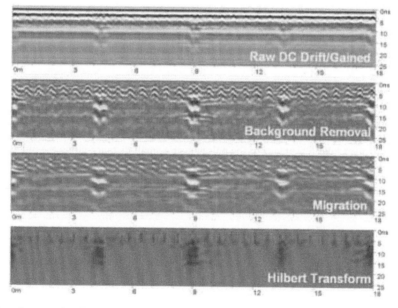

Figure 7c.   An example of a radargram that has undergone several signal processes.

parts of the pulse which are in phase, are constructively added up, whereas features which are not coming from the same reflector will essentially be destructively reduced by addition of out of phase components of the pulse. The migration process does not always work perfectly well since we often do not know the 3D velocity field very well at all. However, this operation is currently usually applied assuming a single constant velocity for the entire site. Such estimates are reasonably good for homogenous structures, and less so for heterogeneous ones. An example of migration on the model in Fig. 5a in which only a single velocity is assumed is shown in Fig. 7b. In this example we can see that the reflections from the hyperbola on the right are collapsed into a single point source reflection. The reflections on the left are not collapses but diminished and spread out

237

because the incorrect velocity is not being used in the migration for this area. Again, in general we never know the complete velocity field so migration results will not always be optimum.

5. Hilbert Transform: The Hilbert transform is used to calculate the envelope of the radargram pulse. The envelop of the pulse (computed using a Fast Fourier Transform in which we shift the negative frequencies by 90 degrees and then perform a Inverse Fourier Transform), can be considered as an operation that gives us a plot that connects the (+) peak amplitude responses of the signal. The Hilbert transform radargrams has no – minus values and is a rectified signal and it is often better for visualizing areas of strong reflection.

An example of a radargram showing a raw radargram that has undergone several of the processes discussed here is shown in Figure 7c. The bottom radargram has undergone several of the signal processes of gain, DC drift removal, background removal, migration, and the Hilbert Transform. This radargram is actually a profile over a set of rebar in concrete.

One can imagine that this final processed radargram looks much closer to the real world structure than the image given in the simple raw radargram (in the top diagram).

4   IMAGE PROCESSING

Once our radargram have been processed the next step is to create a useful image from this data. One of the most useful image for archaeological applications is the time slice image. Time slices represent maps of the amplitudes of the recorded reflections across a site at a specified time or depth (Figure 8). The following procedures are recommended to create the most appealing and useful images for archaeological applications.

5   RECOMMENDED IMAGE CONSTRUCTION PROCEDURE (FOLLOWING GOODMAN ET AL, 1995)

Desample radargrams spatially along the profile line to closely match the profile spacing. Desampling simply involves taking averages of the pulses over a small distance along the radargram. For instance, if the profile spacing is a half meter, it useful to make averages along the line that closely matches this spacing, e.g. 50 cm. One can also make averages at a smaller interval along the line

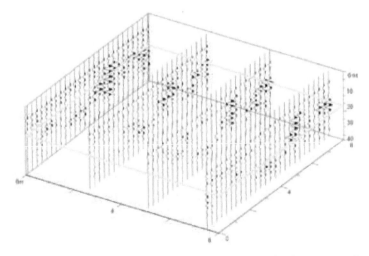

Figure 8.   Time slices represent maps of the amplitudes of the recorded reflections across a site at a specified time or depth.

maybe 25 centimeter or so, but making average every 2 or 3 centimeters for instance gives us too much information along profile. The problem being is that we have very little information between profiles and we need to determine the "blank" spaces between adjacent profiles. If one used all the discreet information along the line, when interpolating these kinds of sampled data, the time slice maps will appear striated and the profile direction that the data was collected will become obvious and lead to more inclusion of noise in the image. Next, with the spatial averages computed, use interpolation algorithms to fill in the spaces between profiles to generate a solid and continuous mapping of amplitudes across a site. There are various methods for interpolating between a set of data, but the two most popular are called inverse distance and krigging. Inverse distance uses a simple mathematical average of nearby points that are to be used in the estimate of the interpolated point, based on the distance to the point to be interpolated. Usually the weighting is set to be proportional to the inverse square of the distance of a known point to the point to be estimated. Krigging is much more complicated interpolation method and it involves solving an inverse matrix to minimize the error between data and interpolated points. Krigging can normally give a little higher resolution in the interpolated maps then inverse distance maps. However, one should note that higher resolution maps may often not show larger reflection features which a more smoothed grid map like inverse distance interpolated maps may show.

Once the images are interpolated, other image processes may be necessary to remove noises in the developed time slices. Shown in the bottom of Plate 20 is a time slice image indicating several kinds of inherent noises—staggering and mosaic noises. Staggering is the apparent periodic shifting of anomalies between adjacent lines in the grid. Staggering is caused by performing a GPR survey using a zigzag survey where lines are also collected in the reverse direction to speed up

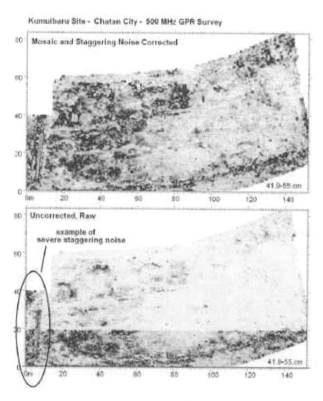

Plate 20.   Shown in bottom diagram is the staggering noise time slice which also has a sharp change in reflection above the line y = 20 m. (See colour plate section)

239

the field collection. If there is a small buffer delay in the GPR control unit or some navigation lag between the positions of radar scans on the ground, then anomalies will get this staggering effect and not be lined up. To correct this problem, a small constant offset is added or subtracted from the time slice points for radargrams collected in the reverse direction and gridding is reapplied.

Another image noise often discovered is an overall reflection difference between different areas making up a larger survey site. These noises, referred to as mosaic noises, must be removed in order to better see the underlying continuous reflection structures across a site. Mosaic noises can be caused by a variety of factors including changes in the soil moisture content between survey days, temperature changes, equipment setting changes or real geologic or subsurface changes in soils across a site. Shown in bottom diagram of Plate 20 is the same staggering noise time slice which also has a sharp change in reflection above the line y = 20 m. The sharp change in reflection for this site was due to rain which saturated the ground and the survey was discontinued until the conditions improved. Nonetheless, these changes in reflection, attributed to different moisture contents across the site, can be fixed using a several methods. The best method usually involves manually adjusting data transforms to get a good match between the different areas. Regaining of areas using 0 mean grid or 0 mean line calculations can also be used to automatically balance the different reflection areas.

6   GPR IMAGING AT ARCHAEOLOGICAL SITES

With the basics in signal processing and image processing presented, the remainder of this report is dedicated to showing examples of GPR imaging at archaeological sites.

Included in this portion of the report are a few examples of GPR surveys made at Roman, Japanese burial mounds, and Native American Indian sites. At Japanese burial mounds an advanced topic which includes static corrections for antenna tilt are introduced. The use of GPS navigation to map unmarked burials at sacred Native American Indian cemeteries is also shown.

Villa of Traianos, Italy:

On of the first GPR surveys made in 1998 in conjunction with the Consiglio Nacionale delle Ricerche in Rome and the Nara National Cultural Properties Research Institute in Japan was the apparent bathhouse to the Villa of Traianos. Using a 500 MHz antenna GPR lines were collected at

Figure 9.   Location of the Villa of Traianos site and an example of GPR calculated time-slice showing buried foundations of the destroyed villa. Several mushroom shape anomalies could clearly be identified in a time slice map near 30 ns.

Plate 21. Ikime Kofun Burial #7, Miyazaki City, Japan. Radargrams and 3D topo warped time slices along with excavation photos of the Ikime Burial #7 are shown. (See colour plate section)

50 cm intervals across the site. Several mushroom shape anomalies could clearly be identified in a time slice map near 30 ns (Figure 9). These structures were immediately interpreted by archaeologists to be dipping pools in the bathhouse. Adjacent to one of the smaller dipping pools identified, an atrium and possibly 4 base stone supports could be estimated from the time slice image. Additional walls and corridors within the bathhouse could also be easily seen within the image.

Ikime Kofun Burial #7, Miyazaki City, Japan:

Since 1992, the Geophysical Archaeometry Laboratory has been surveying many Kofun Period (300–700 AD) burial mounds for the Saitobaru Archaeological Musuem and the city of Miyazaki. One of the difficulties in doing GPR surveys in addition to making sure that plane level navigation is recorded for positioning, is to correct for the effects of tilt in the recorded radargrams as the antenna is pulled over topography that can vary by as much as 30–40 degrees in locations. Correcting for the antenna tilt was implemented into GPR-SLICE v5.0 Software in the summer of 2005. Shown in Plate 21 are radargrams, 3D topo warped time slices along with excavation photos of the Ikime Burial #7. Correcting for antenna tilt involves propagating the assumed vertical ray of the antenna into its correct position beneath the ground using a measured velocity model. Various data corrections to the radargrams, such as sweeping a small angle around the projected vertical ray is necessary to fill in binary voids in the radargram as the tilt-corrected radargrams are being computer generated (Goodman et al, 2006).

For this particular late 4th century burial, a side burial tunnel called a chikashiki yokoana was determined from the GPR survey and later excavated. The tunnel burial was found to be filled in with a darker volcanic soil. In this case, the strong reflections from the chamber were from strong soil contrasts and not from void reflections as was first thought.

Jena Choctaw Cemetery, Kisatchi National Forest, Louisiana:

The current state-of-the-art with GPR is to employ GPS navigation in the survey. In 2003, one of the first successful GPR-GPS surveys at a Native American Indian cemetery was made. The survey sponsored by the United States Forest Service, was initiated at the request of the Jena Choctaw band of Indians. The modern cemetery was missing many gravestones because of maintenance neglect at the site. The purpose of the survey was to relocate unmarked burial so that the cemetery could continue to be used without any problems. Using a Trimble Pro XR GPS navigation system, differential GPS reading which were accurately synced to the radar scans and recorded every 32 scans during the data collection. This corresponded to GPS readings being made every 1 second. The navigation accuracy for this equipment is listed as sub-meter. Using GPS navigation 3D interpolated volumes of the spatially averaged squared amplitude of the radar reflections were generated. The data were then displayed using isosurface rendering in which a chosen reflection level within the volume is illuminated and given shading based on the attitude of the surface being shown. In this example, several small longitudinal reflections can be seen which are from known burials, as well as several reflections which are identified as unmarked burials (Figure 10).

GPR-GPS navigation can also suffer from staggering effects if the GPS readings experience a phase lag with the radar scan number. Shifting of the GPS readings in time or a constant number of scans can be used to eliminate GPS scan lag staggering noises.

San Juan Cathedral, Puerto Rico:

GPR is being used at many historical building for a variety of applications. Evaluating the integrity of columns or walls for fractures, looking for subsurface utilities, or examining the integrity of foundations are among some of the applications that GPR can be a very effective tool. GPR is also being used to discover crypts beneath church floors. In 2006 in conjunction with the Polytechnic University of Puerto Rico, GPR surveys were made at the San Juan Cathedral. Time slices made just in front of the Cathedral alter revealed a rectangular anomaly which is believed to be a subsurface crypt (Fig. 11). Several reflection arms on each side of the main crypt reflection indicated possible corridors leading to adjacent crypts underneath columns to the main church floor.

Similar results were made in 2004 at the Iglesia San Jose Church in Puerto Rico which is the 2nd oldest standing structure in the New World, having been constructed by the Spanish in 1562. Several known and one unknown crypt were easily rediscovered using GPR. GPR at this site was also able to effectively outline areas where water infiltration beneath the floor existed. The GPR results are being used to help restore this church which was designated as a World Heritage site and is thus under extensive restoration and protection.

Nanao City Castle, Ishikawa Ken, Japan:

One of the problems with imaging sites with GPR and using the typical simple 2D time slice displays is that many subsurface features of interest are not level built in the ground. In addition,

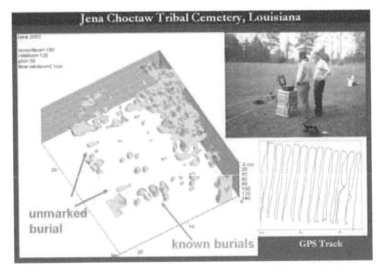

Figure 10. Jena Choctaw Cemetery, Kisatchie National Forest, Louisiana. Example of several small longitudinal reflections can be seen from known burials, as well as several reflections which are identified as unmarked burials.

Figure 11. San Juan Cathedral, Puerto Rico. Time slices made just in front of the Cathedral alter revealed a rectangular anomaly which is believed to be a subsurface crypt.

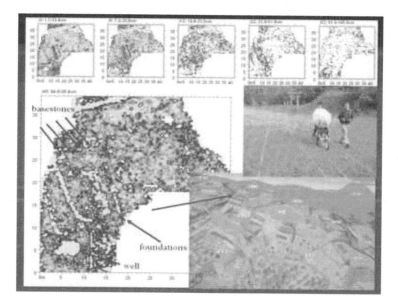

Plate 22. Nanao City Castle, Ishikawa Ken, Japan. Example of overlay analysis. 2D time slices from the complete 3D dataset are first normalized and the relative-strongest-reflectors from each map down to the desired depth range are chosen and then overlaid onto a single map. (See colour plate section)

variable overburden thicknesses of the top soil can create undulating reflectors which are in fact reflections from structures which are horizontally level. Often archaeologists need to have a 2D image that contains all the relevant information regarding a site on a single map. A complete 3D time slice dataset when just previewed in either an animation or at discreet thin time levels, is often not sufficient for the human eye to synthesize and remember the pertinent reflections at each level and make a complete picture of the subsurface structure. Isosurface rendering is also often not sufficient in creating useful images since often only a single isosurface is illuminated, and variable reflections from a continuous structure can go unnoticed on these displays. To implement what is called overlay analysis, the 2D time slices from the complete 3D dataset are first normalized and the relative-strongest-reflectors from each map down to the desired depth range are chosen. These components (displayed in the top series of maps in Plate 22) are not very useful in their separate display, but are very useful when they are "overlaid" on top of one another as well as in an animation form. Overlay analysis has the advantage over simple thick time slices in that the weighting at each level in the overlay can be individually controlled with quick data transforms. The site shown is for the Nanao City Castle, which is from the medieval period. Various structures including base stones for pillar supports and other foundations only became completely visible by overlaying reflections from very shallow levels near 15 centimeters, all the way down to about 1 meter.

7 DISCUSSION

GPR for remote sensing of archaeological sites is fast becoming the most versatile geophysical tool that can help to reduce the need for destructive archaeological excavation. GPR data however are not trivial and to properly process the data and extract the necessary information to solve a problem is not always straight forward. Sometimes filtering can help and other times it can hurt. Simple 2D or 3D time slices or isosurface renders may not be sufficient to solve imaging problems and the user of GPR may have to be creative and apply overlay analysis to extract the necessary information.

With the beginning developments of 2D antenna arrays, and eventually 3D antenna arrays, some of the unknown in creating useful 3D images from GPR will be eliminated. However, until the "ultra radar" device is created, monostatic—single channel survey data will be the most common data collection with GPR. From running simulations for single channel GPR, we learned that one must be careful in the interpretation of radargrams as multiple reflections and the effects of the broad radiation patterns of these kinds of antenna can cause very complicated radar patterns that look nothing like the reflecting subsurface structures.

Nonetheless, for large archaeological sites such that contain Roman buildings for example, or Japanese burial mounds or many other large target features, GPR imaging can provide the necessary images to accurately map and discover lost archaeological remains.

REFERENCES

Goodman, D., Nishimura, Y., Hongo, H. and Noriaki, N., 2006. Correcting for topography and the tilt of the GPR antenna, Archaeological Prospection, 13: 157–161.
Goodman, D., Nishimura, Y. and Rogers, J.D., 1995. GPR time slices in archaeological prospection: Archaeological Prospection, 2: 85–89.
Goodman, D., 1994. Ground-penetrating radar simulation in engineering and archaeology: GEOPHYSICS, 59: 224–232.

Seeing the Unseen – Campana & Piro (eds)
© *2009 Taylor & Francis Group, London, ISBN 978-0-415-44721-8*

Ground-penetrating radar for landscape archaeology:
Method and applications

L.B. Conyers

Department of Anthropology, University of Denver, Colorado, USA

ABSTRACT: Ground-penetrating radar mapping allows for a three-dimensional analysis of archaeological features within the context of landscape studies. The method's ability to measure the intensity of radar reflections from as deep as 5 meters in the ground can produce images and maps of buried features not visible on the surface. A study was conducted in the desert of the American Southwest to study the buried remains of ceremonial architecture within one valley in southern Utah. In this area large circular depressions on the ground were thought to be the remains of ceremonial buildings called kivas, indicating a connection of the people that lived there to a powerful and influential city to the south called Chaco. The ground-penetrating radar analysis of these features, however, showed them to be small family-sized kivas with associated roomblocks, which does not support these people's strong connection to the city to the south. When these buildings were mapped and then placed within the river valley, it was determined that each was likely the product of a single family or extended family, who probably lived by subsistence agriculture. This study shows the applicability of using three-dimensional GPR analysis to place the built-environment within its landscape in order to test ideas about and explain social factors and connections that were in place during prehistoric times.

1 INTRODUCTION

Ground-penetrating radar is a near-surface geophysical technique that allows archaeologists to discover and map buried archaeological features for landscape analysis in ways not possible using traditional field methods. The method consists of measuring the elapsed time between when pulses of radar energy are transmitted from a surface antenna, reflected from buried discontinuities, and then received back at the surface. When the distribution and orientation of those subsurface reflections can be related to certain aspects of archaeological sites such as the presence of architecture, use areas or other associated cultural features, high definition three-dimensional maps and images of buried archaeological remains can be produced. Ground-penetrating radar is a geophysical technique that is most effective with buried sites where artifacts and features of interest are located within 2–3 meters of the surface, but has occasionally been used for more deeply buried deposits. A growing community of archaeologists has been incorporating ground-penetrating radar (GPR) as a routine field procedure for landscape analysis (Conyers 2004a; Conyers and Goodman 1997; Gaffney and Gater 2003). Their maps and images act as primary data that can be used to guide the placement of excavations, define sensitive areas containing cultural remains to avoid and place archaeological sites within a broader environmental context and study human interaction with, and adaptation to, ancient landscapes (Kvamme 2003). Ground-penetrating radar data are acquired by reflecting distinct pulses of radar energy from a surface antenna, reflecting them off buried objects, features or bedding contacts in the ground, and detected those reflections back at a receiving antenna. As radar pulses are being transmitted through various materials on their way to the buried target feature, their velocity will change, depending on the physical and chemical properties of the material through which they are traveling (Conyers 2004a: 45). Each velocity change generates a reflected wave, which travel back to the surface. The velocity of radar energy in the

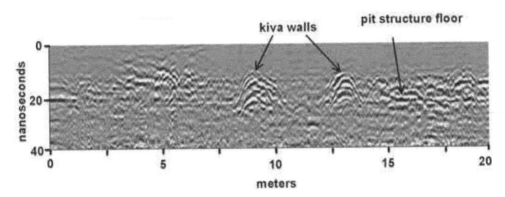

Figure 1. GPR reflection profile showing kiva walls and the floor of a pit structure from the Comb Wash area, southeastern Utah, USA.

ground is also important because when the travel times of the energy pulses are measured and their velocity through the ground is known, distance (or depth in the ground) can be accurately measured (Conyers 2004a: 99), producing a three-dimensional data set. Most typically in archaeological GPR radar antennas are moved along the ground in transects and two-dimensional profiles of a large number of reflections at various depths are created, producing profiles of subsurface stratigraphy and buried archaeological features along lines (Fig. 1).

When data are acquired in a closely-spaced series of transects within a grid, and reflections are correlated and processed, an accurate three-dimensional picture of buried features and associated stratigraphy can be constructed (Conyers 2004a: 148). This can be done visually by analyzing each profile, or with the aid of computer software that can create maps of many thousands of reflection amplitudes from all profiles in a grid. Ground-penetrating radar surveys allow for a relatively wide aerial coverage in a short period of time, with excellent subsurface resolution of both buried archaeological materials and associated geological stratigraphy. This three-dimensional resolution is what gives GPR an advantage over other near-surface methods with respect to buried archaeological feature resolution. Authors of papers to proceedings have to type these in a form suitable for direct photographic reproduction by the publisher. In order to ensure uniform style throughout the volume, all the papers have to be prepared strictly according to the instructions set below. A laser printer should be used to print the text. The publisher will reduce the camera-ready copy to 75% and print it in black only. For the convenience of the authors template files for MS Word 6.0 (and higher) are provided.

2   THE GPR METHOD

The success of GPR surveys is to a great extent dependent on soil and sediment mineralogy, clay content, ground moisture, depth of burial, surface topography and vegetation (Conyers 2004a: 28). It is not a geophysical method that can be immediately applied to any geographic or archaeological setting, although with thoughtful modifications in acquisition and data processing methodology, GPR can be adapted to many differing site conditions. In the past it has usually been assumed by most GPR practitioners that the method would only be successful in areas where soils and underlying sediment are dry (Annan and Davis 1992). Although radar wave penetration, and the ability to reflect energy back to the surface, is often enhanced in a dry environment, recent work has demonstrated that dryness is not necessarily a prerequisite for GPR surveys and even very wet environments are suitable, as long as the medium is not electrically conductive (Conyers 2004b).

The GPR method involves the transmission of high frequency electromagnetic radio (radar) pulses into the earth and measuring the time elapsed between transmission, reflection off a buried discontinuity and reception back at a surface radar antenna. A pulse of radar energy is generated on a dipole transmitting antenna that is placed on, or near, the ground surface. The resulting wave of electromagnetic energy propagates downward into the ground where some energy can be reflected back to the surface at discontinuities. The discontinuities where reflections occur are usually created by changes in electrical properties of the sediment or soil, lithologic changes, differences in bulk density at stratigraphic interfaces and most important water content variations. Reflection can also occur at interfaces between anomalous archaeological features and the surrounding soil or sediment. Void spaces in the ground, which may be encountered in burials, tombs, or tunnels will also generate significant radar reflections due to a significant change in radar wave velocity.

The depth to which radar energy can penetrate and the amount of definition that can be expected in the subsurface is partially controlled by the frequency of the radar energy transmitted. Radar energy frequency controls both the wavelength of the propagating wave and the amount of weakening, or attenuation, of the waves in the ground. Standard GPR antennas used in archaeology propagate radar energy that varies in band width from about 10 megahertz (MHz) to 1200 MHz. Antennas usually come in standard frequencies, with each antenna having one center-frequency, but actually producing radar energy that ranges around that center by about two octaves (one half and two times the center frequency).

The most efficient method in subsurface GPR mapping is to establish a grid across a survey area prior to acquiring data. Usually rectangular grids are established with a transect spacing of one meter or less. Rectangular grids produce data that are easier to process and interpret. Other shapes of grid acquisition patterns may be necessary because of surface topography or other obstructions. Data from non-rectilinear surveys is just as useful as those acquired in rectangular shaped grids, although more field time may be necessary in surveying, and reflection data must be manipulated differently during computer processing and interpretation for reflection amplitude analysis.

The two-way travel time and the amplitude and wavelength of the reflected radar waves derived from the pulses are amplified, processed and recorded for immediate viewing and later post-acquisition processing and display. During field data acquisition the radar transmission process is repeated many times a second as the antennas are pulled along the ground surface in transects. Distance along each line is recorded for accurate placement of all reflections within a surveyed grid. In this fashion two-dimensional profiles, which approximate vertical "slices" through the earth, are created along each grid line (Fig. 1).

Radar energy becomes both dispersed and attenuated as it radiates into the ground. When portions of the original transmitted signal are reflected back toward the surface they will suffer additional attenuation by the material through which they pass, before finally being recorded at the surface. Therefore to be detected as reflections, important subsurface interfaces must not only have sufficient electrical contrast at their boundary but also must be located at a shallow enough depth where sufficient radar energy is still available for reflection. As radar energy is propagated to increasing depths, and the signal becomes weaker as it spreads out over more surface area and absorbed by the ground, making less available for reflection. For every site the maximum depth of resolution will vary with the geologic conditions and the equipment being used. Post-acquisition data filtering and other data amplification techniques (termed range-gaining) can sometimes be applied to reflection data after acquisition that will enhance some very low amplitude reflections in order to make them more visible.

Many ground-penetrating radar novices envision the propagating radar pattern as a narrow "pencil" shaped beam that is focused directly down from the antenna. In fact, GPR waves radiated from standard commercial antennas radiate energy into the ground in an elliptical cone with the apex of the cone at the center of the transmitting antenna (Conyers 2004a: 62). This elliptical cone of transmission occurs because the electrical field produced by the antenna is generated parallel to its long axis and is therefore usually radiating into the ground perpendicular to the direction of antenna movement along the ground surface. The radiation pattern is generated from a horizontal electric dipole antenna to which elements are sometimes added that effectively reduce upward radiation, called shields. Sometimes the only shielding mechanism is a radar absorbing surface placed above the antenna to neutralize upward radiating energy. Because of cost and portability considerations (size and weight), the use of more complex radar antennas that might be able to focus energy more efficiently into the ground in a more narrow beam has to date been limited in archaeology.

Some antennas, especially those in the low frequency range from 10 to 100 MHz, are often not shielded and will therefore radiate radar energy in all directions. Using unshielded antennas can generate reflections from a nearby person pulling the radar antenna, or from any other objects nearby such as trees or buildings. Discrimination of individual targets, especially those of interest in the subsurface, can be difficult if these types of antennas are used. However, if the unwanted reflections generated from unshielded antennas can be identified, they can be easily filtered-out later. If reflections are recorded from randomly located trees, surface obstructions, or people moving about randomly near the antenna, they are more difficult to discriminate from important subsurface reflections and interpretation of the data is much more difficult.

One of the most important variables in GPR surveys is the selection of antennas with the correct operating frequency for the depth necessary and the resolution of the features of interest (Conyers 2004a: 64). Proper antenna frequency selection can in most cases make the difference between success and failure in a GPR survey and must be planned for in advance. In general the greater the necessary depth of investigation, the lower the antenna frequency should be used. But lower frequency antennas are much larger, heavier and more difficult to transport to and within the field than high frequency antennas. For instance a 100 MHz antenna is about 2 meters long. It is not only difficult to transport to and from the field, but must usually be moved along transect lines using some form of wheeled vehicle or sled. In contrast, antennas greater than 400 MHz are usually 50 centimeters or smaller in maximum dimension, weigh very little, and can easily fit into a suitcase.

Low frequency antennas (10–120 MHz) generate long wave-length radar energy that can penetrate up to 50 meters in certain conditions, but are capable of resolving only very large subsurface features. In pure ice, antennas of this frequency have been known to transmit radar energy many kilometers. In contrast the maximum depth of penetration of a 900 MHz antenna is about one meter or less in typical soils, but its generated reflections can resolve features down to a few centimeters in dimension (Conyers 2004a: 47). A trade-off therefore exists between depth of penetration and subsurface resolution. The depth of penetration and the subsurface resolution is actually highly variable, depending on many site-specific factors such as overburden composition, porosity and the amount of retained moisture. If large amounts of electrically-conductive clay, are present, then attenuation of the radar energy with depth will occur very rapidly, irrespective of radar energy frequency. Attenuation can also occur if sediment or soils are saturated with salty water, especially sea water.

The ability to resolve buried features is mostly determined by frequency and therefore the wavelengths of the radar energy being transmitted into the ground. The wavelength necessary for resolution varies depending on whether a three-dimensional object or an undulating surface is being investigated. For GPR to resolve three-dimensional objects, reflections from at least two surfaces, usually a top and bottom interface, need to be distinct. Resolution of a single buried planar surface, however, needs only one distinct reflection and therefore wavelength is not as important in its resolution.

Radar energy that is reflected off a buried subsurface interface that slopes away from a surface transmitting antenna will be reflected away from the receiving antenna and will be lost. This sloping interface would therefore go unnoticed in reflection profiles. A buried surface with this orientation would only be visible if an additional traverse were located in an orientation where that the same buried interface is sloping toward the surface antennas. This is one reason why it is important to always acquire lines of reflection data within a closely spaced surface grid, and sometimes in transects perpendicular to each other.

Some features in the subsurface may be described as "point targets", while other are more similar to planar surfaces. Planar surfaces can be stratigraphic and soil horizons or large flat archaeological features such as floors. Point targets are features such as walls, tunnels, voids, artifacts or any other non-planar object. Depending on a planar surface's thickness, reflectivity, orientation and depth of burial it is potentially visible with any frequency data, constrained only by the conditions discussed above. Point sources, however, often have little surface area with which to reflect radar energy and therefore are usually difficult to identify and map. They are sometimes indistinguishable from the surrounding material, many times being visible only as small reflection hyperbolas visible on one profile within a grid (Fig. 1).

In most geological and archaeological settings the materials through which radar waves pass may contain many small discontinuities that reflect energy, which can only be described as clutter (if they are not the target of the survey). Resolution of clutter is totally dependent on the wavelength of the radar energy being propagated. If both the features to be resolved and the discontinuities producing the clutter are on the order of one wavelength in size, then the reflection profiles will appear to contain only clutter and there can be no discrimination between the two. Clutter can also be produced by large discontinuities, such as cobbles and boulders, but only when a lower frequency antenna that produces a long wavelength is used. In all cases the features to be resolved, if not a large planar surface, should be much larger than the clutter, and greater than one wavelength of the propagating energy in dimension (Conyers 1004a: 65).

The raw reflection data collected by GPR is nothing more than a collection of many individual traces along two-dimensional transects within a grid. Each reflection trace contains a series of waves that vary in amplitude depending on the amount and intensity of energy reflection that occurred at buried interfaces. When these traces are plotted sequentially in standard two-dimensional profiles the specific amplitudes within individual traces that contain important reflection information are sometimes difficult to visualize and interpret. Rarely is the standard interpretation of GPR data, which consists of viewing each profile and then mapping important reflections and other anomalies sufficient, especially when the buried features and stratigraphy are complex. In areas where buried materials are difficult to discern, different processing and interpretation methods, one of which is amplitude analysis, must be used. In the past when GPR reflection data were collected that had no discernable reflections or recognizable anomalies of any sort the survey was usually declared a failure and little if any interpretation was conducted. With the advent of more powerful computers and sophisticated software programs that can manipulate large sets of digital data, important subsurface information in the form of amplitude changes within the reflected waves has been extracted from these types of GPR data (Conyers 2004a: 148).

An analysis of the spatial distribution of the amplitudes of reflected waves is important because it is an indicator of subsurface changes in lithology and other physical properties. The higher the contrasting velocity at a buried interface, the greater the amplitude of the reflected wave. If amplitude changes can be related to important buried features and stratigraphy, the location of higher or lower amplitudes at specific depths can be used to reconstruct the subsurface in three-dimensions. Areas of low amplitude waves indicate uniform matrix material or soils while those of high amplitude denote areas of high subsurface contrast such as buried archaeological features, voids or important stratigraphic changes. In order to be correctly interpreted, amplitude differences must be analyzed in discrete slices that examine only the strength of reflections within specific depths in the ground. Each slice consists of the spatial distribution of all reflected wave amplitudes at various depths, which are indicative of these changes in sediments, soils and buried materials.

Amplitude slices need not be constructed horizontally or even in equal time intervals. They can vary in thickness and orientation, depending on the questions being asked (Conyers and Goodman 1997). Surface topography and the subsurface orientation of features and stratigraphy of a site may sometimes necessitate the construction of slices that are neither uniform in thickness nor horizontal. To compute horizontal amplitude slices the computer compares amplitude variations within traces that were recorded within a defined time window (that can become depth-windows if velocities are known). When this is done both positive and negative amplitudes of reflections are compared to the norm of all amplitudes within that window. No differentiation is usually made between positive or negative amplitudes in these analyses; only the magnitude of amplitude deviation from the norm. Low amplitude variations within any one slice denote little subsurface reflection and therefore indicate the presence of fairly homogeneous material. High amplitudes indicate significant subsurface discontinuities, in many cases detecting the presence of buried features. An abrupt change between an area of low and high amplitude can be very significant and may indicate the presence of a major buried interface between two media. Degrees of amplitude variation in each time-slice can be assigned arbitrary colors or shades of gray along a nominal scale. Usually there are no specific amplitude units assigned to these color or tonal changes.

Using three-dimensional GPR reflection data, buried features can be rendered into isosurface images, meaning that the interfaces producing the reflections are placed in a three-dimensional picture and a pattern or color is assigned to specific amplitudes in order for them to be visible (Conyers et al. 2002; Conyers 2004a: 163; Goodman et al. 2004; Leckebusch 2003). In programs that produce these types of images certain amplitudes (usually the highest ones) can be patterned or colored while others are made transparent. Computer-generated light sources, to simulate rays of the sun, can then be used to shade and shadow the rendered features in order to enhance them, and the features can be rotated and shaded until a desired image results.

3 EXAMPLE OF THREE-DIMENSIONAL GPR MAPPING FOR LANDSCAPE

One area of landscape analysis success with GPR is the high altitude desert areas of Utah, USA, which contains abundant buried archaeological remains, including pit houses, kivas (semi-subterranean circular pit features used for ceremonial activities) and storage pits (Conyers and Cameron 1998). In this area whole valleys might contain buried archaeological features that are all but invisible on the surface, aside from scattered pottery sherds. The climate and geological processes active in this area have produced an abundance of dry sandy wind-blown sediment that often covers and obscures the underlying archaeological features (Conyers and Osburn 2006).

Traditional archaeological exploration and mapping methods in this area that have been used for the discovery of buried sites includes visual identification of artifacts in surface surveys, random test pit excavation and the spatial analysis of subtle topographic features, all of which might indicate the presence of buried architecture. These methods are extremely haphazard and random, often leading to mis-identification or non-identification of many features. In order to test the GPR method for archaeological landscape analysis in this area, a number of tests were performed in one valley, called Comb Wash in southeastern Utah, USA (Fig. 2).

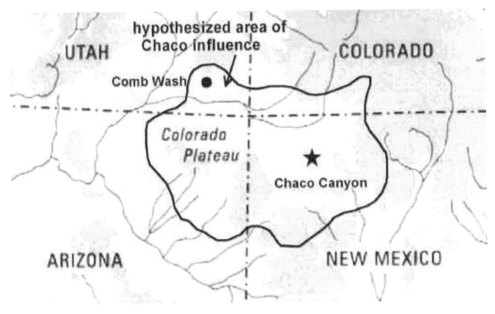

Figure 2.   Base map showing the Comb Wash study area in relationship to the dominant are of Chaco Canyon, New Mexico, USA.

Surface analysis and minor testing over the last 3 decades indicated that this area contained an abundance of buried features thought to be the product of influence from a dominant culture in the area, called Chaco (Lekson 2006). Initial interpretations indicated that most of the Comb Wash area was inhabited by people that were in direct contact with Chaco (located hundreds of kilometers to the southeast) in New Mexico (Cameron 2001, Hurst 2000).

The Chaco period of influence in this region of the American Southwest, beginning about A.D. 900, is characterized by widespread and distinctive architectural styles that have been linked to a shared ideology (Lekson 2006). It was during this time that many of the most impressive buildings with complex architecture and massive stone construction were constructed. These large structures are characterized by above-ground room blocks (called pueblos) and associated semi-subterranean circular pit structures termed kivas, which were used for ceremonial and other activities. Often during Chaco times these kivas and pueblos were "over-built" presumably to impress others, and are often referred to with the moniker "great" by archaeologists who study this region. Kivas that were very large and indisputably Chaco in origin or influence are therefore termed Great Kivas. Archaeologists have proposed many hypotheses about the way Chaco leaders might have exerted influence on the surrounding region, but there is no consensus among scholars as to why or how this was accomplished (Lekson 2006).

Usually when Great Kivas and other large buildings from the Chaco period are found at sites on the margin of the Chaco influence they are referred to as Chaco-outliers, and economic and religious connections with the Chaco center are hypothesized. Some scholars have proposed that the outliers represent military strong posts, or that the people living there were subsumed under a tribute and redistributive system controlled by Chaco (Lekson 2006). Throughout the linear valley at Comb Wash, Utah four large circular depressions were hypothesized to be buried Great Kivas. The valley was therefore hypothesized to have been a regional center of Chaco integration, which was integrally tied to the larger Chaco center far to the southeast.

To test these ideas, GPR data were collected in large grids over these depressions in 2003. Previous work in the vicinity had shown that the GPR method could produce images of buried kivas with good resolution (Conyers and Cameron 1998). The excellent resolution of these buried features in GPR reflection profiles is a function of the distinct interfaces between the stone walls and compacted earth and masonry floors with the sandy matrix, producing distinct radar reflections (Figure 1).

Four GPR grids were collected on the Comb Wash depressions found along the valley floor with antenna transects spaced at 50 centimeters. The GSSI SIR-2000 control system with 400 MHz center frequency antennas was used with 50 reflection traces collected per meter in time windows ranging from 30 to 50 nanoseconds. Reflection profiles show distinct vertical walls and floors of pit features (Figure 1). In all four grids amplitude slice-maps were constructed in 5 nanosecond (two-way time) slices, each of which is approximately 30 centimeters thickness in depth (Figs. 3–5).

It was anticipated that large circular amplitude features of Great Kivas would be imaged using this data processing method, mimicking the size of the surface depressions and following on the ideas that this valley was well integrated with Chaco, based on the size of these buried kivas. These surface depressions range in diameter from 10–15 meters, which is the usual size of Chaco period Great Kivas elsewhere. Instead the GPR reflection amplitude slice-maps yielded a much different picture of these buried sites. Two of the four sites (Sites 1 and 2) showed much smaller circular pit house features ranging in diameter from 5 to 7 meters (Figures 3 and 4). At these two sites the GPR maps also showed a palimpsest of multiple superimposed pit features, indicating at least two, but potentially more periods of construction and modification in this one area. Site 3 showed only one pit feature constructed into bedrock, also about 6 meters in diameter (Figure 5). The fourth test site yielded no features that could be identified as architectural whatsoever in the amplitude maps and the depression presumably is not archaeological in origin.

To test the origin and age of the resulting GPR amplitude features at Sites 1 and 2, augers and standard open excavations were conducted. Vertical stone walls were uncovered in the locations where the highest reflection amplitudes were mapped, definitively showing that these are

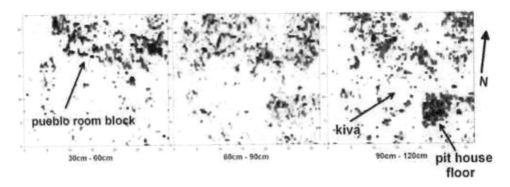

Figure 3. Amplitude slice-maps of Site 1 and Comb Wash, Utah showing near-surface room block with deeper pit house and kiva structures.

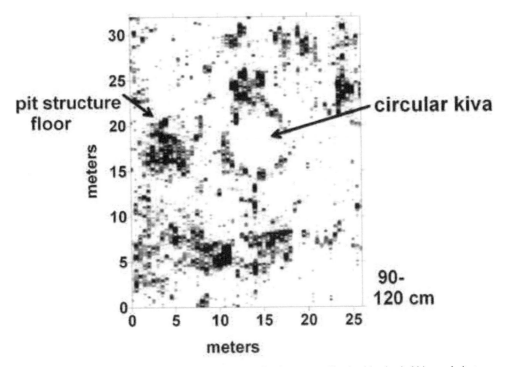

Figure 4. Amplitude slice-map of Site 2, showing similar features as Site 1 with a buried kiva and pit structure floor.

masonry-lined structures are much like other excavated Great Kivas, but much smaller. Ceramic artifacts encountered in association with the floors of these structures date to both before, during and after the period of presumed Chaco influence in the area. These excavations also confirmed multiple phases of construction at these sites, which had been hypothesized from the GPR amplitude maps. At two of the sites at least 2 kiva and pit structure building and subsequent abandonment episodes over many centuries were demonstrated. Their small size and the abundance of everyday cooking artifacts and other utilitarian tools supports the hypothesis that these were kivas used for multiple functions and not just ceremonies, as would have occurred in Chaco Great Kivas.

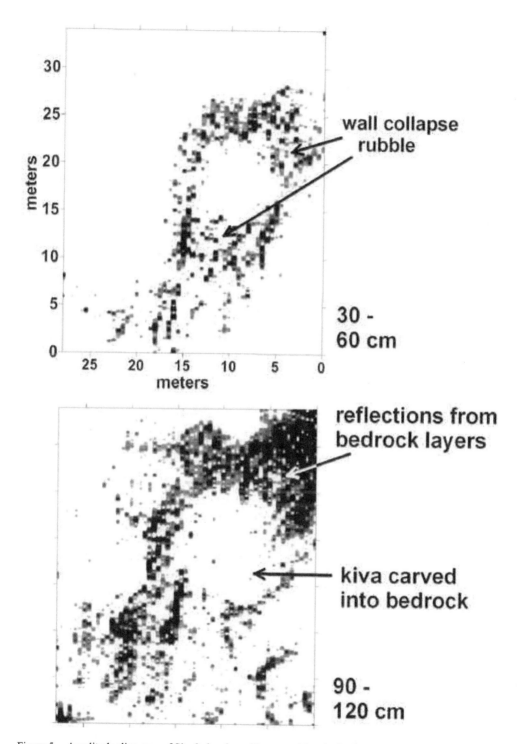

Figure 5.   Amplitude slice-map of Site 3 showing a kiva carved into bedrock.

In addition to kivas and other pit features, GPR reflection data mapped the remains of small room blocks to the north of two of the underground kivas in the upper 50 cm slice (Figure 3). The remains of these larger stone features were partially visible on the ground surface as rubble piles, which had been assumed, prior to GPR imaging, to be the surface remains of buried antechambers connected to the Great Kivas, which had been described elsewhere in the area (Hurst 2000). The GPR maps, however, showed no architectural elements preserved below the upper 30–60 centimeters in these areas, and the spatial patterning of the reflections in the upper slices show instead the foundations of above ground buildings that were composed of between 6 and 8 rooms each (Figure 3). These buildings likely served the habitation and storage functions for this small group of people, built to the north of the kivas in order to block the cold winter winds.

Site 3 GPR mapping illustrated a kiva cut into bedrock whose walls had partially collapsed long after abandonment (Figure 5). No room block was seen in the GPR maps at this site perhaps because its stone building materials were originally set on bedrock and had been recycled and reused elsewhere long ago. The kiva at Site 3 was also 6–7 meters in diameter, indicative of a small group size.

In the case of the Comb Wash GPR analysis, four features that had been assumed to be the product of Chaco influence were determined to be simple family dwellings. In the context of this area's integration with the powerful cultural center to the southeast, GPR mapping along with information from a few excavations, provided the definitive tools that refuted this long held interpretation. When the buried features mapped by GPR were placed within an overall landscape context it was seen that Comb Wash was little more than a simple agrarian community that may have been peripherally influenced by Chaco, but hardly dominated by this powerful entity. The GPR method, selectively applied to what were considered to be the most important buried sites in the overall landscape, provided this new interpretation.

4   CONCLUSION

Ground-penetrating radar has the unique ability of near-surface geophysical methods to produce three-dimensional maps and images of buried architecture and other associated cultural and natural features for landscape analysis. Using high-definition two-dimensional reflection profiles, three-dimensional maps of amplitude changes can define physical and chemical changes in the ground that are related to archaeological buried materials of importance. When these data and maps are used to test ideas about human adaptation to landscapes, they offer a powerful and time-efficient way to study ancient human behavior, social organization and other important archaeological concepts.

In the processing of GPR reflection data for landscape analysis, maps and images must be generated and integrated with information obtained from other buried cultural artifacts to provide age and context for the mapped sites. This can be done by placing these cultural data from excavations within horizontal amplitude maps that produce images of only certain amplitudes within a three-dimensional volume of radar reflections. In all cases, the results of these amplitude images must be differentiated from the surrounding geological layers. When these multiple datasets are interpreted archaeologically, they can provide a powerful tool for the integration of archaeological sites within a landscape context.

REFERENCES

Annan, A.P. and J.L. Davis, 1992. *Design and development of a digital ground penetrating radar system*. In J.A. Pilon (editor), *Ground penetrating radar*. Geological Survey of Canada, Paper 90–4: 49–55.
Cameron, Catherine M., 2001, The Northern San Juan Region in the Post-Chaco era. In Bluff/Comb Wash Project Research Design. Department of Anthropology, University of Colorado, Boulder.
Conyers, Lawrence B., 2004a, Ground-penetrating Radar for Archaeology. AltaMira Press, Walnut Creek, California.

Conyers, Lawrence B., 2004b, Moisture and soil differences as related to the spatial accuracy of amplitude maps at two archaeological test sites. Proceedings of the Tenth International Conference on Ground Penetrating Radar, Delft, The Netherlands, June 21–24, 2004.

Conyers, Lawrence B. and Catherine M. Cameron, 1998, Finding buried archaeological features in the American Southwest: New ground-penetrating radar techniques and three-dimensional computer mapping. Journal of Field Archaeology 25 (4): 417–430.

Conyers, Lawrence B. and Dean Goodman, 1997, Ground-penetrating Radar: An Introduction for Archaeologists. AltaMira Press, Walnut Creek, California.

Conyers, Lawrence B., Ernenwein, Eileen G. and Leigh-Ann Bedal, 2002, Ground-penetrating radar (GPR) mapping as a method for planning excavation strategies, Petra, Jordan. E-tiquity Number 1 http://e-tiquity.saa.org/%7Eetiquity/title1.html.

Conyers, Lawrence B. and Tiffany Osburn, 2006, GPR Mapping to test anthropological hypotheses: A study from Comb Wash, Utah, American Southwest. Proceedings of the 11th International Conference on Ground-penetrating Radar, June 19–22, 2006, Columbus, Ohio, USA.

Davis, J.L. and A.P. Annan, 1992, Applications of ground penetrating radar to mining, groundwater, and geotechnical projects: selected case histories. In Pilon, J.S., Editor, Ground Penetrating Radar. Geological Survey of Canada paper 90–4, Ottawa: 49–56.

Gaffney, Chris and John Gater, 2003, Revealing the Buried Past: Geophysics for Archaeologists. Tempus, Stroud, Gloucestershire.

Goodman, Dean and Piro, Salvatore, Nishimura, Yasushi, Patterson, Helen and Vince Gaffney, 2004, Discovery of a 1st century AD Roman amphitheatre and other structures at the Forum Novum by GPR. Journal of Environmental and Engineering Geophysics 9: 35–42.

Hurst, Winston B., 2000, Chaco outlier or backwoods pretender? A provincial Great House at Edge of Cedars Ruin, Utah. In Great House Communities Across the Chacoan Landscape. Edited by J. Kantner and N.M. Mahoney. Anthropological Papers of the University of Arizona, n. 64, University of Arizona Press, Tucson, pp. 63–78.

Kvamme Kenneth L., 2003, Geophysical surveys as landscape archaeology. American Antiquity 63 (3): 435–457.

Leckebusch, J., 2003, Ground-penetrating radar: A modern three-dimensional prospection method. Archaeological Prospection 10: 213–240.

Lekson, Steven H., 2006, The Archaeology of Chaco Canyon: An Eleventh-Century Pueblo Regional Center. School of American Research Press, Santa Fe, New Mexico. The above material should be with the editor before the deadline for submission. Any material received too late will not be published. Send the material by airmail or by courier well packed and in time. Be sure that all pages are included in the parcel.

Seeing the Unseen – Campana & Piro (eds)
© *2009 Taylor & Francis Group, London, ISBN 978-0-415-44721-8*

Comparative geophysical survey results in Japan: Focusing on kiln and building remains

Y. Nishimura

National Research Institute for Cultural Properties, Nara, Japan

ABSTRACT: Regarding the archaeological prospection, some different environmental factors, involved for Japan and Southeast Asia than for America and Europe, namely, the soils as represented by rice paddies, must be considered. Because of the need to be flooded in the early stages of rice agriculture, paddy soil has a hard, even-grained bed underlying the cultivated soil. The combined bed and cultivated layers have a thickness of more than 30 cm, and generally contain large deposits of iron and manganese. The presence of paddy soil exerts great influence on, and serves as a hindrance to, archaeological prospection.

In geophysical prospection, such as electric resistivity surveying, at times the energy of the electric current does not reach the deep strata because of the paddy layer. As a result, signals detectable at the surface are extremely weak and the signal to noise ratio is poor. In magnetic surveying, the highly magnetized bed of the paddy layer greatly suppresses the degree of magnetic anomaly exhibited by underlying features. In Ground Penetrating Radar (GPR) surveying, the attenuation of microwave energy is severe, and the effective survey depth becomes shallow.

During the geophysical surveys conducted in Japan, attention must also be paid to the regional soil characteristics. Excluding the central portion of the archipelago, areas mainly to the north and south are covered with dark soils derived from volcanic ash, and in these regions the soil magnetism is extremely high. The underlying soil is usually a yellow loam. On the other hand, in alluvial areas having a yellow or yellow-brown ground surface, the soil is not high in magnetism but is clayish in nature. Accordingly, the contrast in electric resistivity between archaeological features and the surrounding soil is low. The microwave energy of GPR is greatly attenuated in this layer, and the effective survey depth is shallow.

This paper introduces examples of geophysical prospections in these different types of soil zones in Japan. Next, comments will be made regarding the results of prospection of postholes, in other words, building features, in which the detection of extremely weak signals is required.

1   INTRODUCTION

With regard to archaeological prospection, there are different environmental factors involved for Japan and Southeast Asia than for America and Europe, namely, the soils as represented by rice paddies. Because of the need to be flooded in the early stages of rice agriculture, paddy soil has a hard, even-grained bed underlying the cultivated soil. The combined bed and cultivated layers have a thickness of more than 30 cm, and generally contain large deposits of iron and manganese. The presence of paddy soil exerts great influence on, and serves as a hindrance to, archaeological prospection.

In aerial photo interpretation, examples of archaeological features being detected from soil or crop marks are extremely rare. This is because differences between the features and the surrounding soil are not readily reflected at the surface level, due to the existence of paddy soil. The paddy layer acts as an impermeable membrane, preventing the differences between the features and the surrounding soil from influencing overlying strata.

In geophysical prospection, such as electric resistivity surveying, at times the energy of the electric current does not reach the deep strata because of the paddy layer. As a result, signals detectable at the surface are extremely weak and the signal to noise ratio is poor. In magnetic surveying, the highly magnetized bed of the paddy layer greatly suppresses the degree of magnetic anomaly exhibited by underlying features. And in ground penetrating radar (GPR) surveying, the attenuation of microwave energy is severe, and the effective survey depth becomes shallow.

In addition, in prospection conducted in Japan, attention must also be paid to regional soil characteristics. Excluding the central portion of the archipelago, areas mainly to the north and south are covered with dark soils derived from volcanic ash, and in these regions the soil magnetism is extremely high. The underlying soil is usually a yellow loam. On the other hand, in alluvial areas having a yellow or yellow-brown ground surface, the soil is not high in magnetism but is clayish in nature. Accordingly, the contrast in electric resistivity between archaeological features and the surrounding soil is low. The microwave energy of GPR is greatly attenuated in this layer, and the effective survey depth is shallow.

The following introduces examples of geophysical prospection in these different types of soil zones in Japan. Next, comments will be made regarding the results of prospection of postholes, in other words, building features, in which the detection of extremely weak signals is required. (Fig. 1)

Figure 1.   General view of tunnel style kiln.

2 PROSPECTION OF KILN SITES: RESULTS FROM TWO TYPES OF SOIL ZONES THE GPR METHOD

2.1 *A volcanic-ash derived soil zone: goshogawara kiln site, aomori prefecture*

This kiln site lies in Aomori Prefecture, the northernmost part of Honshu, and is located in a zone of volcanic-ash derived soil. On a grass-covered slope, currently maintained as part of a park, and over an area 27 m in both the east-west and north-south directions, magnetic prospection was conducted and using a fluxgate magnetometer and a proton magnetometer, electric resistivity prospection with a RM15 resistance meter, GPR with a 400 MHz antenna, and electromagnetic method prospection with a EM38 conductivity meter.

From examples previously discovered in the vicinity, the Sue ware kiln remains were presumed to have a tunnel-shaped structure of around 6 m in length.

2.1.1 *Magnetic prospection*

Measurements were taken using a FM18 fluxgate magnetometer at grid points spaced 1 m apart in both directions. It is difficult to point out the position of the kiln with precision from the results. This is because bands of magnetic anomalies are prominent, which appear to be highly magnetic volcanic-ash soil that that has been deposited along troughs or depressions, making it difficult to discern anomalies that could belong to a kiln.

2.1.2 *Electric resistivity prospection*

Electric prospection used a RM15 resistance meter with a pole-pole configuration in which a spacing of 1 m was maintained for the mobile electrodes. Measurements were again taken at grid points spaced 1 m apart in both directions. From the results, an isolated area of high resistivity is seen slightly west from the center of the survey area. This is inferred to be the kiln. It was noted with this method that resistivity is low over the western half of the survey area, which comprises the uphill portion of the slope. This is presumed to show that the surface layer was stripped away when the grounds were prepared for the park.

2.1.3 *Ground penetrating radar prospection*

A GPR survey was conducted using a SIR-2P unit from GSSI with a 400 MHz antenna (30 cm square), taking scans in a north-south direction along lines spaced 50 cm apart. When the data were processed as horizontal distribution (time slice) maps, a strong isolated anomaly was observed at the same position as seen in the resistivity survey, which was therefore inferred to be the kiln.

2.1.4 *Electromagnetic prospection*

Electromagnetic method prospection was conducted with a EM38 conductivity meter from Geonics in a vertical mode, with measurements taken both parallel and at a right angles to the direction of the line of measurement, and the average taken as the result. Measurements were taken as previously at grid points spaced 1 m apart in both directions. From the results, an anomaly is seen to concentrate to the east of, or in other words in the downhill direction from, the locus inferred as the kiln from the resistivity and GPR surveys; this result is thought to reflect conditions at deeper strata (this was pointed out to me during a lecture in the Field School by Dr. Albert Hesse and Dr. Michel Dabas).

As a result of applying and comparing the various methods of prospection described above, the magnetic survey data were reexamined in reference to the resistivity and GPR results, and the position and scale of the kiln were estimated. Archaeological excavation was conducted on the basis of these prospection results, and the kiln was ascertained as lying between 20–60 cm beneath the ground surface, and as approximately 6.5 m long with a maximum width of 1.8 m.

The prospection conducted at this site has significance for its demonstration of possible difficulties, stemming from the soil in which the remains lie, even for kilns that have been magnetized by heat, and are generally thought easily detected by magnetic prospection. These results, including the greater than expected efficacy of resistivity and GPR prospection, should provide useful insights when prospection is conducted under similar soil conditions in the future. (Plate 23).

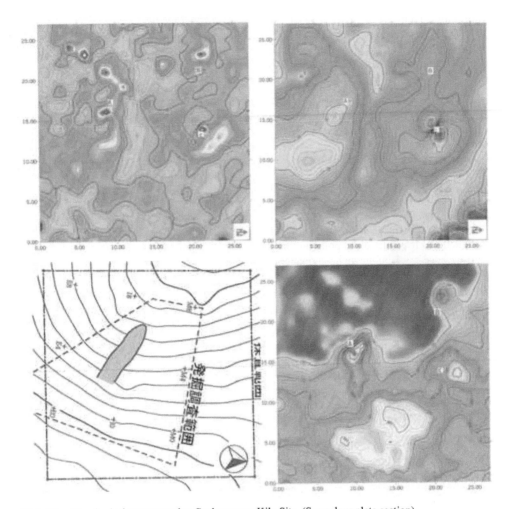

Plate 23.   Geophysical survey results: Goshogawara Kiln Site. (See colour plate section)

2.2   *A non-volcanic soil zone: sabukaze kiln site, okayama prefecture*

This site lies in Okayama Prefecture, in the western part of Honshu, and is located in a zone covered with soil not derived from volcanic-ash, but created from the erosion of granite. Prospection with a proton magnetometer and test excavation were conducted at this site in 1978, and the remains of four kilns plus features thought to be a workshop were detected. In order to prepare the locale as a historic site, the local Board of Education planned a re-examination covering 180 m², an area larger than that of the 1978 survey, for the 2003 and 2004 fiscal years. This was taken as an opportunity to conduct comparative prospection with magnetic, resistivity, and GPR surveys at the site of the No. 2 kiln, which remained unexcavated. One purpose was to clarify the difference with the results given above from a volcanic-ash derived soil zone.

Prospection results, taken from a survey area 19 m east-west by 14 m north-south at the site of the No. 2 kiln, are compared and examined below. The kiln at this site also had a inclined tunnel-shaped structure that utilized a slope.

2.2.1   *Magnetic prospection*

In the magnetic prospection, in the results from both the FM18 fluxgate and G856 proton magnetometers, the magnetic anomaly thought to indicate the kiln appears to be over 10 m in length. As

260

this was thought a bit large for a Sue ware kiln, a 7 m length on the uphill end of the slope, which showed a greater level of anomaly with the fluxgate magnetometer, was inferred to be the body of the kiln. Further, as a faint anomaly extends northward from its northern end, it was thought that perhaps the chimney remained as well.

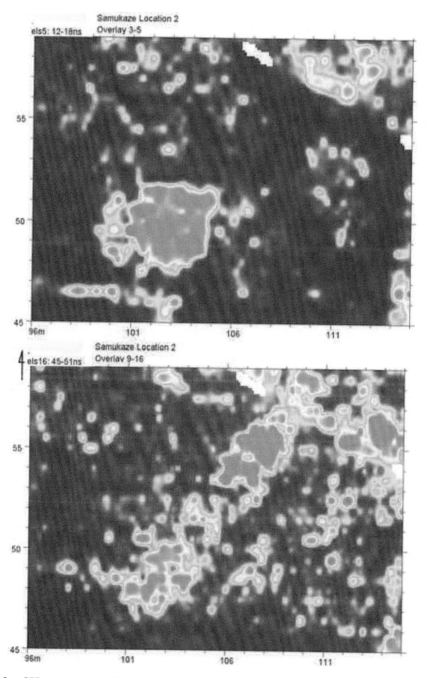

Figure 2. GPR survey results, Upper and lower layers: Samukaze Kiln Site.

2.2.2 Electric resistivity prospection

The measurements were made using a Wenner arrangement, with the interval between the electrodes widened at 0.5 m increments from 0.5 m to 2 m. Among the horizontal representations of the data, in that made for an electrode interval of 1.5 m, areas of high resistivity are divided into two regions on the upper and lower parts of the slope, similar to the situation seen in the GPR survey. If the entire span of the high resistivity areas is seen as the body of the kiln, its length would be interpretable as about 8 m, but it was noted that over the southern half a portion of relatively high resistivity, thought to represent some structure other than the kiln body, spreads out toward the west.

2.2.3 Ground penetrating radar prospection

From the horizontal distribution maps produced with the processed data, the body of the kiln is not detectable at shallow depths up to 15.2 ns, but at the portion thought to represent the stoking area or ash pile in front of the kiln, there is an anomaly of about 2–3 m diameter. The position accords with the southern half of the anomaly seen in the magnetic prospection. This was thought to be possibly a more recent rubbish pit, but as it can be observed in profile (Line 50) to begin from a depth of about 30 cm, it was also considered possibly a feature related to the Sue kiln.

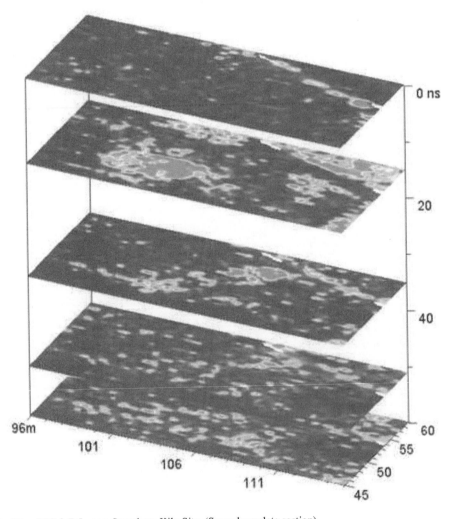

Plate 24.   GPR 3-D Image: Samukaze Kiln Site. (See colour plate section)

Examining the horizontal distribution map for deeper positions (35–41.4 ns), the body of the kiln appears to extend upward into the side of the slope on the one hand, but continues also, after a short gap, in the downhill direction (Fig. 2). If this entire area is taken as the extent of the kiln, its overall length is on the order of 10 m.

From the results of the test excavation conducted after the prospection, the kiln was inferred to be slightly less than 2 m wide and about 10 m in total length. The round anomaly detected in each type of prospection was found to be large amounts of Sue and other materials, gathered during recent agricultural development of the area and deposited in a depression, postdating the kiln, caused by the collapse of its ceiling at a point near the fire box.

In this case, the magnetic anomaly indicating the body of the kiln was clear, serving as a typical example of the utility of magnetic prospection with regard to features that have been subjected to heat. The kiln was also identified with resistivity and GPR prospection, but while the presence and scale of soil areas having different properties could be pointed out in these surveys, their nature as features subjected to heat could not be determined. Only by using several types of prospection methods, and learning thereby the physical and chemical nature of the anomaly, does it become possible to infer it as a kiln.

In the GPR survey, it was possible to show separately the features from the upper and lower strata. (Plate 24) The methods of data processing and display peculiar to this type of prospection also enable three-dimensional display; as this enables learning in detail the historic progression of change in land utilization, GPR may be regarded as holding a unique position among the prospection methods currently in use (At the 6th International Conference on Archaeological Prospection held in Rome in September, 2005, there were several presentations in which results from not only GPR, but also resistivity and magnetic prospection, were analyzed and displayed three-dimensionally. Detailed three-dimensional display similar to that for GPR will probably be developed in the near future for other prospection methods as well).

3 PROSPECTION OF BUILDING FEATURES: DETECTING POSTHOLES

For ancient buildings erected on pillars that stand directly in holes dug in the ground, detection with any type of prospection is difficult. As the dirt removed when the posthole was dug is replaced in the hole within a short period of time, the soil within and outside the posthole is fundamentally the same, providing little contrast for the feature with its surroundings. I have previously introduced the first example in Japan in which embedded-pillar buildings were detected with GPR. The target of the prospection was the main storehouse area of an ancient regional government office complex, where scans were taken in an east-west direction along lines spaced 50 cm apart, over a survey sector 92 m east-west and 135 m north-south, using a SIR-2P unit from GSSI with a 400 MHz antenna. As a result, five storehouse buildings lined up with their long axes oriented to the north, plus a ditch serving as a western limit to the precinct, were detected. Each building was comprised of twenty rectangular postholes slightly under 1 m, and from the configuration of full-sized pillars running four in the east-west axis by five going north-south, and taken together with the results of previous test excavations at the site, they were inferred to represent the main storehouses. (Fig. 3)

3.1 *Comparative examination using various prospection methods*

Features comprised of postholes were detected as just described with GPR, but as it had not previously been possible to detect such remains in Japan regardless of the prospection method, it was decided to examine further the type of soil conditions under which these results were obtained.

For the purpose of this examination, one of the buildings detected with GPR was selected, and resistivity, magnetic, and radar (with 400 and 200 MHz antennas) surveys were first conducted in the area where it stood from the ground surface. Then the surface strata were removed down to the level where the outlines of the postholes became visible, and resistivity and magnetic prospection were conducted again.

263

Figure 3.　Post-hole excavation.

3.1.1　*Resistivity and magnetic prospection*

In the resistivity survey, measurements were taken at the ground surface using a RM15 resistance meter with a pole-pole configuration, a span of 1.5 m for the mobile electrodes, and grid points spaced 0.5 m apart. The postholes began from a shallow depth of about 30 cm below the surface, and the electrode span was decided from the conventional knowledge that these features would have at least 70–80 cm in depth. Measurements at the level where the features were exposed were taken with an electrode span of 0.5 m. In the end, electric resistivity prospection proved unable to detect the existence of the postholes from either the ground surface or from the level at which the features were exposed.

In magnetic prospection using a FM-18 fluxgate magnetometer, grid points spaced 0.5 m apart were also used for measurements taken at both the surface and the level at which the features were exposed. From the results, while it cannot be said that all of the postholes were detected from the ground surface, their presence was nevertheless sufficiently detectable. In measurements from the level at which the features were exposed, all of the postholes were clearly discernible individually, and it was thus learned at this site that magnetic prospection is an effective method for the prospection of postholes.

From the results of the resistivity and magnetic surveys, it can be said that in order to differentiate at this site between the soil within and outside of postholes, it is more effective to pay attention to differences in magnetism than electrical resistance. For postholes detected in test excavations, the features were filled with soil presenting a dark brown color, and it is inferred that this soil is highly magnetic. Further, whereas the yellow-brown soil surrounding the dark-brown feature soil appears clearly different to the naked eye, it was not possible to differentiate them in terms of the factor of electric resistance.

The greater magnetism on the inner part of the postholes also has bearing on the reflection of microwaves during radar prospection, and it is this high level of magnetism that is thought to make a deciding contribution to the results (Fig. 4–5).

3.1.2 Ground penetrating radar prospection

As stated above, postholes were successfully detected from the ground surface using a 400 MHz antenna at this site (Fig. 6–7), but one focus of interest in the current comparative examination

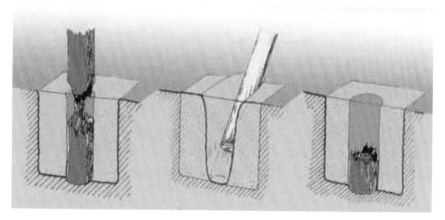

Figure 4. Variety of post-hole remains.

Figure 5. Image of building.

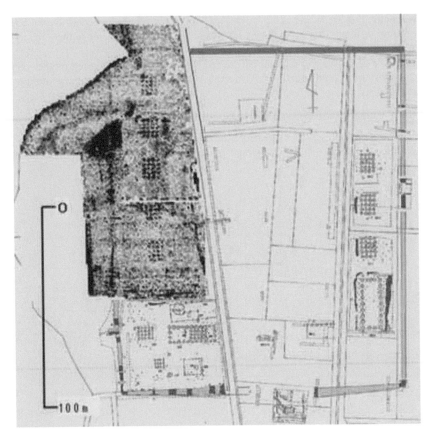

Figure 6. Complex of ancient local government office and GPR survey result: Shimotakahashi Local Government Office Site.

was to see what kind of results could be obtained with a 200 MHz antenna. As previously reported (Yasushi Nishimura, "A Trial GPR Survey for Detecting Post Hole Buildings: Target Identification in Low Contrasted Soil Structures." Presented at "Recent Work in Archaeological Geophysics," Geological Society and English Heritage, London, 2002.12), with the larger and lower frequency 200 MHz antenna, for which the wave length is accordingly long, it has been regarded difficult to detect features like the postholes at this site (Plate 25).

Beginning in 2006, however, the use of staggering compensation is being proposed in the program created by Dr. Dean Goodman (Dean Goodman et al., GPR Data Analysis for Large Scale Area Surveys at Futenma MCAS, Irebaru, Hanzanbaru and Kumuibaru Sites in Okinawa, Abstract-Annual meeting 2006 for Archaeological Prospection Society of Japan, pp. 40–41, 2006.06), utilized for the time slice analysis, and as a result of applying this compensation, it was found that in over half of the cases the postholes were detectable with the 200 MHz antenna. Accordingly I would like to take this opportunity to revise my earlier observations.

The phenomenon of stagger noise is caused by a time difference for signals picked up with the receiving antenna and recorded in the control unit; the term refers to gaps between the loci where signals are recorded bidirectionally, produced in surveys where the antenna travels in alternating directions, in other words in zigzag fashion. It might also be called the recording time lag noise, or the recording gap noise.

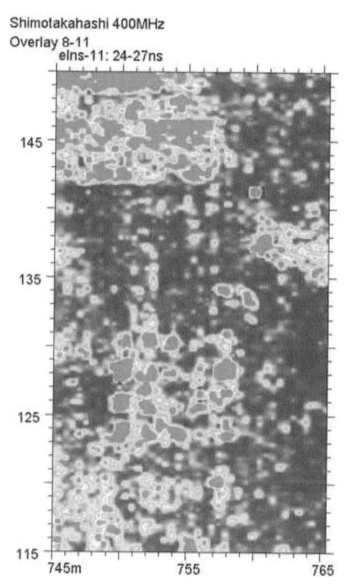

Figure 7.   Details of building (400 MHz antenna): Shimotakahashi Local Government Office Site.

This compensation has not been considered important in GPR prospection until now, but as can be seen from this example, when anomalous reflections which should be displayed as produced by a single posthole end up in a distorted arrangement, the actual form of the anomaly cannot be accurately shown. By compensating for this noise, as stated above, results very close to those of a 400 MHz antenna are obtainable with a 200 MHz antenna as well. In can be said from this that because the penetration of a lower frequency antenna is deeper, that the range over which GPR can be effectively applied should increase in the future.

Also, considering the resolution of the 200 MHz antenna, one reason why it was possible to detect postholes may be the shortening of the wavelength caused by attenuation of the signal within the soil, but in this regard we must await the results of more theoretical examinations.

267

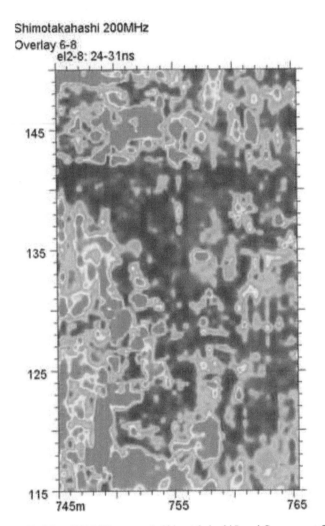

Plate 25.   Details of building (200 MHz antenna): Shimotakahashi Local Government Office Site.

ACKNOWLEDGMENT

For the opportunity to introduce applications geophysical prospection in Japan to the XVth International Summer School in Archaeology, I am grateful to the organizing committee's Dr. S. Campana, of the University of Sienna, and to Dr. S. Piro, of ITABC-CNR.

REFERENCES

A.A. 2005. Proceedings Extended Abstracts of 6th International Conference on Archaeological Prospection, Ed. S. Piro, ISBN 88-902028-0-7, CNR, Roma, Italy.
NISHIMURA Y., 2002. A Trial GPR Survey for Detecting Post Hole Buildings: Target Identification in Low Contrasted Soil Structures. In "Recent Work in Archaeological Geophysics," Geological Society and English Heritage, London, 2002.12.
GOODMAN D., 2006. GPR Data Analysis for Large Scale Area Surveys at Futenma MCAS, Irebaru, Hanzanbaru and Kumuibaru Sites in Okinawa, Abstract-Annual meeting 2006 for Archaeological Prospection Society of Japan, pp. 40–41, 2006.06.

Field Work Section

Test site background and outcome of the Aiali project

S. Campana

Department of Archaeology and History of Arts, University of Siena, Italy

S. Piro

Institute of Technologies Applied to Cultural Heritage – National Research Council,
Monterotondo Sc.(Roma), Italy

The place name Aiali is sited on lowland between the medieval town of Grosseto and the Roman town of Roselle in central Italy (Fig. 1). The site was detected from the air during the Aerial

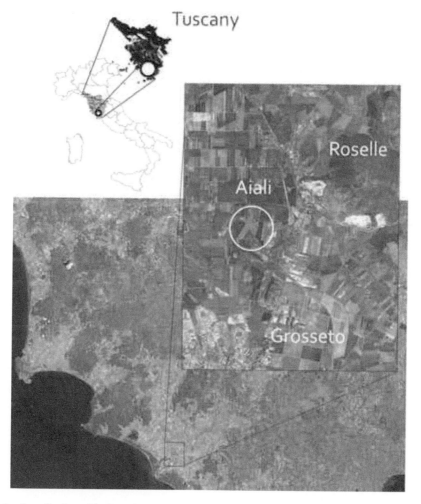

Figure 1. Localization of the investigated area.

Plate 26.   Ground view of the test site. (See colour plate section)

Archaeology Research School organised by the University of Siena in 2001 (Campana et al. 2006). Aerial survey allowed us to recognize an area within which the growth of the wheat varied in such a way as to reveal an articulated group of traces that made up the plan of a complex of structures interpreted as a Roman villa, 4 hectares in extent. In the following years Aiali has become the most important test site for the Laboratory of Landscape Archaeology and Remote Sensing at University of Siena (Plate 26). Since 2001 we have collected, processed and interpreted many different kinds of data:

- Quickbird-2 satellite imagery
- historical and recent vertical coverage (from 1954 to 2005)
- oblique air photographs in various years, seasons and lighting condition (2001, 2004, 2007)
- field-walking survey (grid & replicated collection)
- geophysical survey (magnetometry, GPR, EM, ERT, ARP©)
- DGPS survey to produce detailed DTM
- … in the next future ARP (fast geoelectrical data capture sys, LiDAR)

One intention of the Aiali project is to apply the highest available level and intensity of archaeo-logical prospection methods on a large, complex and stratified site, producing material from the from Etruscan, Roman and Medieval periods.

The archaeological objective and outcome of the Aiali project has to take account of the criti-cal impact of the kinds of information that are available for recording: to use his own words "in assessing the potential or interpretation of a landscape it is at least as important to know what may *not* be visible as to appreciate what *is* visible". At the same time Aiali forms the starting point for a more wide-ranging approach to the study of the landscape between Grosseto and Roselle.

Remote sensing: Aerial photography and satellite imagery

S. Campana

Department of Archaeology and History of Arts, University of Siena, Italy

1   AERIAL SURVEY AND OBLIQUE PHOTOGRAPHY

As noted in the introduction, the site was first detected in early summer 2001 during the aerial archaeology research school through aerial survey exercise. In the following years we did not have the opportunity to survey the site from the air because of logistical problems until the spring of 2004 (Campana et al. 2006). In that year, from the end of May to the middle of June, throughout the crop-ripening season, the site was repeatedly monitored from the air to record the aerial visibility of the cropmarks, using flights at intervals of between two and four days to document their development. This procedure allowed the clear identification of new traces that had not been visible when the site was first discovered in 2001 (Fig. 1). Photography was also continued into the later months of the summer. The new evidence included traces of an abandoned river-course and two new structures in the north-western part of the field, adjacent to the main complex of buildings.

The site was also monitored from the air in 2005 but the maize crop proved unhelpful. A similar situation applied in 2006, when the field was not under cultivation. Crop rotation returned the field to grain in 2007 and this has been the opportunity for the recovery of new information. The new aerial survey, as the previous, was repeated between May and June to follow the development of the crop-ripening and to get the best aerial visibility. The result has been very interesting because

Figure 1.   Oblique air photographs taken at different dates in 2004. From top left: 25 June, 30 June, 4 July, 9 July, 13 July, 25 September.

we achieve new and unexpected outcome. The photographs show quite clearly a new feature in the north-eastern portion (Fig. 2, *top left*) of the field moreover other evidence are visible, unfortunately not so distinctly, in the north-western side of the field. At this stage of the research we can recognize three different areas of the site characterized from evidence with different orientation, probably related with three building complex.

2 VERTICAL AIR PHOTOGRAPHY

As usual we started our work by examining the oldest available aerial photographs, in this case from the national coverage of 1954. Unfortunately no features were visible on this historical flight because the area was at that time used for olive cultivation (Fig. 3). Even though the land-use changed to grain cultivation between the 1950s and 1970s we did not find any features on vertical photographs of 1976, 1996, 2001, 2002, 2005 and 2007.

3 QUICKBIRD-2 SATELLITE IMAGERY

At the end of spring 2002 we planned the acquisition of 70 Km^2 of Quickbird-2 imagery around Aiali. We ordered the multispectral and the panchromatic data. Unfortunately Digital Globe acquired the data relatively later than we had hoped and requested, on 13 June, right at the end on the intended time window. Even though satellite imagery in this case failed to make a meaningful contribution at Aiali itself, we may consider that is related to serendipity. Without a multitemporal approach aiming a long time research strategy the input of satellite imagery (but also, more generally, airborne remote sensing data) will not be effective for intra-site analysis but only somewhat for large scale landscape studies. In this case, for instance, on the same image set it is possible to see many other archaeological features, some of them very close to Aiali. One such site lies only 800 m to the south-east, where field-walking survey has revealed the existence of a large medieval mound (Fig. 4).

Figure 2.   Oblique air photographs taken at different dates in 2007. Top: 12 May, from top left: 12 May, 25 May, 26 May, 9 July.

Figure 3. Above, 1954 vertical photograph with the field used for olive cultivation. Centre, 1996 vertical; despite the changed land-use no features are visible. Bottom, 2001 vertical; still nothing is visible but it is interesting to note that, as in the satellite imagery, it is possible to see features relating to the Brancaleta site, 800 m southeast of Aiali.

Figure 4.   Detail of the Aiali site and surroundings of the quickbird-2 imagery multispectral and panchromatic. In the area of the test-site (Aiali) no features are visible whereas on the south a huge ditch is clearly visible.

Seeing the Unseen – Campana & Piro (eds)
© 2009 Taylor & Francis Group, London, ISBN 978-0-415-44721-8

Geophysical surveys

S. Campana
Department of Archaeology and History of Arts, University of Siena, Italy

S. Piro
Institute of Technologies Applied to Cultural Heritage – National Research Council, Monterotondo Sc.(Roma), Italy

The XV International Summer School at Grosseto has been organized with three days of theoretical lessons and four days into the field collecting geophysical measurement. This school has been organized with the aim to show and to use directly on the field the main important geophysical methods on a real archaeological situation.

The Aiali test-site, with dimension of about 4 hectares in extent, has been divided in four squared test-areas of 50×50 m (Fig. 1).

Figure 1.   Sample areas (A-B-C-D) superimposed over the oblique air photography.

Each test area has been assigned to one student's group and has been investigated, during the different days, employing four different geophysical methods: Magnetometry (fluxgate and Overhouser gradiometers, Cesium optical pumping magnetometer), Ground Penetration Radar (GPR), inductive ElectroMagnetic (EM), Electrical Resistivity Tomography (ERT). This means that each student's group has had the opportunity to employ all different geophysical methods in the same area for the location of the same archaeological features (Plate 27).

The following chapters show the field work achieved during the Summer School in the Aiali test-site employing the highest available level and intensity of archaeological prospection methods on a large, complex and stratified site, producing material from the Etruscan, Roman and Medieval periods.

For the purpose of magnetic prospecting a highly automated system consisting of a 4-Foersterprobes (Ferex DLG) on a chart with GPS and a Overhauser magnetometer (GSM-19) were applied in the same grid. At the time of the summer school there was no caesium-magnetometer

Plate 27.   Geophysical survey at Aiali-test site. From top left: GPS SIR 3000; GSSI Terravision (Geostudi Astier); Helmut Becker using the magnetometer GEM Overhouser; Foerster DLG Kartograph multi-probe gradiometer; field lecture; survey with the electomagnetometer EM-31; setting up the electomagnetometer EM-38 before survey; EM-38 survey; Gianfranco Morelli processing the ERT data; Michel Dabas setting up the geoelectrical instrument (syscal-pro); Michel Dabas explaining field data acquisition to a group of students; differential GPS and total station survey; Helmut Becker Scintrex Smartmag SM4G-special system in various sensor configurations (portable and trolley); Automatic Resistivity Profiler (ARP© Terranova). (See colour plate section)

system available. Therefore the whole area had been remeasured some weeks later with a Scintrex Smartmag SM4G-special in various sensor configurations again in the same grid and under the same surface conditions for comparison (the magnetometer was made available for the test by Schweitzer-GPI company).

During the field surveys, site D was surveyed firstly with GPR equipped with 270 Mhz bistatic antenna with constant offset. The obtained results indicated that the data were quite noisily and characterised by low resolution. This last effect could probabily depend by the wavelength of the signal respect to the averaged dimensions of the buried archaeological structures. For this reason the GPR surveys at sites A, B and C were made employing 400 Mhz bistatic antenna with costant offset.

One year later the area has been measured with the Automatic Resistivity Profiler (ARP©-Terranova University of Paris).

Seeing the Unseen – Campana & Piro (eds)
© 2009 Taylor & Francis Group, London, ISBN 978-0-415-44721-8

Fluxgate, overhouser and caesium-magnetometry

H. Becker
Becker Archaeological Prospection, Eurasburg, Germany

S. Campana
Department of Archaeology and History of Arts, University of Siena, Italy

T. Himmler
Foerster Group, Germany

I. Nicolosi
Institute of Geophysics and Vulcanology, Rome, Italy

For the purpose of magnetic prospecting a highly automated system consisting of a 4-Foerster-probes (Ferex DLG) on a chart with GPS and a Overhauser magnetometer (GSM-19) were applied in the same grid. At the time of the summer school there was no caesium-magnetometer system available. Therefore the whole area had been remeasured some weeks later with a Scintrex Smartmag SM4G-special in various sensor configurations again in the same grid and under the same surface conditions for comparison (the magnetometer was made available for the test by Schweitzer-GPI company).

Most impressive to everybody was the performance of the Foerster-4-probe system with high resolution differential GPS on a chart. The limits of fluxgate-magnetometry are known and became also evident at the site. In general multi-probe fluxgate magnetometry is the ideal instrument for near surface structures with a high magnetization contrast. At Roselle-Aiali this was the case for the main building of the Roman villa. But the sensitivity of the Foerster probes and the vertical gradient configuration were not suitable for the detection of the deeper structures. However the vertical gradient configuration resulted in a perfect reduction of the very strong noise of the high voltage power line nearby (some squares were measured directly under the phone line). The survey system consisted of 4 vertical gradient fluxgate probes with sensor spacing of 65 cm. The lower sensor was guided 25 cm above ground. The FEREX data logger offers a sampling rate of up to 100 Hz for each of the included 4 channels. In the present case 10 Hz had been selected. A measuring range of 10,000 nT is offered, with a sensitivity of 0.1 nT. One interface of the FEREX was used to connect a high resolution DGPS for positioning. The system offers a navigation aid by means a heading indicator on the built in screen, that keeps the user on straight tracks during data sampling (Fig. 1).

A part of the Roselle-Aiali archaeological area has been surveyed with a GEM System GSM 19 Overhauser magnetometer. The survey has been done once without GPS using as reference tradition grids and after that equipped with an internal low cost GPS board, in order to evaluate the performance of the positioning system of this instrument compared to the more accurate acquisition by means of ground grid definition.

The survey instrument, a single sensor scalar magnetometer, has been configured with a measurement sampling rate of 2 Hz maintaining a distance sensor-ground of about 20 cm; this magnetometer has a sensitivity of 0.01 nT with a maximum measurable spatial gradient of 10,000 nT/m.

During the survey the magnetometer internal clock is synchronized with the gps signal in order to precisely attribute a spatial datum to the magnetic field values. The internal GPS board is a 12 channel EGNOS enabled system with a resolution of less than 1.5 meters (http://www.gemsys. ca/). During the acquisition the GPS coverage has been quite good with not less then 8 satellites and EGNOS correction signal available.

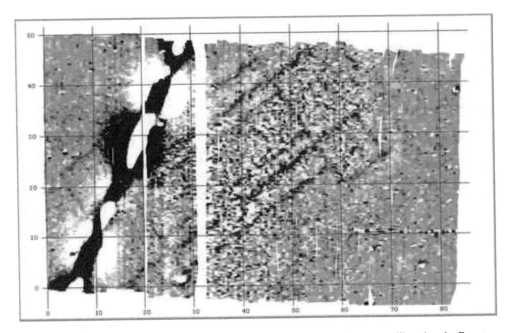

Figure 1. Magnetic map of the data collected on the main building of the roman villa using the Foerster-4-probe system.

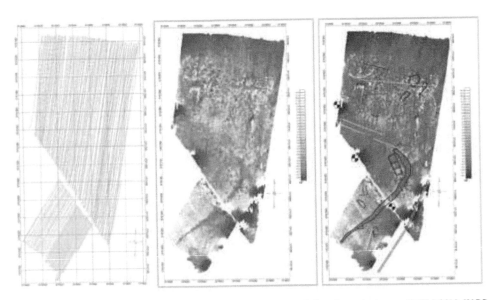

Figure 2. From the left side: surveyed profile lines at Roselle-Aiali archaeogical area. (UTM 32N, WGS 84); magnetic anomaly map of Roselle-Aiali archaeological area; interpretation of magnetic data.

power line tower;   pipe;   possible subsurface archaeological structures;
linear anomalies;   negative magnetization contrast area (paleochannel? Road?)

To monitor the magnetic field temporal variations, a second magnetometer has been used in a fixed station nearby the survey area (base station); a GEM System GSM 19 T proton precession magnetometer with a sampling rate of 1 Hz with an internal GPS board to synchronize the two magnetometers clocks has been used, assuming that the temporal variations of the base station and of the survey area were the same. After fixing a reference time, the temporal variations measured by the base station unit has been removed from the data acquired by the survey unit, so that the residual magnetic field represents the spatial variations component only.

The global surveyed area is represented in figure 2 with two colored lines groups indicating two different acquisitions days; the "blue lines" group (area 1) surveyed area, with main lines direction of N10°, is 2 hectares wide, the "red lines" one (area 2), with mean direction N 38°, is 0.4 hectares wide. The first group has been acquired in 4 hours of continuous measurement, the second in 2 hours. The total survey lines length is about 22 km with 43000 measures; the mean line spacing for the first group is 1.25 meters with a spatial density of 1.4 measures/m², for the second group the mean line spacing is 1.1 meters for station density of 3 measures/m².

In order to map the magnetic anomaly values, a regular data distribution (mesh) of magnetic stations has been realized applying a gridding algorithm to the real measurements; the grid spacing for both areas is 0.5 meters interpolated.

In figure 2 it is reported the resulting magnetic anomaly map displaying the data mesh by means of a grey graded scale. The chosen grey scale linear dynamic range is +/− 15 nT that represents the best interval to visualize the measured magnetic anomalies.

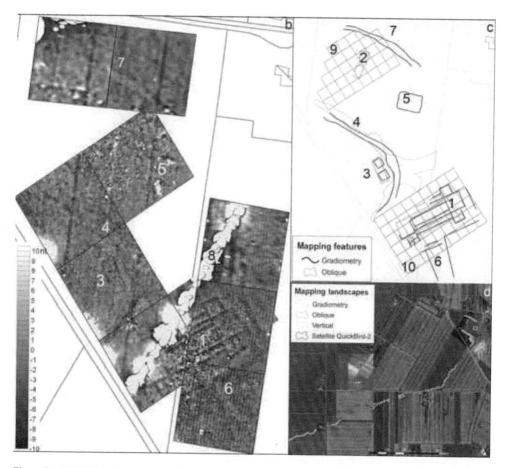

Figure 3.   GEM Overhouser magnetic map.

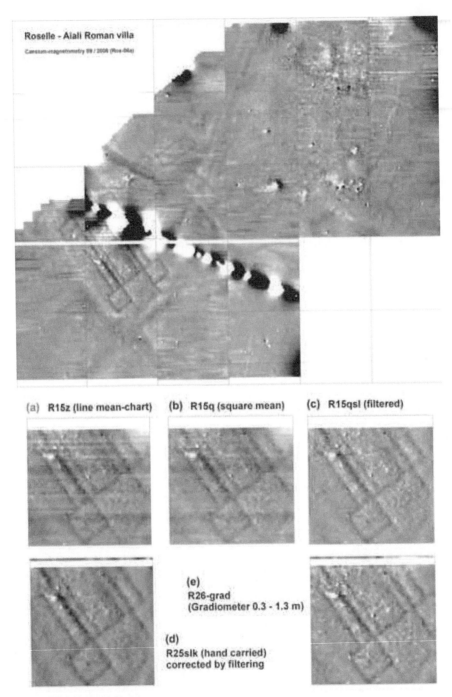

Figure 4. Roselle-Aiali. Caesium-magnetometry with Scintrex Smartmag SM4G-special with duo-sensor configuration on a chart. Sensitivity 30 pT at 10 Hz cycle. Raster 0.10 × 0.5 m (interpol. 0.25 × 0.25 m), dynamics +/– 25 nT, 40 m-grid. Roselle-Aiali. Caesium-magnetometry of a part of the main building with various sensor configurations. (a) measurement on a chart, reduction of the time dependent anomalies by the reduction on the line mean value, (b) same by square mean value, (c) reduction of the long wavelength anomalies on the square mean value, (d) measurement with the hand carried frame and (e) gradiometer measurement with 0.3 m and 1.3 m above ground.

The survey without GPS allow to achieve better results. It proved possible to collect 2 hectares of gradiometer data at intervals of 50 cm along profiles 1 m apart. The magnetic map show a series of magnetic anomalies which closely replicate the traces visible on the oblique air photographs (Fig. 3b, nos. 1,3,4). In addition to confirming, very precisely, the evidence seen from the air, the magnetic survey added a series of anomalies that fill in many of the gaps in the main building complex (Fig. 3b, no. 1 and Fig. 3c, no. 1).

In conclusion the Overhauser magnetometer seemed to be a quite good system for archaeological prospection independent of the sensor configuration (vertical gradient measurements, variometer = correction by a base station and non-compensated duo-sensor configuration were tried). The main disadvantages for archaeological prospection should be recognized in the too low measuring speed and the heavy weight.

The caesium-magnetometry with Smartmag SM4G-special was applied for the whole field in the duo-sensor configuration on a chart with half automated positioning by the rotation of the wheel for non compensated total field measurements. Spacing of the lines was set to 0.5 m and sampling rate to 0.1 Hz (10 measurements per second, corresponding to 10–15 cm sample spacing). Later in data processing and visualization of the magnetograms the raster was interpolated to 0.25×0.25 cm. For the reduction of the high frequency noise a bandpass filter in the magnetometer processor was set to 5 Hz.

Resampling and the reduction of the diurnal geomagnetic variations by line-mean and square mean value were processed by RESAM2. Further data processing and visualization of the magnetograms were achieved by Geoplot 3, Archaeo-Surveyor, Surfer 8 and Photoshop. In the raw data the long wave length noise of the high voltage power line is strongly visible. But this could be cancelled by desloping filters in Archaeo-Surveyor. Unfortunately the main building of the Roman villa had been cut by a modern pipe line with steel flanges, which gave very strong magnetic disturbances, which could only partly removed by filtering techniques (Fig. 4).

Part of the main building had been remeasured by Smartmag caesium-magnetometer with different methods for comparison. Figure 4a, b, c shows the measurement on the chart with long wave length reduction by line mean (a), square mean (b) and complete reduction by deslope filtering on the line (c). Figure 4d gives the example of the measurement with a hand carried frame with the distance triggering on the line only every 5 m by switching. The high frequency noise is minor, which is due to a more quite movement of the sensor compared with the rather bouncy push-pulling method of the chart over a field with lots of stones at the surface. Figure 4e shows the result of a hand carried vertical gradient measurement over 1 m with the lower sensor at 0.3 m and the upper sensor at 1.3 m above ground. Despite of the optimal reduction of the long wave length anomalies of the power line we loose by the gradiometer setup of the sensors the information of the deep structures under the surface and in speed the factor of 2. Considering the essential importance of the 3 "S" in archaeological prospection—Speed, Spatial resolution and Sensitivity—one should forget gradiometer—or variometer measurements and not waste the second sensor for minor quality of the magnetogram of an archaeological site.

REFERENCES

Becker H. 1995. From Nanotesla to Picotesla—a new window for magnetic prospecting in archaeology. *Archaeological Prospection*, 2, 217–228.

Becker H. 1999. Duo and quadro-sensor configuration for high speed/high resolution magnetic prospecting with caesium magnetometer. *Archaeological Prospection*. Proceedings of the 3rd International Conference on Archaeological Prospection. Fassbinder JWE, Irlinger E (eds). Munich.

Campana S., Piro S., Felici C., Ghisleni. 2006. *From Space to Place: the Aiali project (Tuscany-Italy)*, in Campana S., Forte M. (eds.), *From Space To Place*.

Tabbagh J. 2003. Total Field Magnetic Prospection: Are Vertical Gradiometer Measurements Preferable to Single Sensor Survey? *Archaeological Prospection* 11: 75–82.

Seeing the Unseen – Campana & Piro (eds)
© *2009 Taylor & Francis Group, London, ISBN 978-0-415-44721-8*

Electrical surveys

S. Campana
Department of Archaeology and History of Arts, University of Siena, Italy

M. Dabas
Département de Géophysique Appliquée, UMR Sisyphe,
Université Pierre et Marie Curie—Paris VI, France

G. Morelli
Geostudi Astier, Livorno, Italy

During the International Summer School electrical measurements have been collected employing Syscal Pro system. This instrument was very useful to train the students on the field but unfortunately also extremely slow compared to the size of the sample areas and the resolution needed for archaeological tasks. The problem lies in the fact that we didn't collect enough measurements during the School to get data about the samples and to compare the results with the other geophysical methods. To go over this issue, in autumn 2007 the site has been surveyed emploing the Automatic Resistivity Profiler (ARP©). This system has been developed in France by the GEOCARTA company, a CNRS France spin-off (Dabas, this volume). The principle of ARP© is rather simple, relying on the standard galvanic electrical method, used widely for different applications since its discovery by Marcel and Conrad Schlumberger in the 1930s. The ARP© system was first designed for agricultural applications in 2001. It was not until 2004, however, that the system was released for archaeological surveying, because of the need to enhance its positional and measurement accuracy. The surveyed area involves the whole of the Aiali site and its surroundings to a total extent of about 12 hectares. All the raw data have been processed by a 1D median filter along transect first and then interpolated by a spline bicubic process on a squared mesh (40 cm × 40 cm for investigation depths 0–50 cm & 0–100 cm but 50 cm × 50 cm for the deeper one (0–170 cm). In this occasion we show the maps of apparent electrical resistivity at different depths of investigation using a black & white colour scale (Figs. 1 and 2). We decided to publish the maps of the whole site but we should outline that at this stage of the research on the Aiali site it would be unfair to do any parallel between the ARP result and the other methods (excluding the magnetic map of Helmut Becker) outside the sample areas. Fluxgate and Overhouser magnetic systems and GPR surveyed only the areas inside the samples. Anyway we have to admit that the results of ARP are extremely interesting. The ARP© data makes it possible to add new and extremely significant information to the western side of the surveyed area but also to integrate information on known features (this volume, part II, Putting everything together…). The survey also produced a high resolution DTM of the investigated area.

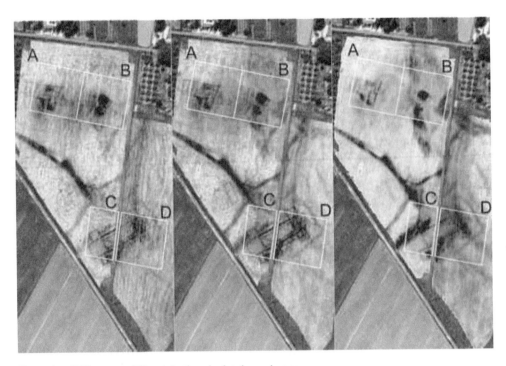

Figure 1.   ARP maps at different depth and related sample area.

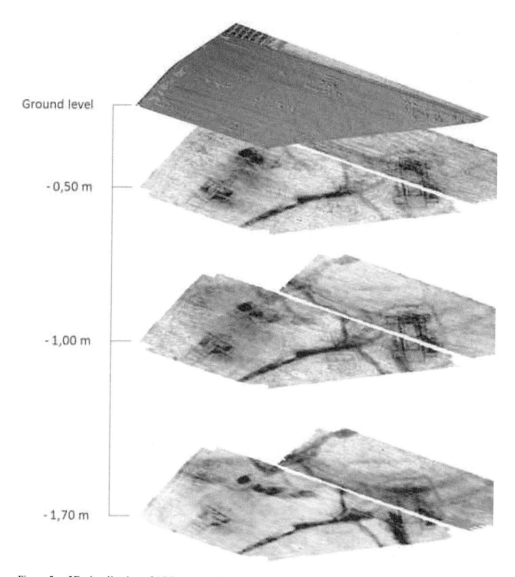

Ground level

- 0,50 m

- 1,00 m

- 1,70 m

Figure 2.   3D visualization of ARP maps at different depths.

Seeing the Unseen – Campana & Piro (eds)
© 2009 Taylor & Francis Group, London, ISBN 978-0-415-44721-8

Electromagnetic survey (EM38 Geonics ltd)

A. Tabbagh
Université Pierre et Marie Curie, Paris, France

Four different zones of the site were surveyed (named A, B, C and D) by the different groups of students with a distance between successive profiles of approximately 1 m. Another survey of the whole western parcel had been also achieved with a wider distance between the profiles (approximatively 5 m). While the operator was walking his position was recorded by GPS and his route can be drawn afterward. The routes are presented in figure 1. The survey of one zone took about 30 mn.

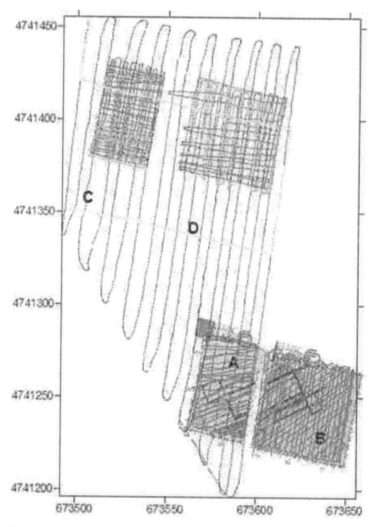

Figure 1. Routes of operators in the field.

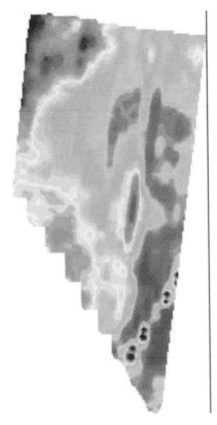

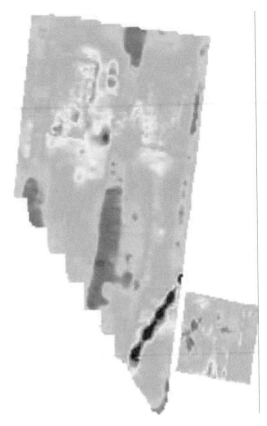

Figure 2.   Conductivity map (HCP configuration).

Figure 3.   Susceptibility map (HCP configuration).

The EM38 apparatus (two parallel coils, 1 m apart, 14.6 kHz) measures both the apparent electrical conductivity (quadrature out-of-phase component) and the apparent magnetic susceptibility (in-phase component). When used in the HCP (vertical board) configuration its depth of investigation is about 1.5 m for the conductivity measurements but it is very low, around 0.4 m, for the susceptibility ones. Contrary to the whole parcel that had been surveyed in HCP configuration, the VCP configuration (horizontal board) was preferred for the four areas surveys to have a better depth of investigation for susceptibility measurements (about 0.6 m). The in-phase measurements exhibit a high temperature drift, which must be corrected by fitting a cubic curve to the data and then by subtracting it. To determine the true susceptibility value one must check the true zero value by handling the instrument at a 1 m height (or more) above the ground.

The conductivity map of the whole parcel is presented in figure 2 and the susceptibility map in figure 3.

The conductivity ranges between 2 and 20 mS/m (50 Ω.m) and is centred at about 10 mS/m (100 Ω.m, in green value on the figures). The susceptibility values range from 10 10^{-5} SI to 170 10^{-5} SI and are centred at about 70 10^{-5} SI. In both maps the pipe corresponds to anomalous values.

The four-area maps are presented in figures 4 and 5. The apparent conductivity is significantly higher than in HCP configuration, which corresponds to the fact that the surface layer is more conductive than deeper layers. The measurement mesh is too wide to observe the pattern of the building and a smaller mesh would be more informative on such site. The effect of the pipe is larger on conductivity maps, where it approximatively reaches 15 m in the perpendicular direction, than on susceptibility maps.

Figure 4.   Conductivity map of A, B, C and D areas (VCP configuration).

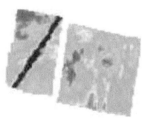

Figure 5.   Susceptibility map of A, B, C and D areas (VCP configuration).

Ground Penetrating Radar (GPR) surveys at Aiali (Grosseto)

D. Goodman
Geophyscial Archaeometry Laboratory, Woodland Hills, CA, USA

S. Piro
Institute of Technologies Applied to Cultural Heritage – National Research Council, Monterotondo Sc.(Roma), Italy

1 INTRODUCTION

GPR surveys were conducted at four sites as part of XVth. Summer School *Geophysics for Landscape Archaeology* of the University of Siena at Grosseto in July of 2006. The main purpose of the GPR was to introduce the GPR method to students in archaeological geophysics. The sites surveyed were located in Grosseto. Crop marks visible on aerial photographs as well as from ground surface remains, indicate that a destroyed Roman villa is present on part of the survey site. The possibility to further the information below the ground surface on the villa using GPR was a secondary objective for the study.

For sites A, B, and C, a SIR 3000 (GSSI) employing a 400 MHz antenna was used. Data were recorded at 50 cm profile spacing. Scan density along the radar track was set to 40 scans/m over the ground using a survey wheel for navigation. The recording time window was 50 ns, which corresponds to about 1.5 m penetration depth with this smaller antenna. Site D was surveyed with a 270 MHz antenna using a time window of 80 ns and an approximate depth of penetration of 2.4 m. A nominal microwave velocity of about 7 cm/ns was determined from fitting hyperbolas to the raw field data. This was used in estimating a penetration depth from the GPR survey.

During the 3 day field survey, Site D was actually surveyed first with GPR. The survey results which were processed later that evening indicated that the 270 MHz data were quite noisy and there was very little useful information in the resultant images. For this reason, the following surveys at C, and at sites A and B on the final day were all completed with the smaller antenna, which eventually proved to show much better results.

All the GPR data were processed in GPR-SLICE v5.0 Ground Penetrating Radar Imaging Software (www.GPR-SURVEY.com). The basic radargram signal processing steps included:

- Post processing pulse regaining
- DC drift removal
- Data resampling
- Background filter

Once the basic raw radargrams were conditioned, the radar pulses were then averaged spatially and in time. Every 25 centimeters along the profile and over a 6 ns time window, the squared amplitude of the radar pulses were averaged. Consecutive time windows for computing time slices were generated at about 50% overlap between adjacent levels. Using these averages, interpolated and solid 3D volumes of reflections amplitudes were generated.

2 GPR RESULTS

GPR time slice images for site AB are shown in Figures 1, 2, Plate 28–29. Shallow time slices indicate that the reflections to the tops of wall of the destroyed villa are quite shallow and begin

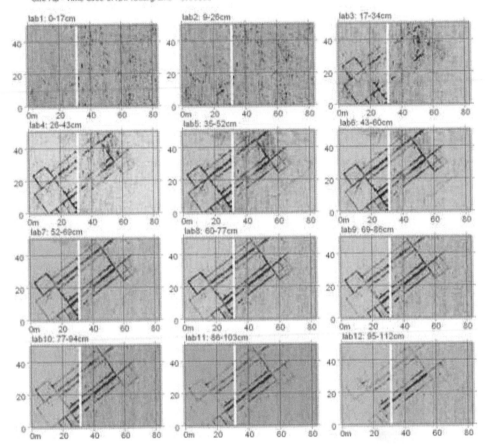

Site AB - Time Slice of raw radargrams - Grosetto

Figure 1.   400 MHz GPR time slices of raw radargram collected at site A and B in Aiali (Grosseto).

to appear on the third slice that is from 17–34 centimeters below the ground surface (Fig. 1). A utility pipe that crosses the site can be identified by a faint linear reflection in a deeper slice from 43–60 cm (Figs. 1 and 2). A fence diagram (Plate 28) and an isosurface amplitude render (Plate-29) help to show the 3D attitude and possible shapes of the remnant wall structures of the destroyed villa.

The survey at site C indicates the presence of some Roman buildings beneath this area (Figs. 3, 4). Unfortunately, because the GPR results are less definitive on the building structure designs in this location, may indicate that the ancient buildings have less integrity and are much less intact then buildings from the Roman villa which are in an adjacent field. In addition to the main building located in the center, a faint rectangular structure can be seen on the left side of the surveyed site (Fig. 4).

The 270 MHz images created for site D indicate that this area probably does not have any intact Roman buildings that can be detected with GPR. Time slices images computed to a depth of 2.25 meter do not show any well defined structures that could be associated with any subsurface buildings. This may suggest that either there were never any building foundations buried in this location, or, that the degree of destruction from modern farming has completely stripped away any remnants of the Roman foundations. Nonetheless, there does appear to be very very faint/broken rectilinear reflections which may be Roman buildings, but the GPR images were not clear enough to accurately resolve these noisy structures.

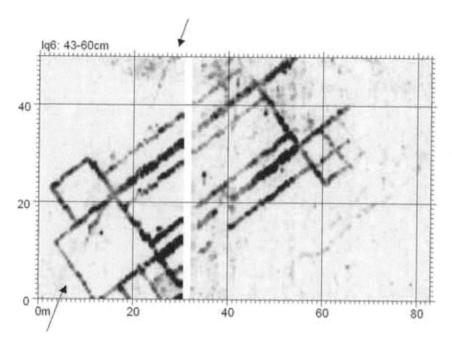

Figure 2.   400 MHz GPR time slice between 43–60 cm depth at site AB at Aiali (Grosseto). The faint reflection of a utility crossing the destroy villa is identified with arrows.

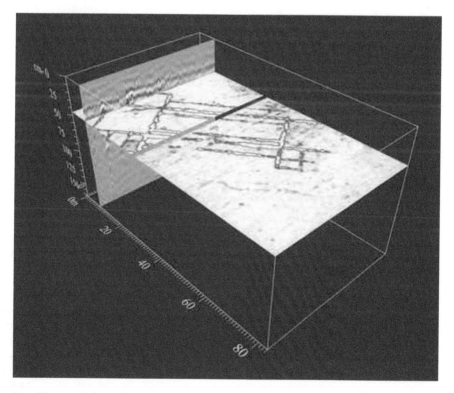

Plate 28.   3D example fence plot generated for site AB at Aiali (Grosseto). (See colour plate section)

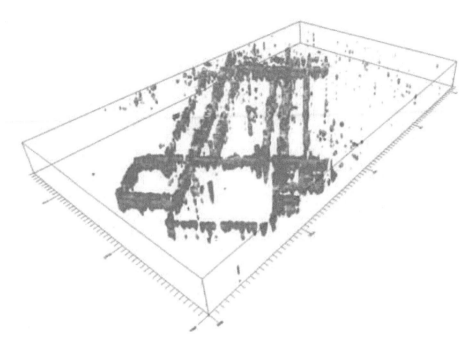

Plate 29. 3D isosurface render for the villa located below site A and B. (See colour plate section)

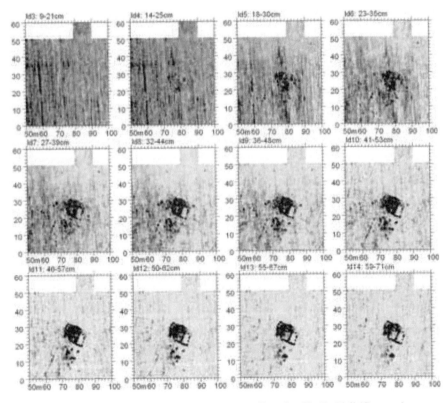

Figure 3.   400 MHz GPR time slices of raw radargrams collected at Site D, Aiali (Grosseto).

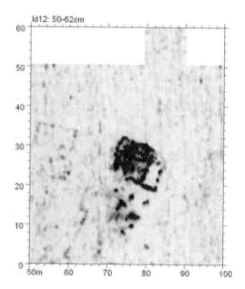

Figure 4.   400 MHz time slice between 50–62 cm at site D, Grosetto.

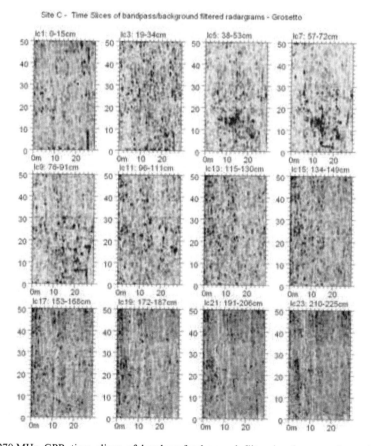

Figure 5.   270 MHz GPR time slices of bandpass/background filtered radargrams from site C, Aiali (Grosseto).

301

REFERENCES

Conyers L.B., Goodman D., 1997. Ground Penetrating Radar. An introduction for archaeologists. AltaMira Press, Walnut Creek, California, (ISBN 0-7619-8927-7).

Gaffney V., Patterson H., Piro S., Goodman D., Nishimura Y., 2004. Multimethodological approach to study and characterise Forum Novum (Vescovio, Central Italy). *Archaeological Prospection*, Vol. 11, pp. 201–212.

Goodman D., Piro S., Nishimura Y., Patterson H., Gaffney V., 2004a. Discovery of a 1st century Roman Amphitheater and Town by GPR. *Journal of Environmental and Engineering Geophysics*, Vol. 9, issue 1, pp. 35–41.

Goodman D., Schneider K., Piro S., Nishimura Y. Pantel A.G., 2007. Ground Penetrating Radar Advances in Subsurfaces Imaging for Archaeology. In "Remote Sensing in Archaeology", Ed. J. Wiseman and F. El-Baz. Chapter 15, pp. 367–386.

Neubauer W., Eder-Hinterleitner A., Seren S., Melichar P., 2002. Georadar in the Roman Civil town Carnuntum, Austria. An approach for archaeological interpretation of GPR data. *Archaeological Prospection*, 9: 135–156.

Piro S., Goodman D., Nishimura Y., 2003. The study and characterisation of Emperor Traiano's Villa (Altopiani di Arcinazzo – Roma) using high-resolution integrated geophysical surveys. *Archaeological Prospection*, 10, pp. 1–25.

Seeing the Unseen – Campana & Piro (eds)
© *2009 Taylor & Francis Group, London, ISBN 978-0-415-44721-8*

Field walking survey, artifact collection and analysis: Remarks of site development from late republican period to middle ages

E. Vaccaro & M. Ghisleni
Department of Archaeology and History of Arts, University of Siena at Grosseto, Italy

S. Campana
Department of Archaeology and History of Arts, University of Siena, Italy

1 FIELDWORK

During the course of 2004 we carried out field-walking survey and surface collection. We realized suddenly some peculiarities of the Aiali site. The high density of artefacts scatter influenced our capability to recognize the boundaries of different areas and we had also to take in to account the large chronological range of the evidence. However we identified, not without uncertainness, the areas of highest artifact scatter density and we did the GPS survey (Fig. 1).

After that we deiced to collect anything because from our point of view the only way to try to understand such a complex site was to approach the artefacts collection within a grid system (Mattingly 2000). The grid has been planned into the GIS and the polygons (10 by 10 meters) exported to the DGPS to allow the satellite navigation in to the field to build the physical grid (Fig. 2).

We can anticipate that ground survey and the study of the collected material has confirmed the archaeological character and interpretation of the site as seen in the air, demonstrating a good level of correspondence between the aerial evidence and concentrations of archaeological finds. In the next chapter it should be clear the great advantage of the grid collection for the archaeological understanding of the site, allowing the development of distributional maps based on different attributes (chronology, function, etc.).

Stefano Campana

2 SOME REMARKS ON THE AIALI SITE PLAN

Integrating all the data from the multi-methodological Aiali project (aerial and geophysical intensive surveys in particular) we are now able to present a preliminary reconstruction of the topographical structure of this site, which was occupied from Roman to medieval times (Figs. 3–4).

Several very interesting elements ensue from the analysis of the plan of both the main building complexes identified, which are undoubtedly related to two different periods of construction, as we can see from the different alignments.

A large building, of an overall size of about 1800 m² (37 × 49 mt), was found in the northern part of the surveyed area. This complex was organized around a rectangular courtyard orientated NE-SW: some rooms opened off the central courtyard, that perhaps was characterized by a peristyle, mostly located in the north-eastern part, while no clearly visible trace of rooms can be seen in the south-western part.

The plan of this building seems to be that of a small to medium size *villa rustica*, whose residential area, in the light of surface materials collected during some intensive field walking surveys, seems to have lacked monumental features.

Even more interesting is a second building complex with an E-W orientation, situated at a distance of 120 mt from the first. This has a very compact plan, characterized by a rectangular

303

Figure 1.   GPS survey of the highest density of artifact scatters.

structure (38 × 10 mt) at the centre and two symmetrical square rooms (about 16 × 16 mt) on the short sides. Two long and narrow corridors run along the rectangular structure, while four roughly quadrangular structures (about 10 × 10 mt) are located on the corners. This complex, like the first one, is about 1800 m² in surface. Three small buildings, located to the NW of this second site, have the same alignment and dimensions as the four corner structures.

It is quite evident that the different orientation of the complexes shows that they belong to two distinct building phases, but it is difficult to establish the relative chronology of both sites with only the planimetric data to go on.

A particularly interesting element of complex 2 is represented by the four corner structures, which could be identified as towers, in the light of comparison with late antique "fortified" residential buildings in Italy and in Tracia as well[1].

In the case of Aiali the four structures, which could be part, together with the rest of the villa, of the same building project, cannot be dated without an archaeological excavation. It seems certain that these rooms/towers were built at the same time because of their position and dimensions, but there is no clear evidence whether these were part of the original plan of the villa or were added later.

Villas with corner towers were already present during the early imperial period, although this design pattern is more frequent during late antiquity, not only as an effect of the militarization of the countryside, but also as a socio-economical marker for the owner[2].

---

[1] See in particular the villa of Palazzolo, close to Ravenna, attributed to a goth owner and both the villas of Orlandovtsi and Pernik in Tracia, Sfameni 2005, p. 615, Figs. 7–9.
[2] *Ibidem*, p. 613.

Figure 2.   Gridded area (10 meters gray squares) and artifact collection grid (10 meters white squares) super-imposed over the geophysical sample areas (A, B, C, D).

This kind of tower, after all, did not always have a military and defensive function, but they often served aesthetic purposes, as in the case of the villa of La Olmeda, in *Hispania*, that was monumentalized in the 4th century AD[3].

The complex 2 of Aiali, unique in Tuscany, seems to have reached its maximum development during the mid imperial period, at least on the basis of surface pottery analysis[4], and could be similar to the seigniorial villas, mentioned in the mid to late imperial written sources with the term *praetoria*, because they consisted of isolated residential buildings surrounded by farm buildings as the military *praetoria* were surrounded by barracks, stables and grain stores[5]. The topographical organization of complex 2 would fit in the planimetric patterns of late antique *villae-praetoria*, since it is characterized by a residential building at the centre, surrounded by minor structures. Perhaps this villa of the *ager Rusellanus* was the dwelling of a *possessor* and the administrative centre for the management of the agricultural resources within a large estate.

Mariaelena Ghisleni-Emanuele Vaccaro

3   AIALI IN THE LONG PERIOD: THE CERAMIC EVIDENCE

Field walking survey at the site of Aiali has been carried out by means of DGPS virtual grids to allow a precise topographic localisation of finds. Two grids, made of squares each having a

[3] Chavarría Arnau 2007, pp. 216–219.
[4] See *infra* Ghisleni.
[5] Carandini 1989, p. 102

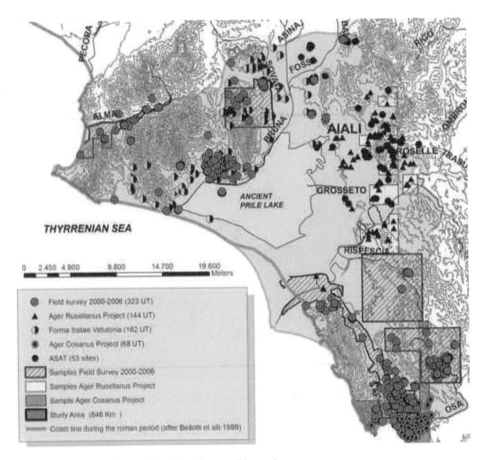

Figure 3.   Aiali in the study area in Southern Tuscany (Vaccaro).

side of ten meters, for a total extension of 0,86 hectares, about 25% of the total size of the site (about 4 hectares), were used. All the material available on the surface was collected and afterward selected: the pottery was sorted out and quantified for total number of fragments and weight. Only typologizable fragments (1100), have been put in the analytical data sheets, taking into account the method of the minimum (1044) and maximum number of forms.

From late republican to mid imperial periods (200 BC–300 AD): an hypothesis
on the topographic articulation of the site

We have analyzed the distribution of the main classes of fine ware, such as Black Glaze Ware (BGW), *terra sigillata* italica, and African Red Slip Ware "A" (ARS), into both grids to define the articulation of the site from the late republic period to 3rd century AD.

A total number of 11 fragments of BGW was collected, a small quantity compared to the many fragments of other classes of fine ware. It is possible that, since Aiali was occupied for such a long period from late republican up to medieval times, the evidence of this class of fine ware related to the older phase of the site was effaced in surface assemblage, but it is also possible that this small number of fragments ensues from a smaller number of people on the site in the late republican period in comparison to the following phases (Fig. 5).

These 11 fragments are located in seven squares, six of which are situated in the same grid placed in the north-western part of the site, just on complex 1. In the wider grid located on

306

Figure 4.   Aiali and Rachalete in the light of aerial and geophysical surveys (Vaccaro).

complex 2, well explored in its spatial articulation thanks to the repeated analysis of remote sensing, only one fragment is documented. This data would indicate a limited occupation of the site in the late republican period, probably concentrated in the north-western part of the site. Perhaps this is evidence of a small to medium farm site.

We should emphasize that the higher quantity of BGW on complex 1, and in general of each ceramic class, in comparison with the second grid could ensue from different post-depositional events. In fact, on the basis of the integration between intensive geophysical surveys and surface collection, we can argue that the archaeological deposit of complex 2 may be more preserved than that of complex 1. It is possible that the frequent ploughings, because of the presence of firmer stone buildings, have done less damage to the site in the southern part. On the other hand, the archaeological deposit in the north-western area would have been damaged more, as the higher number of pottery on the surface could testify.

The total number of collected fragments of *terra sigillata italica* is 339: 50 shreds are documented in the grid placed on complex 2, whereas 289 are attested in the other grid, where they are equally distributed in almost all squares (Fig. 6).

The comparison between quantitative distribution of BGW and *terra sigillata italica* would seem to reveal a notable growth in population size of the site from the 1st century AD on, when the area of complex 2 seems to have been permanently occupied.

However, the site had its greatest development between the 2nd and mid 3rd centuries AD, as proved by the very considerable number of fragments of A production in ARS documented by 993 sherds (Fig. 7). Within the overall amount of fragments related to this production we recognized a minimum number of 179 typologizable forms, which were put in the analytical data base (See appendix 1).

Figure 5.   Black Glaze Ware at Aiali (Ghisleni).

841 fragments of ARS "A" come from the first grid where have a systematic distribution: evidence of a notable density of population on this area. A similar pattern is documented on complex 2 where the total number of ARS "A" is higher than that of *terra sigillata italica* and the distribution over the grid is wider as well.

In this case, the ceramic data allows us to reconstruct an interesting development of the site in early and most of all mid imperial periods. However, we can not exclude that an extensive excavation could bring some more evidence about late republican phase, that now seems to have been so evanescent.

Mariaelena Ghisleni

Late antiquity (300–600 AD)

Although the analytic study of the surface pottery collected over the site is still in progress, it is possible to propose some observations on the transformation of the settlement from the mid imperial period to late antiquity. At the current state of research, useful and thorough information on all the tableware documented on the site from the late 1st to the 6th century AD is available: African Red Slip Ware (ARSW), Colour Coated Ware (CCW) and *sigillata tarda dell'Italia centro-settentrionale*.

ARS is very well documented with A, C and D productions, covering the whole aforementioned period, so that it represents an important guide for the reconstruction of the changing occupation of the site in the long term. The percentage evaluation of the three productions of ARS offers a rough preliminary element of discussion. The A production is documented with at least 993 fragments, while C and D respectively with 25 and 32 (Fig. 8). After assigning each fragment to an ARS production, the analytical study has focused on all the diagnostic sherds which could be

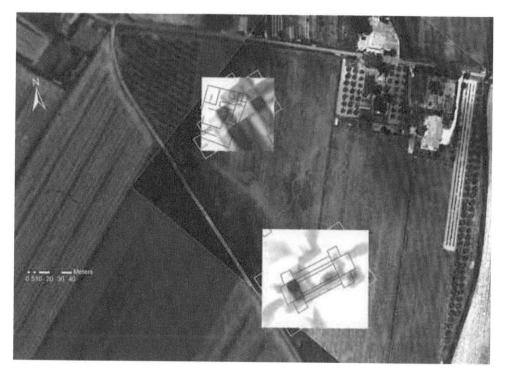

Figure 6. Density map of Terra Sigillata Italica at Aiali through the use of Inverse Distance Weighted interpolation (Ghisleni).

related to a typology; through this approach it is possible to assume that at least 179 minimum number of individuals (MNI) belong to A production, 16 to C and finally 20 to D (Graph 1).

Patterns of ARS supply at Aiali seems to have been unusual compared to the general trend of Southern Tuscany, so it needs to be examined in greater depth and the transformation of the site in the long period must be carefully considered[6].

In fact, the reference frame of ARS distribution over the *ager Rusellanus*, and broadening the perspective, over the whole coastal and sub-coastal area between the Osa stream and the Alma river (a large part of the Grosseto Province), shows quite a clear trend to which the accurate data from Aiali can be compared[7] (Graph 2).

As in the case of Aiali, it is possible to observe in the whole study area a widespread circulation of ARS from the beginning of the 2nd century AD. This is a consequence of three different but correlated factors: the increase of food supply (grain and oil) from Tunisia to Rome within the *annona* system, the commercial redistribution of surplus Tunisian goods (foods and craft products) from the Rome market toward other market places of the Western Mediterranean, and finally the central role of north-African *navicularii*, capable of promoting their own exchanges beyond Roman fiscal control[8].

---

[6] For a quantitative study of ARS see Fentress-Perkins 1988.

[7] This large sample area has been the subject of intensive and systematic archaeological surveys and excavations coordinated by myself since 2000, first for my degree thesis and then for my Ph.D. In July 2007 the Ph.D thesis, focused on settlement patterns and circulation of pottery in this area from the 4th up to the 10th century AD, was presented at the University of Siena.

[8] For these aspects see Panella 1993, pp. 619–654; Reynolds 1995, pp. 106–136; McCormick 2001, pp. 83–103; Wickham 2005, pp. 708–713.

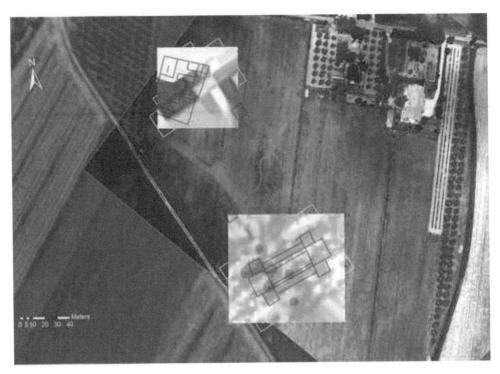

Figure 7. Density map of ARS "A" at Aiali through the use of Inverse Distance Weighted interpolation (Ghisleni).

Both in the study area and at Aiali the exponential growth of ARS reached its peak in the first half of the 3rd century. Subsequently the total amount of ARS in circulation decreased greatly, partly because of the general trend of economic decline that affected the Italic Peninsula during the second half of the 3rd and also because of the decline in the production of ARS A and the inability of the new C production, developed in Central Tunisia, to fill the gap left by the vessels in A promptly[9].

4th century percentages of ARS are very different between Aiali and the other settlements of the study area. In fact from the second half of the 4th ARS was circulating again in large quantities in most of the sites that continued to be occupied in the late Roman period. The general trend was quite positive at least until the beginning of the 6th century, although there were two phases of decline coinciding with the second quarter of the 5th, possibly an immediate effect of the Vandal conquest of Carthage in 439 AD and the interruption of the fiscal route with Rome, and, at the end of the same century, because of the fall of the Roman empire.

At Aiali, on the other hand, the volume of ARS does not provide any evidence of a revival of imported fine ware during the 4th century. Although occupation of the site continued up until the 6th, as we will explain shortly, the supply of ARS persisted at the low levels of the second half of the 3rd, which are not comparable to the very high values of the 2nd and the first half of the 3rd century.

If we were to interpret this data without considering all the information from the systematic study of late antique table ware produced on a regional/sub-regional scale we would be led to believe there was a crisis of the settlement and a marked reduction of the occupied site from the mid 3rd century on.

---

[9] For a similar trend in *Hispania* see Reynolds 2005, pp. 374–376.

Figure 8.   ARS "C" at Aiali (Vaccaro).

Actually ARS dating from the 4th to 6th is still documented at Aiali, but the total number of vessels, as we have seen above, is very scarce (see in particular Appendix 2), and the distribution over the surface site seems to be more reduced compared to the systematic distribution of A products (Fig. 9).

We may note, however, that although the percentage of late ARS is very low, a few forms dating to 5th and 6th century are attested at the two main building complexes.

Specifically, within the grid matching the villa with corner rooms/towers two flanged bowls (Hayes 91 generic and Baradez 1961, Plate-II, n. 7, the latter previously unknown in this area) are documented and a small bowl (Hayes 99 generic), while in the grid over the villa with the central courtyard at least four late antique forms have been found: a cup (Hayes 99A), two flanged bowls (Hayes 91 A/B) and finally a very late large dish (Atlante I, Plate-XLVI, n. 6), whose presence in Southern Tuscany is paralleled only at the dig close to the northern hill of Roselle.

Obviously, synthesis data furnished by the study of surface pottery needs to be very carefully assessed, but we can speculate on the interpretative value to assign to the percentage difference among the three main ARS productions. So the issue that we have to explain is whether the anomaly would be the massive supply of the older ARS production at Aiali or if the marked drop in imported fine ware from the second half of the 3rd century on reflects a radical decrease in the ability of the late antique Aiali dwellers to purchase Mediterranean goods in significant quantities.

Comparing the relative percentages of ARS at Aiali in the long term with those from all the other Roman to late antique rural settlements in the study area, we can see that Aiali is an exception because of its such large amount of A production. In fact even where the first production of ARS is prevalent compared to C and D, the relative proportion is never so high in favour of A. In the light of surface analysis, until an extensive dig at Aiali can be undertaken, we can only assume that this complex settlement was characterized by a marked demographic and economic growth between the late 1st and the mid 3rd century AD, quite different from the trends of other rural sites

311

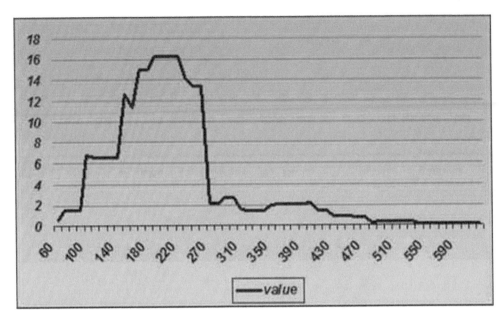

Graph 1.   ARS supply per year at Aiali.

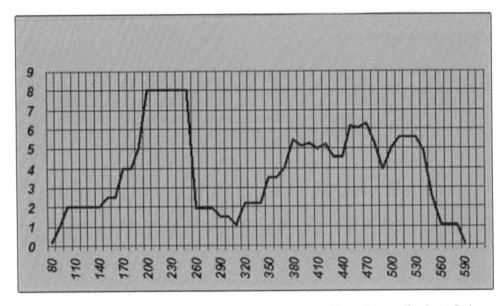

Graph 2.   ARS supply per year in the study area (Alma basin, mid and lower Bruna valley, lower Ombrone basin and Osa valley). ARS data from Aiali are not included in the graph.

in the area. The economic vigour and the close link with Mediterranean trades is also shown by the large amount of Tunisian cooking ware on the site.

We must also note the appearance, from the 4th century on, of a large quantity of regional alternative tableware, in particular CCW, that seems to have been sufficient to counterbalance the ARS drop. These regional craft products were widely distributed over all the *ager Rusellanus* as well as the recently studied area of Southern Tuscany.

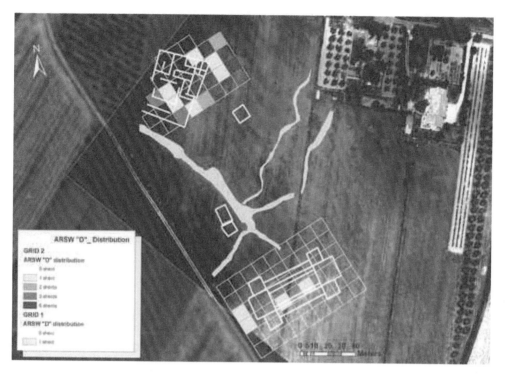

Figure 9. ARS "D" at Aiali (Vaccaro).

In this regard, it is interesting to compare the proportion between Tunisian fineware and regional tableware at Aiali to several 4th to early 5th century ceramic assemblages from the ancient town of Roselle, whose study is now in progress. In this urban centre, in fact, the consumption of tableware is characterized by a strong preponderance of regional CCW (44,8%) and *sigillata tarda dell'Italia centro-settentrionale* (43,5%) compared to ARS of the same period (about 11,5%).

Overall, the marked increase of regional classes with an *engobe* and sometimes with a real slip decoration could have been determined by a widespread demand for pseudo-fineware available at a lower price than that of the overseas craft products.

ARS continued to be attested especially in the coastal sites during all the 5th century up to the beginning of the 6th, but the totale number of imported vessels was much lower than that of the tableware produced by regional and sub-regional workshops. In the course of the 6th century, the devastating effects of the Greco-Gothic war on the socio-economic framework of a very large part of Italy, including this sample area, produced a marked reduction of circulating Mediterranean goods, to such an extent that ARS became a sort of middle range luxury good, which almost only the privileged social classes could afford[10].

This kind of goods continued to reach Aiali between the 5th and the middle of the 6th century, but probably only occasionally. More information regarding dates and events in the settlement can be deduced from the analysis of the volume and distribution of Italic tableware.

---

[10] For this sample area see Vaccaro *forthcoming* with related bibliography; more in general, for the Western see Zanini 1996 and Wickham 2005, pp. 708–713.

It is worth noting that at least 7 forms of *sigillata tarda dell'Italia centro settentrionale*[11] are documented at Aiali; this ceramic class is well documented in the late Roman ceramic assemblages of the nearby urban site of Roselle (only 3,8 km to the E of Aiali) but uncommon in rural settlements (Fig. 10).

Identification of this specific class within the very articulated Italic productions of tableware of late antiquity is no easy matter. In the case of Aiali we had the opportunity to compare the fabrics of some red slipped diagnostic sherds to those of many vessels from 4th to early 5th century ceramic assemblages found in the digging close to the northern hill of Roselle, whose assignment to *sigillata tarda dell'Italia centro settentrionale* was guaranteed by a very typical overpainted decoration on the base in red/orange or brown colour, creating a motif with concentric circles, sometimes joined by *tremoli a raggera*.

However, at Aiali, CCW is the most frequent tableware from the 4th up to 6th century AD; in fact it is attested by 652 fragments and at least a minimum number of 100 classifiable individuals[12] (Fig. 11).

We can argue that the decrease of ARS from the mid 3rd century on, so pronunced if it is compared to the exceptional values of the previous period, should not be automatically considered as evidence of decline of this site and reduction of the occupied area, especially in the light of the systematic distribution and high quantities of regional late Roman table ware over all the parts of the settlement, intenisvely investigated through the use of virtual collection grids.

Both regional and extra-regional workshops produced a very articulated formal repertorire that included open forms imitating some ARS prototypes and many open and closed forms which were inspired by formal patterns of independent tradition, increasing the table ware range to a very notable degree, adding some new functional vessels not often found in the Tunisian repertoire and only occasionally exported from there[13].

Among the more widely distributed forms imitating ARS archetypes, we should mention several dishes with more and less incurved rim, sometimes with a triangular section, clearly inspired by the shape of Hayes 61A and Hayes 61B, several flanged bowls deriving from Hayes 91 later versions, in particular Hayes 91C, with a less developed flange below the rim, and finally a few bowls similar to Hayes 99.

For other forms it is more difficult to recognize a formal link with ARS prototypes, as in the case of many bowls with a more or less horizontal flat rim which sometimes seem to be related to Hayes 58 and Hayes 59, sometimes to Hayes 67.

The same concept is arguable in the case of several large bowls with a vertical and indistinct or slightly incurved rim. These vessels, in fact, show such a simple and functional shape that they could be the fruit of a regional tradition unrelated to Tunisian formal influence; however, we can also suggest a slight link with two types of flat based dishes in ARS, such as Hayes 62 and Hayes 64.

---

[11] For this ceramic class see Fontana 2005; about the presence of this class at several urban sites in *Tuscia* see: Michelucci 1985 (Roselle), Bianchi-Palermo 1990a, pp. 158–168 (Fiesole); Tortorella 1991, pp. 103–114 (Arezzo). For the first distinction of *sigillata medioadriatica*, a mid to late Roman *sigillata* produced in central-eastern areas of Italy, see Brecciaroli Taborelli 1978.

[12] The dates of CCW documented at Aiali is furnished by the morphological comparison with the data from well excavated urban sites in Tuscany, first of all that close to the northern hill of Roselle (which I am studying at present) and the *Domus dei Mosaici* at Roselle (Michelucci 1985), and by use of the chrono-typology of CCW found in the urban excavations at Siena (Cantini 2005), Fiesole (Bianchi-Palermo 1990b) and Arezzo (Tortorella 1991). Finally, we made a comparison between CCW from Aiali and that collected on many late Roman rural settlements of the northern part of Siena countryside (Valenti 1996).

[13] High values of closed forms in ARS are documented in 5th, 6th and mid 7th centuries contexts excavated by the British Mission at Carthage (see Fulford-Peacock 1984, pp. 84–87). In the coastal study area from the Osa stream to the Pecora river basin, closed forms in ARS are very unusual: in fact, only two individuals of the late antique bottle *Fulford closed forms* 2 are respectively documented at the sea-port dump of *Portus Scabris* and at the digging close to the northern hill of Roselle.

Figure 10.   Sigillata chiara tarda dell'Italia centro-settentrionale at Aiali (Vaccaro).

In short, this last hypothetic link is not ascertainable at the moment owing to the lack of whole profile vessels, showing the shape of both the bases and the rims; the latter in CCW are systematically smaller than those of similar forms in ARS.

The regional production of many large basins, documented by many typological rim variants—the rim is always thickened but the upper part of it can vary from flat to rounded—, seems to have been completely independent from Tunisian fine ware influence.

In the surface pottery assemblage of Aiali, this very functional and deep form, probably used to consume liquid and semi-liquid foods, is the most frequent found within the CCW class.

Some interesting data is available from the analysis of CCW distribution in relation to that of ARS over both grids.

In the n°1 grid, CCW is documented in at least 27 cells of a total 54, exceeding A production of ARS, that is attested in 22 cells. Although in both grids the total amount of fragments of Tunisian fine ware is greater than CCW, the wide distribution of late Roman regional table ware over the surface ceramic assemblage of Aiali is very interesting for the reconstruction of late antique settlement patterns. In fact, this data may be evidence that late Roman settlement size coincided with that of early and mid imperial periods.

Distribution values in the n°2 grid are very similar to those of n°1; here CCW is documented in 27 cells out of 32, while A production of ARS was collected in 28. The high density of late Roman CCW sherds over the western part of the second grid, corresponding precisely to the notable density of fragments of D production ARS in the same area, is particularly important.

In conclusion, in the light of ceramic data now available, the site seems to have maintained its size, without any evident reduction in the long period from the early imperial to late antique periods, perhaps up to the beginning of the 6th century; subsequently evidence for settlement size continuity becomes weaker and the villa was probably abandoned during the second half of the 6th or at the beginning of the 7th century at the latest.

315

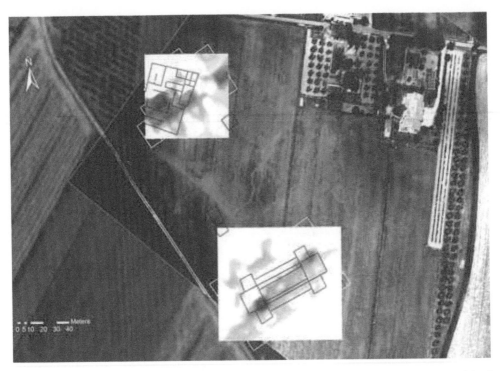

Figure 11. Density map of Colour Coated Ware at Aiali through the use of Inverse Distance Weighted interpolation (Vaccaro).

Obviously a systematic analysis of surface ceramics is not sufficient to give us detailed and reliable information on long-term functional transformations of the site. Only an open-area excavation could tell us what functional changes or introduction of new building techniques in perishable materials occurred on the site, throughout the long period of its occupation, and whether they follow the same patterns of transformation found at many excavated late Roman villas in Italy from the 3rd and 4th centuries on[14]. What can be said is that the articulated site of Aiali probably played a central role in Roman rural settlement hierarchy of the *ager Rusellanus* and in late antiquity as well, with a continuing supply of Mediterranean goods up until that period—particularly noting the presence of Late roman Tunisian amphorae such as Keay XXV, LV and LXII types—but with also increasingly strong links with regional production centres from the 4th century on.

These commercial relationships were probably aided to a certain degree by this site's proximity to the town of Roselle, whose market was still at the centre of a complex network of regional trades of foods and craft products, in late Roman period.

Emanuele Vaccaro

The medieval phase of Aiali: an hypothesis based on surface ceramics

From the late 6th century on, as we have seen above, surface ceramic sequence at Aiali is no longer useful as evidence of continuity of the site in the early middle ages. This pattern seems to have been typical of all the main villas of the study area between the Osa stream and the Alma river

---

[14] For the changes in late antique villas see Francovich-Hodges 2003; Lewit 2003; Arthur 2004; Brogiolo-Chavarría Arnau-Valenti (edd) 2005; Lewit 2005; Wickham 2005.

Figure 12. Density map of medieval pottery at Aiali through the use of Inverse Distance Weighted interpolation (Vaccaro).

basin, which held a central role in rural settlement network from the early imperial period through to late antiquity. This change is probably due to the new choices which preferred agglomerated settlements to dispersed ones, defining a rural landscape characterized by new villages on the hilltops and on the plains[15].

However the case of Aiali is quite different from most of the other villas of the *ager Rusellanus*. In fact, the gap in surface material culture of the site ends in 9th century, when the presence of some ceramic forms dating from late 9th/beginning of 10th to 11th centuries shows evidence for a renewed occupation[16].

The systematic analysis of the very rich and articulated pottery assemblage of Aiali allows us to recognize at least a minimum number of 30 forms dating to the medieval period. The use of virtual collection grids has been very useful to understand the distribution of medieval pottery and the degree to which the Roman and late antique settlement surface was reused. Although the total number of medieval diagnostic sherds is low, the fact that these are documented in many cells of both grids is very significant (Figs. 10–11). In fact, this may be evidence of a similar settlement size between the Roman to late antique and medieval periods; although the occupation density

---

[15] For the early medieval rural settlements in Tuscany see Francovich-Hodges 2003; Francovich 2004; Valenti 2004 e Valenti 2005; for the lower Ombrone valley see Vaccaro 2005.

[16] The dates of medieval surface pottery at Aiali are furnished by morphological comparison with ceramics from stratigraphic excavations of some early medieval villages in central-southern Tuscany, such as Poggio Cavolo (see Vaccaro-Salvadori 2006), Campiglia Marittima (see Grassi 2006), Grosseto (see Valdambrini 2005) and Montarrenti (see Cantini 2003).

Figure 13.   Classes of medieval pottery (late 9th-11th/12th centuries) at Aiali (Vaccaro).

was very dissimilar. After a long *iatus* in the archaeological sequence, dating from the beginning of the 7th up until late 9th century AD, a new settlement emerged at Aiali, whose patterns are not definable without an extensive dig.

The small number of classifiable forms prevents us from reconstructing the articulation of pottery sets during the medieval occupation of the site. However is interesting to look for analogies and differences between medieval ceramics from Aiali and from other contemporary rural sites of the study area, and in particular 10th to 11th centuries pottery assemblages at the hilltop village of Poggio Cavolo. An important point is that all the fabrics identified in medieval pottery from Aiali are the same as those we have already documented at Poggio Cavolo.

However, the similarities in medieval material culture of these sites are not restricted to this point, they also concern formal patterns. The small distance between the sites, about 7 km, obviously explains the presence in both sites of the same forms, produced in sub-regional workshops and distributed on a small local scale. Medieval pottery assemblages at Aiali reflect the simplification of pottery production and circulation typical of the early medieval period, when the total number of pottery classes and forms was greatly reduced. Three main ceramic classes (*acroma depurata*, *acroma semidepurata* and kitchen ware) are documented at Aiali and in most of the other rural sites of the study area during this period, while glazed pottery, such as sparse glaze ware produced in Tuscany and *Latium*, and red painted ware are rare. Mediterranean amphorae, dating from the 10th to 12th centuries, are occasionally found at three sites but not in the Aiali ceramic context.

Among the most frequent functional forms for domestic purposes are closed vessels, such as handled jugs and jars, produced both in *acroma depurata* and in *acroma semidepurata*, and a fairly uncommon basin with rather a broad flat rim. Two minimum numbers of closed forms in red painted ware are documented as well; on the basis of the fabric these were produced in regional

318

workshops[17]. Only one of these is typologizable: it is a jug with vertical rim and a thickened strap handle attached at the neck.

Kitchen ware as well as domestic pottery is characterized by a strong prevalence of closed forms: cooking pots are very well documented, while the presence of open forms is reduced to a bowl/lid and a *testo*.

Emanuele Vaccaro

4 A FRAMING HYPOTHESIS OF THE MEDIEVAL PHASE OF AIALI IN THE CONTEXT OF THE RURAL SETTLEMENT IN THE PLAIN BETWEEN GROSSETO AND ROSELLE

New intensive field surveys along the lower Ombrone and Bruna valleys have revealed the development, at the end of the early middle ages, of several new rural settlements on the plain, that reused more ancient sites abandoned or reduced in size from the late 6th/early 7th century on; all of these shared a common position close to the major Roman road that wound around the Prile salt lake[18]. Medieval demographic trends over the small area between the sites of Aiali and *Casoni del Terzo*, a few kilometres to the N from Grosseto, are particularly interesting.

Toponymical data allows us to locate the papal *curtis* of *Flacianum, cum caio suo qui dicitur tertio*, which was mentioned in the estates list drawn up by cardinal Deusdedit during the 80s of the 11th century, in the countryside close to *Casoni del Terzo*; the part of the list concerning this section refers to property events at the beginning of the 8th century AD[19].

This text, listing some papal properties placed around Prile lake from Castiglione della Pescaia on, mentions immediately after *Flacianum* a *villa magna et fossa, que vocatur Flexu*.

Archaeological records from recent and previous filed surveys[20], undertaken by the University of Siena over the plain between Grosseto and Roselle, give us a picture of flourishing socio-economic trends from the late 9th to 12th centuries AD.

Unfortunately, the archaeological data now available on the area between Aiali and *Casoni del Terzo* fails to identify the rural settlements mentioned in written sources on the field. It is however interesting to point out how surface archaeological record and texts coincide in underlining the articulated settlement network of this area.

The restudy of pottery collected on a Roman farm site at 10 m above the sea level, in the locality of *Casoni del Terzo*, has identified several medieval ceramics, previously unidentified, which can testify to a later reuse of the settlement between the late 9th and 11th centuries. In this case, in the absence of punctual georeferencing of medieval pottery, we can not define the settlement size and decide whether it was permanent or seasonal site.

Of course, the *curtis* of *Flacianum*, although located in the area close to *Casoni del Terzo* on the basis of toponymical evidence, can not be identified as the farm site mentioned above; however we should not exclude that the latter could have been a very small farmstead related to the papal *curtis*.

Two major sites of the plain to the N of Grosseto were probably respectively a medieval settlement that reused the Roman and late antique building complexes of Aiali and a medieval village in the locality of Brancaleta which should probably be interpreted as the *castellare* of *Rachalete*,

---

[17] For the similarities of shapes and fabrics with red painted ware attested in the urban contexts of Pisa, dating to the 10th and 11th centuries, see Abela 2000, pp. 119–120 e Abela 1993, pp. 413–418.

[18] For the Roman and medieval roads of this area see in particular Citter 2005 (ed.).

[19] For the document see Von Glanvell 1905, Lib. III, cap. 191; for the right chronological location of the papal estates mentioned in 191 section see Kurze 2002, pp. 397–414.

[20] A part of the flood plain between Grosseto and Roselle was investigated by Dr.Carlo Citter in the second half of the 1980s, see Citter 1988–1989 and Citter 1996. I restudied the late Roman and medieval pottery collected during that field survey in order to identify settlement phases dating from the 5th to 10th centuries AD, in the light of late antique and early medieval ceramic sequences now available.

mentioned for the first time in a boundary document of 1262[21] (Fig. 4). The sites are at a distance of slightly less than 400 m apart, and both lie along the *Aemilia Scauri* Roman road[22].

As we have said, ceramic evidence at Aiali reveals a medieval reuse of Roman and late antique buildings of the site, after it was abandoned for about three centuries. The most interesting information comes from the analysis of medieval pottery distribution, which covers both the building complexes visible on the magnetic map. Widely scattered distribution of medieval fragments seems to be evidence for a similarity between Roman to late antique topographical layout and settlement size and that of the medieval period; population density seems to have been greatly reduced. The seeming correspondence between the Roman and medieval inhabited surfaces at Aiali, such as in the case of a few rural settlements of the study area, probably depended on the search for reusable stone buildings and stone materials to be reused for new dwellings. The strong link with a Roman road, that still served local rural settlements in early middle ages, was probably also a strong factor. Surface ceramic assemblage at Aiali fails to provide any evidence for a continuity of the site between late antiquity and the late 9th/beginning 10th century AD. However, the papal *villa magna* mentioned in early 8th century written source could be proof of the presence of an early medieval farm site close to Aiali, at the centre of a papal estate[23].

Since the *curtis* of *Flacianum*, as the toponymy reveals, was placed close to *Casoni del Terzo*, the *villa magna*, mentioned immediately afterwards, is likely to be situated in close proximity to Aiali or exactly on this site, since it represents one of the main monumental Roman settlements in the hinterland of Roselle.

Possible evidence of the closeness or even the topographic correspondence between the early medieval papal *villa* and the Roman building complex could be found by comparison between the archaeological field record and written sources. In fact, intensive aerial and geophysical surveys carried out on the site since 2001, permit not only the identification of Roman buildings but also several traces of an abandoned river-course, respected by the layout of ancient structures. This stream bends sharply, almost at right angles, opposite the villa with corner towers, while two adjacent rectangular structures lie close to the angle. The 8th century text refers to a *villa magna et fossa que vocatur Flexu*, that is a large settlement and a ditch called *Flexu* because of its meandering course. This clear correspondence could be more than a mere hint, perhaps suggesting the presence of an 8th century phase at Aiali.

This hypothesis, however, must be confirmed by extensive archaeological excavation, capable of discovering an early medieval occupation of the site that has not been revealed by the multimethodological non destructive project. There is a possibility that in such a rich and articulated pottery assemblage, as that of Aiali, the huge amount of Roman and late antique ceramics have covered up fainter traces of 8th century pottery.

The nucleated settlement of *Rachalete* lies to the SE of Aiali, in an area indicated on the cadastral map on a scale of 1 to 5000 with the toponym of *Brancaleta*, which clearly derives from the name of the *castellare* first mentioned in 1262. This document attests that the site, reduced to a *castellare*, that is a settlement partially abandoned or already depopulated[24], was owned by the bishop of Grosseto.

It is possible that, as Roberto Farinelli has argued[25], *Rachalete* had lost its own castle district before 1179, since it was not mentioned in a property trade between the bishop of Grosseto and

---

[21] This site was discovered recently during the archaeological field surveys carried out by LAP&T in fall 2006; the study of material culture collected on the site is still in progress, so we have at our disposal only a small amount of preliminary data. Information about this site is offered in Campana 2006, pp. 131–132. For the written sources mentioning the site see Farinelli 2000, p. 160, Tab. 1.II.e.

[22] For the hypothetical reconstruction of this road see Citter 2005, p. 73, Plate 13.

[23] See above.

[24] For the meaning of *castellare* in 12th up to 14th centuries written sources of southern Tuscany see Farinelli 2000, p. 184.

[25] Farinelli *forthcoming*, record n.12920.

Ildebrandino VII Aldobrandeschi. In this document, in fact, the estate related to the castle of *Montecurliano*, only 2,5 km to the E from *Rachalete*, is described through the use of property boundaries of the neighbouring castles of Istia and Roselle, but not through those of *Rachalete*. Besides, a preliminary chronological sequence provided by a preliminary observation of surface pottery seems to refer to a settlement occupied for a short time between the late 9th/beginning 10th to 12th centuries AD.

This village on the plain, with a micro-relief so similar to those of castle mounds, was probably characterized by rectangular-shaped enclosing walls, as a trace, clearly visible in satellite imageries and in a vertical aerial photo taken in 2001, has revealed[26]. This structure surrounds an area of about 3700 m²: a size that is very similar to the fortified summits of many hilltop villages of the study area, such as Poggio Cavolo (circa 4600 m²), Calvello (3800 m²), Castel Maus (circa 2400 m²), Castellaccio del Prile (circa 3400 m²) and Montepescali (4775 m²)[27], just to mention some important sites. So this settlement can be interpreted as a medium size village surrounded by defences holding a substantial rural population, as the large amount of medieval pottery reveals. This settlement together with Aiali and several other small rural sites, which have been little by little identified by the restudy of previously collected surface pottery and new intensive field surveys, constituted an articulated settlement network over the plain between Grosseto and Roselle in the medieval period.

APPENDIX 1

ARSW –A–*		
Hayes 2	3 (MNI)	Second half 1st
Hayes 3 B	2 (MNI)	75–150 AD
Hayes 3 C	10 (MNI)	First half 2nd
Hayes 6 (generic)	2 (MNI)	Late 1st-second half 2nd
Hayes 6 A	2 (MNI)	Late 1st-early 2nd
Hayes 6 B	6 (MNI)	Second half 2nd
Hayes 7 A	2 (MNI)	70-second half 2nd
Hayes 7 B	2 (MNI)	First half 2nd
Hayes 8 A	21 (MNI)	90- mid 2nd
Hayes 8 B	9 (MNI)	3rd
Hayes 9 A	17 (MNI)	100–160
Hayes 9 B	4 (MNI)	Second half 2nd
Hayes 10	1 (MNI)	150–300
Hayes 14 A	29 (MNI)	175–250
Hayes 14 B	7 (MNI)	Late 2nd-first half 3rd
Hayes 14 C	13 (MNI)	Late 2nd-first half 3rd
Hayes 14 (generic)	22 (MNI)	Mid 2nd-first half 3rd
Hayes 14/17	1 (MNI)	Second half 2nd-first half 3rd
Hayes 16	6 (MNI)	Second half 2nd-second half 3rd
Hayes 17 A	3 (MNI)	Second half 2nd-first half 3rd
Hayes 26	7 (MNI)	Second half 2nd-early 3rd
Hayes 27	6 (MNI)	Late 2nd-first half 3rd
Hayes 31	2 (MNI)	First half 3rd
Hayes 32	1 (MNI)	First half 3rd
Hayes 34	1 (MNI)	Late 2nd-early 3rd

* Dates from Atlante I

---

[26] For this trace see in particular Campana 2006, p. 132, Figs. 1–2.
[27] For the surface size of the medieval castle of Montepescali see Farinelli 2000, p. 157, Tab. 1.II.b; the sizes of the other settlements were measured by myself through a systematic use of a dGPS during punctual field surveys.

APPENDIX 2

ARSW –C–

Hayes 50 and related variants (230/240–400 +)*	14 (MNI)
Hayes 45A (230/240–320)*	1 (MNI)
Hayes 42 (220–240/250)*	1 (MNI)

ARSW –D–

Hayes 58 (290/300–375)*	1 (MNI)
Hayes 59 (320–420)*	2 (MNI)
Hayes 61A (325–400/420)*	2 (MNI)
Hayes 67 (360–470)*	5 (MNI)
Hayes 91A/B (350–530)°	2 (MNI)
Hayes 91, generic	1 (MNI)
Baradez 1961, Plate-II, n. 7 (400–550)°	1 (MNI)
Mackensen 1993, Taf. 63, n. 18.11, Hayes 93B variant (450–530/550)•	1 (MNI)
Hayes 99A (450–530/550)	1 (MNI)
Hayes 99 (?), generic	1 (MNI)
Atlante I, Plate-XLVI, n. 6 (?) (late Vth/early VIth-550)°	1 (MNI)
Stamped decoration: palm branch (Hayes type 1) and concentric circles (Hayes type 26). Styles A(i) and A(ii). Date 320–350 and 350–420*	1 (MNI)
Stamped decoration: concentric circles (Hayes type 26). Styles A(i) and A(ii). Date 320–350 and 350–420*	1 (MNI)

Dates from:
* Hayes 1972
° Atlante I
• Tortorella 1998

ABBREVIATIONS

Atlante I–*Atlante delle forme ceramiche. I. Ceramica Fine Romana nel Bacino Mediterrane (medio e tardo impero),* Supplemento a Enciclopedia dell'Arte classica e orientale, Roma 1981.

Atlante II–*Atlante delle forme ceramiche. II,* Supplemento a Enciclopedia dell'Arte classica e orientale, Roma 1981.

Hayes–Hayes, J.W., *Late Roman Pottery,* Londra 1972.

Keay–Keay S., *Late Roman amphorae in the western mediteranean. A typology and economic study: the catalan evidence,* B.A.R., i.s., 196, 2, voll., Oxford 1984.

MOREL–Morel J.P., *Céramique campanienne. Les formes,* Roma 1981.

BIBLIOGRAPHY

Abela 2000–Abela E., *Ceramica dipinta a bande rosse (DR), in* Abela E., Berti G., Bruni S. (edd.), *Ricerche archeologiche medievali a Pisa. I. Piazza dei Cavalieri, la campagna di scavo 1993,* Florence, pp. 119–120.

Abela 1993–Abela E., *Ceramica dipinta in rosso (DR),* in Bruni S. (ed.), *Piazza Dante: uno spaccato di storia pisana,* Pontedera, pp. 413–418.

Arthur 2004–Arthur P., *From Vicus to Village: Italian Landscapes, AD 400–1000,* in Christie N. (ed.), *Landscapes of Change. Rural Evolutions in Late Antiquity and the Early Middle Ages,* Aldershot.

Bianchi-Palermo 1990a–Bianchi S., Palermo L., *Terra sigillata chiara italica,* in De Marinis G. (ed.), *Archeologia urbana a Fiesole. Lo scavo di via Marini-via Portigiani,* Florence, pp. 158–168.

Bianchi-Palermo 1990b–Bianchi S., Palermo L., *Ceramica a vernice rossa tarda,* in De Marinis G. (ed.), *Archeologia urbana a Fiesole. Lo scavo di via Marini-via Portigiani,* Florence, pp. 169–188.

Brecciaroli Taborelli 1978–Brecciaroli Taborelli L., *Contributo alla classificazione di una terra sigillata chiara italica*, in "Rivista di studi marchigiani", I1, pp. 1–38.

Brogiolo-Chavarría Arnau 2005–Brogiolo G.P., Chavarría Arnau A., *Aristocrazie e campagne nell'Occidente da Costantino a Carlo Magno*, Florence.

Brogiolo- Chavarría Arnau-Valenti 2005–Brogiolo G.P., Chavarría A., Valenti M. (edd.), *Dopo la fine delle ville: le campagne dal VI al IX secolo*, XI Seminario sul Tardoantico e l'Altomedioevo, Mantua.

Campana 2006–Campana S., *Remote sensing at work*, in Campana S., Piro S., Felici C., Ghisleni M., *From space to place: the Aiali project (Tuscany-Italy)*, in Campana S., Forte M. (edd.), From space to place, 2nd International Conference on Remote Sensing in Archaeology, CNR, Roma, 4–7 dicembre 2006, Oxford, pp. 131–132.

Cantini 2005–Cantini F., *Archeologia urbana a Siena: l'area dell'ospedale di Santa Maria della Scala prima dell'ospedale, altomedioevo*, Florence.

Cantini 2003–Cantini F., *Il Castello di Montarrenti*, Florence.

Carandini 1989–Carandini A., *La villa romana e la piantagione schiavistica, in Storia di Roma, 4. Caratteri e morfologie*, Turin, pp. 101–200.

Chavarría Arnau 2007–Chavarría Arnau A., *El final de las villae en Hispania (siglos IV-VII D.C.)*, Paris

Citter 2005–Citter C. (ed.), *Lo scavo della chiesa di S.Pietro a Grosseto. Nuovi dati sull'origine e lo sviluppo di una città medievale*, Florence.

Citter 1996–Citter, C. (ed.), *Grosseto, Roselle e il Prile. Note per la storia di una città e del territorio circostante*, Mantua.

Citter 1988–1989–Citter C., *La topografia archeologica del territorio di Roselle-Grosseto*, Degree Thesis in Medieval Archaeology, Univeristy of Siena, Prof. R.Francovich.

Farinelli forthcoming–Farinelli R., *I castelli nella Toscana delle "città deboli". Dinamiche del popolamento e del potere rurale nella Toscana meridionale (secoli VII-XIV)*.

Farinelli 2000–Farinelli R., *I castelli nei territori diocesani di Populonia-Massa e Roselle-Grosseto (secc. X-XIV)*, in Francovich R.-Ginatempo M. (edd.), *Castelli. Storia e archeologia del potere nella Toscana medievale*, vol. I, Florence, pp. 141–185.

Fentress-Perkins 1988–Fentress E., Perkins P., *Counting African Red Slip Ware, in L'Africa romana*, Atti del V Convegno di studio (Sassari 1987), Sassari 1988, pp. 205–214.

Fontana 2005–Fontana S., *Le ceramiche da mensa italiche medio-imperiali e tatdo-antiche: imitazioni di prodotti importati e tradizione manifatturiera locale*, in Gandolfi D. (ed.), *La ceramica e i materiali di età romana. Classi, produzioni, commerci e consumi*, Bordighera, pp. 259–278.

Francovich 2004–Francovich R., Villaggi dell'altomedioevo: invisibilità sociale e labilità archeologica, in Valenti M., *L'insediamento altomedievale nelle campagne toscane. Paesaggi, popolamento e villaggi tra VI e X secolo*, Florence, pp. IX-XXII.

Francovich-Hodges 2003–Francovich R., Hodges R., *Villa to Village*, London

Fulford-Peacock 1984–Fulford M.G., Peacock D.P.S., *Excavations at Carthage: The British Mission. The Avenue du President Habib Bourguiba, Salammbo. The Pottery and other ceramic objects from the site*, Sheffield.

Grassi 2006–Grassi F., *La ceramica tra VIII e X secolo nella Toscana meridionale: le tipologie, le funzioni, l'alimentazione*, in Francovich R., Valenti M. (edd.), *IV Congresso Nazionale di Archeologia Medievale*, Florence, pp. 461–467.

Kurze 2002–Kurze W., *Notizie dei Papi Giovanni VII, Gregorio III e Benedetto III nella raccolta dei canoni del Cardinal Deusdedit*, in Kurze W., *Studi toscani. Storia e Archeologia*, Castelfiorentino, pp. 397–414.

Lewit 2005–Lewit T., *Bones in the Bathhouse: re-evaluting the notion of 'squatter occupation' in 5th–7th century villas*, in Brogiolo G.P., Chavarria A. and Valenti M. (edd.), *Dopo la fine delle ville: le campagne dal VI al IX secolo*, XI Seminario sul Tardoantico e l'Altomedioevo, Mantua, pp. 251–262.

Lewit 2003–Lewit T., *'Vanishing villas': what happened to élite rural habitation in the West in the 5th-6th c?*, in "J.R.A.", 16, pp. 260–274.

Mackensen 1993–Mackensen M., *Die spätantiken Sigillata- und Lampentöpfereien von El Mahrine (Nordtunesien). Studien zur nordafrikanischen Feinkeramik des 4. bis 7. Jahrunderts*, Voll. I-II, Munich

McCormick 2001–McCormick M., *Origins of the Eurpean Economy. Communications and Commerce AD 300–900*, Cambridge.

Michelucci 1985–Michelucci M., *La domus dei mosaici*, Montepulciano.

Panella 1993–Panella C., *Merci e scambi nel Mediterraneo tardoantico*, in Carandini A., Cracco Ruggini L., Giardina A. (edd.), *Storia di Roma 3. L'età tardoantica. II. I luoghi e le culture*, Turin, pp. 613–697

Reynolds 1995–Reynolds P., *Trade in the Western Mediterranean, AD 400–700: the ceramic evidence*, BAR International Series 604, Oxford.

Reynolds 2005–Reynolds P., *Hispania in the Late Roman Mediterranean: Ceramics and Trade*, in Bowes K., Kulikowski M. (edd.), *Hispania in Late Antiquity. Current Perspectives*, Leiden, pp. 369–486

Sfameni 2005–Sfameni C., *Le villae-praetoria: i casi di San Giovanni di Ruoti e di Quote San Francesco*, in Volpe G., Turchiano M. (edd.), *Paesaggi e insediamenti rurali in Italia meridionale fra tardoantico e altomedioevo*, Bari, pp. 609–622.

Tortorella 1998–Tortorella S., *La sigillata africana in Italia nel VI e nel VII secolo d.C.: problemi di cronologia e distribuzione*, in Saguì L. (ed.), *Ceramica in Italia VI-VII secolo*, Florence, pp. 41–69

Tortorella 1991–Tortorella S., *La ceramica fine da mensa di età romana*, in Melucco Vaccaro A. (ed.), *Arezzo. Il colle del Pionta. Il contributo archeologico alla storia del primo gruppo cattedrale*, Arezzo, pp. 103–114.

Vaccaro *forthcoming*–Vaccaro E., *Dinamiche insediative e gestione del territorio tra tarda età repubblicana e tarda antichità nella Toscana meridionale: il campione di quattro valli fluviali (valle dell'Alma, media e bassa valle del Bruna, bassa valle dell 'Ombrone, valle dell 'Osa)* in Valenti M. (ed.), *Il V secolo, la Toscana alle soglie del medioevo*.

Vaccaro 2005–Vaccaro E., *.3 Dinamiche insediative; .4 I manufatti ceramici di Podere Serratone: metodi di analisi dei materiali da superficie per lo studio della cultura materiale di un abitato di pianura*, in Campana S., Francovich R., Vaccaro E. (with texts by Frezza B. and Ghisleni M.), *Il popolamento tardoromano e altomedievale nella bassa valle dell'Ombrone. Progetto Carta Archeologica della Provincia di Grosseto*, in "A.M." XXXII, pp. 461–480.

Vaccaro-Salvadori 2006–Vaccaro E., Salvadori H., *Prime analisi sui reperti ceramici e numismatici di X secolo dal villaggio medievale di Poggio Cavolo (GR)*, in Francovich R., Valenti M. (edd.), *IV Congresso Nazionale di Archeologia Medievale*, Florence, pp. 480–484.

Valdambrini 2005–Valdambrini C., *Il materiale ceramico proveniente dallo scavo della chiesa di S.Pietro a Grosseto. Osservazioni preliminari*, in Citter C. (ed.), *Lo scavo della chiesa di S.Pietro a Grosseto. Nuovi dati sull'origine e lo sviluppo di una città medievale*, Florence, pp. 33–47.

Valenti 2005–Valenti M., *La formazione dell'insediamento altomedievale in Toscana. Dallo spessore dei numeri alla costruzione di modelli*, in Brogiolo G.P., Chavarria A., Valenti M. (edd.), *Dopo la fine delle ville: le campagne dal VI al IX secolo*, XI Seminario sul Tardoantico e l'Altomedioevo, Mantua, pp. 193–220.

Valenti 2004–Valenti M., *L'insediamento altomedievale nelle campagne toscane. Paesaggi, popolamento e villaggi tra VI e X secolo*, Florence.

Valenti 1996–Valenti M., *La ceramica comune nel territorio settentrionale senese tra V-inizi X secolo*, in Brogiolo G.P., Gelichi S. (edd.), *Le ceramiche altomedievali (fine VI-X secolo) in Italia settentrionale: produzione e commerci*, VI seminario sul tardoantico e l'altomedioevo in Italia settentrionale, Mantua, pp. 143–169.

Von Glanvell 1905–Von Glanvell W., *Die Kanonensammlung des Kardinals Deusdedit*, Padeborn

Wickham 2005–Wickham, C., *Framing the Early Middle Ages. Europe and the Mediterranean 400–800*, New York.

Zanini 1996–Zanini E., *Ricontando la terra sigillata africana*, in "A.M.", XXIII, pp. 677–688.

Putting everything together: GIS-based data integration and interpretation

S. Campana
Department of Archaeology and History of Arts, University of Siena, Italy

S. Piro
*Institute of Technologies Applied to Cultural Heritage – National Research Council,
Monterotondo Sc.(Roma), Italy*

1. AIALI TEST-SITE

Bearing in mind the preliminary stage of the research it would be premature to attempt a definitive interpretation. At this point, however, we are confident enough to outline a few archaeological and methodological issues.

Firstly, it is necessary to stress the very clear difference in the map between the amorphous concentrations provided by the field-walking surface-collection data and the integrated results of the remote sensing techniques (Plate 30 and Fig. 1). The integration of the various techniques enriches the resulting information both quantitavely and qualitatively. Plate 30 clearly shows how the comparison between the various methods makes it possible in some cases to distinguish the same features while at the same time to recognize elements that are not necessarily revealed by the other survey techniques. In these cases the inherently different characteristics of the various techniques produce a quantitative enrichment in the representation of the buried evidence as seen in the composite map (Plate 30). It is important to appreciate that where the anomalies seem to match one another the 'redundancy' between the different sources of information is more apparent than real. The integration of the magnetic, ERT and GPR methods makes it possible to acquire information on the geometrical pattern, depth and even some of the chemical or physical properties of the buried features, though no single technique reveals all of these characteristics.

In this case, for instance, there are readily distinguishable on the ERT maps three groups of buildings (Plate 30): the first at the west (Plate 30, no. 1), the second at the south-east (Plate 30, no. 2) and the third at the north (Plate 30, no. 3). The availability of the magnetic data for the whole of the field allows us to establish that the large building complexes at the far west (Plate 30, no. 1) and at the north (Plate 30, no. 2) are made of different materials from the large building at the south-east (Plate 30, no. 3). Further differences are revealed by comparison between the GPR, the magnetic and the ERT data for the complex at the north (Plate 30, no. 2). In this case the ERT data, so effective in uncovering the geometry of complexes (Plate 30, no. 1) and (Plate 30, no. 3), does not produce clear results. Similar difficulties can be seen in the magnetometer measurements, while a much stronger response is provided by the GPR signals (Plate 30, interpretation of GPR maps). We might reasonably suggest that the different responses perhaps reflect different structural characteristics in the buried deposits. It is worth observing in this context that the overall plan of the data shows probably evidence of three building complexes, situated close to one another but apparently with quite different plan-forms and variable orientations. If it is difficult, from this discussion, to dispute the obvious contribution of the ground-based and remote-sensing techniques, the importance of the archaeological field data is equally undeniable. As we have seen, the surface material was collected and recorded in this case within a predetermined grid, making it possible to overlay on the remotely sensed data a distribution-map of the finds-concentrations, offering clues to the chronological or even functional phasing of the buildings, etc.

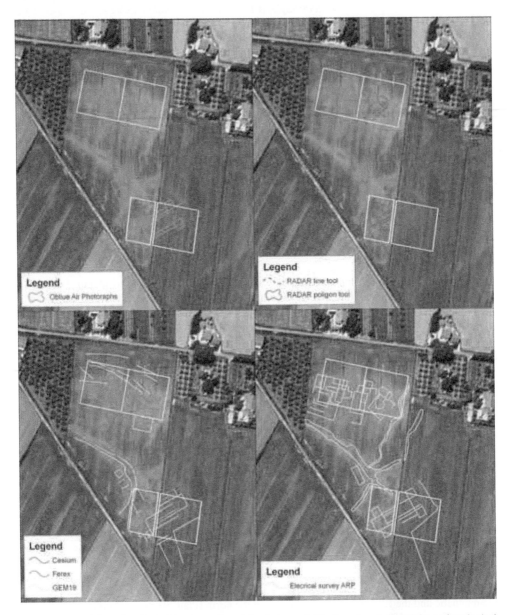

Plate 30. Top left: archaeological interpretation of the features visible from the air. Top right: archaeological interpretation of three different sets of magnetic data—fluxgate, overhouser and caesium. Bottom left: archaeological interpretation of the GPR survey. Bottom right: interpretation of the archaeologyical features visible in the fast electrical imaging data set. (See colour plate section)

The finds-distribution demonstrates the occupation of the site as a whole, and of the three building complexes, throughout the period from the early imperial age to the middle centuries of the medieval period, with an apparent interval in the VIIth and VIIIth centuries (perhaps attributable to problems in the recognition of ceramic material from this particular phase). In the absence of grid collection we could have added little else. The quantitative analysis of the finds-distribution, however, allows us suggest a more intensive occupation during the first imperial age of the building

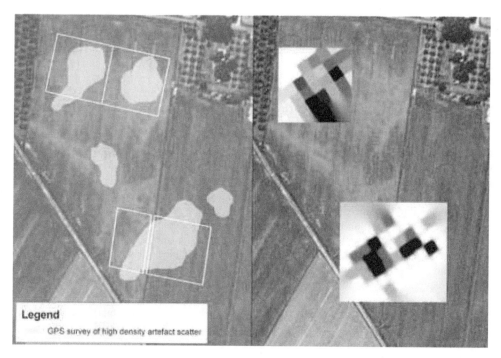

Figure 1. Data gained from field-walking. Left GPS survey of the main density artifact scatters; right, grid-based artifact collection and related representation of finds distribution.

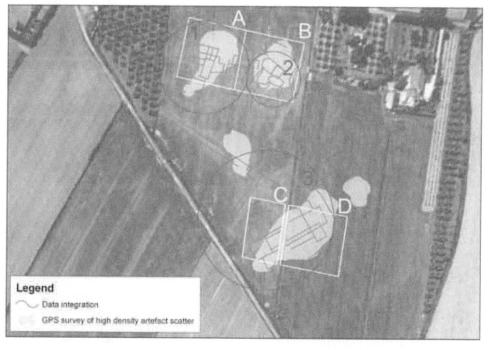

Plate 31. Integrated interpretation. Notice the striking difference between the GPS survey of the higher density of artifact scatters and the integrated remote sensing datasets. (See colour plate section)

complex at the west (Plate 31, no. 1), while from the IInd-IIIrd centuries AD the centre of greatest concentration falls on complex (Plate 31, no. 3) and in the medieval period in zone (Plate 31, no. 2). These speculations seem to be supported by the differing plan-forms of the various complexes as outlined by the remotely-sensed data. The layout of complex (Plate 31, no. 1) has local parallel with buildings of late republican/early imperial age (this volume Vaccaro et al.).

An interesting aspect of the building complex in area (Plate 31, no. 3) is the presence of the four squared structures at the corner which might perhaps be interpreted as towers, of exactly the kind found in some 'fortified' late antiquity residential complexes, whether Italic or Thracian. Without resorting to excavation it is impossible to ascribe a specific chronological attribution to the four towers, which could in reality form part of a single building project along with the rest of the villa. At this stage of the research, it seems fairly clear that the four quadrangular structures, given the their identical dimensions, were built within a single phase, though it is not possible to say for sure whether they formed part of the original layout of the villa or were added at a later date. It is as well to remember that villas with towers are already present within the first imperial age though this kind of plan becomes more widespread in late antiquity, not only, perhaps, in response to militarization within the countryside but also as an indication/symbol of the social rank of the owner (Sfameni 2005). Complex (no. 2) is most clearly defined in the radar data which, amongst a variety of other indications, reveals the only instance on the site of a possible apsidal structure. The association between the prevalence of medieval material in this area and the presence of an 'apse' might lead us to suggest the hypothetical existence of a religious structure.

In conclusion we should recognize that each technique has produced a remarkable increase in both the quantity and quality of the available archaeological information. In particular the use of geophysical surveys represent an advance of great importance in our researches. In addition, to confirm the evidence recovered from the air, the geophysical data has without doubt added new and otherwise unseen evidence. It could also be argued that without field-walking and artifact grid collection it would not have been possible to achieve the necessary feedback to create a convincing archaeological interpretation of the site in terms of its dating and development over time.

2. WHAT ABOUT LANDSCAPES?

We believe, should recognized that on large, complex and stratified archaeological site such as Aiali the contribution of an integrated approach is extraordinarily effective. The contrast between the data gained from field-walking alone and that achieved through the combination of various prospecting methods within a GIS environment is extraordinary, showing clearly—even, we hope, to the most sceptical of archaeologists—the extremely important role of remote sensing in the archaeological process (Plate 30).

The unavoidable question arising from this sample study would probably be: when one needs to know about a buried site, where would archaeology be without remote sensing? But nowadays we believe the question and the challenge should really be: when one needs to know about buried *landscapes*, where would archaeology be without remote sensing? We believe that the next challenge should be to extend this kind of research to the broader landscape in the valley between the town of Grosseto and Roselle (Powlesland 2006; Gaffney and Gaffney 2006). We have just started PhD research in this area, with the aim of collecting about 600 hectares of magnetic data using the Foerster fluxgate magnetic system. The same strategy has been followed during the ARP© survey, connecting Aiali to a location about 800 m away where aerial reconnaissance and field-walking survey allowed us to identify a medieval mound or castle. Between the two sites the ARP© survey shows many features invisible from the ground and in some cases also invisible (so far) from the air through aerial photography. It is easy to understand, however, that it will never be possible to extend this approach to areas of the scale covered by the University's archaeological mapping projects for the region as a whole. For instance, the archaeological map of the province of Grosseto covers an area of 4030 kmsq and that of Tuscany as a whole about 22.990 kmsq.

The archaeological objectives and outcome of the Aiali project will be conditioned by the critical impact of the kinds of information that are available for recording: to use our introductory words 'in assessing the potential or interpretation of a landscape it is at least as important to know what may *not* be visible as to appreciate what *is* visible'. Achieving suitable sample areas for the application of the highest level of research intensity and archaeological visibility will, in our view, produce better returns:

1. Conservation—increased awareness of the archaeological resource as a whole, so as to create more effective and better-adapted policies for landscape monitoring and conservation.
2. Academic issues—the recognition of 'emptiness' (that is, the absence of evidence) as being equally important as the *presence* of archaeological evidence in what we hope will be a new approach to research into the development of settlement patterns and landscape history
3. The future—the search for better visibility in our present 'emptinesses' will hopefully produce new and perhaps different kinds of data, in turn creating new kinds of feedback into the investigative and interpretative process.

The challenge we now face is fundamentally linked to the scale of visualization but also to a changing approach to landscape studies. The huge archive collected in recent decades by the University of Siena and these recent developments in improving the variety and intensity of research

Plate 32.   Landscape between Grosseto and Roselle and related areas where we applied very high intensity and integrated survey methods. Form a landscape-based approach it should be easy to understand the huge lost of information due to the absence of information between "sites". Missing the connective tissue of our landscape we definitively lost evidences of relationships. (See colour plate section)

will not be enough to answer our long-term needs. These strategies will allow us to achieve only 'dots' or islands of knowledge, generally without any connection in between as prof. Dominic Powlesland argued many times (Plate 32). We will probably have to reduce the sample area, moving from regional-scale to a 'real' and continuous landscapes—the local scale (this volume, Campana)—of more limited size, while still applying the broader interpretative and theoretical framework of archaeology in general. We have to accept and respond to the inherent complexity of the record, not only in the research design but also in the composition of the research team, in the methods that we use in the field and in our data processing in the office. We have to develop new strategies aimed at implementing a 'total archaeology' approach to our chosen study area, so as to understand *through real archaeological evidence* the meaning of the 'emptinesses' imposed by limitations of visibility or accessibility within the framework of the landscape.

REFERENCES

Gaffney C. and Gaffney V. 2006. *No further territorial demands: on the importance of scale and visualization within archaeological remote sensing*, in From artefacts to anomalies: Paper inspired by the contribution of Arnold Aspinall (University of Bradford 1–2 December 2006), http://www.brad.ac.uk/archsci/conferences/aspinall/presentations/

Powlesland, D., 2006. *Redefining past landscapes: 30 years of remote sensing in the Vale of Pickering*, in Campana S., Forte M. (eds.), From Space to Place, Proceeding of the IInd International Conference Remote Sensing Archaeology, Rome 4–7 December 2006, Archaeopress, BAR International Series, Cambridge, pp. 197–201.

Sfameni C., 2005, *Le villae-praetoria: i casi di San Giovanni di Ruoti e di Quote San Francesco*, in Volpe G., Turchiano M. (eds.), *Paesaggi e insediamenti rurali in Italia meridionale fra tardoantico e altomedioevo*, Bari, pp. 609–622.

Author index

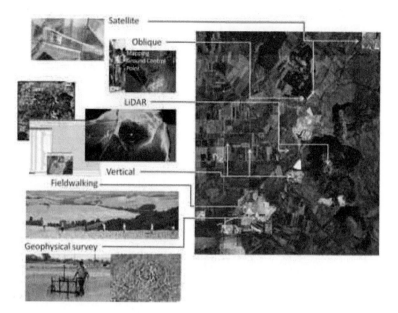

Plate 1. Archaeological mapping. Integration of different detection methods to draw the jigsaw puzzle in which we can measure and position each piece of information while at the same time perceiving the overall picture, whether synchronically or diachronically.

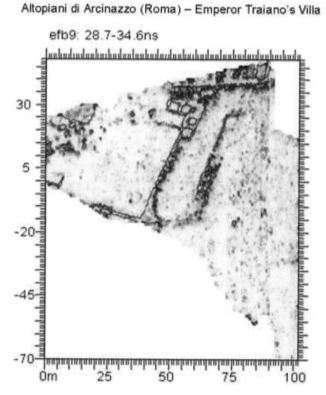

Plate 2. Traiano's Villa, GPR time slice in the time window 29–35 ns (twt). The map shows clear anomalies due to the presence of walls and rooms.

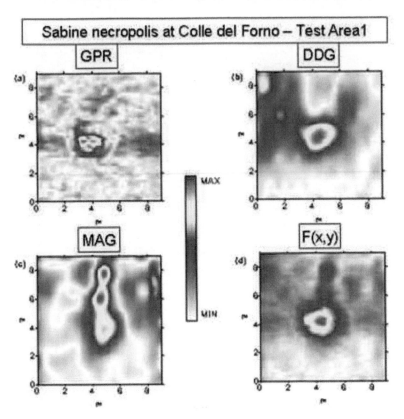

Plate 3. Test_Area1. Comparison of normalised results for the different employed methods: (a) GPR data, (b) dipole-geoelectric data, (c) cross-correlated magnetic data, (d) integration of all results.

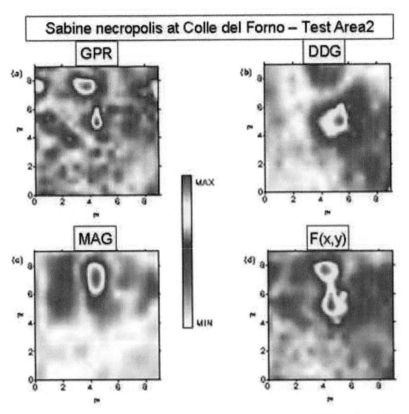

Plate 4. Test_Area2. Comparison of normalised results for the different employed methods. For other description see Fig. 16.

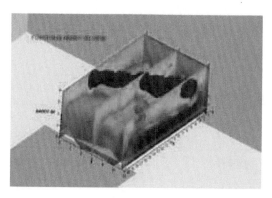

Plate 5. An image produced by inverting resistivity data over a drain at Fountains Abbey (England) (Tsourlos, 1995; Coppack et al., 1992).

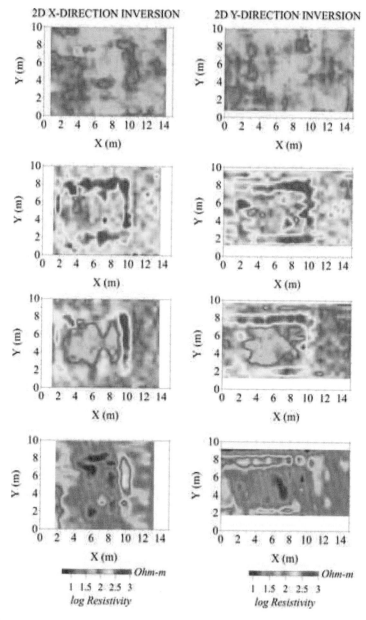

Plate 6. Example of depth slices obtained from tomographic data using the modified RM15™ resistivity meter. The slices are at 0.125, 0.375, 0.625 and 0.875 m depth from top to the bottom. Data from the archaeological site of Sikyon in Greece (Papadopoulos et al., 2006).

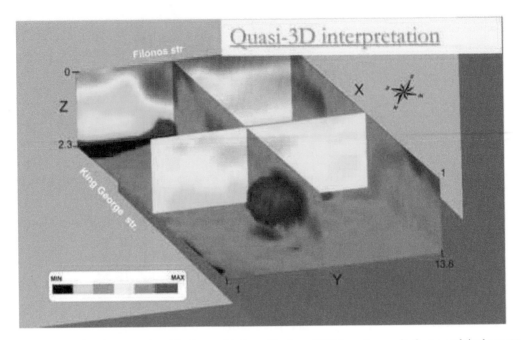

Plate 7.   Pseudo 3-D image produced by the graphical combination of 2-D inversion results from a park in the centre Piraeus. A pronounced 3-D target is detected.

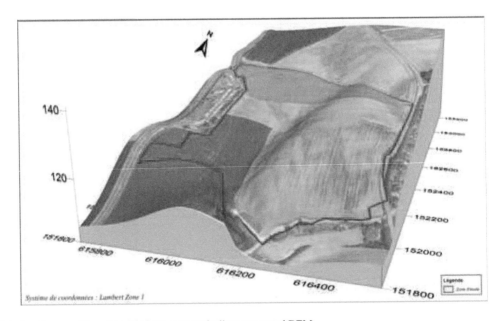

Plate 8.   Superposition of aerial photo over vertically exaggerated DEM.

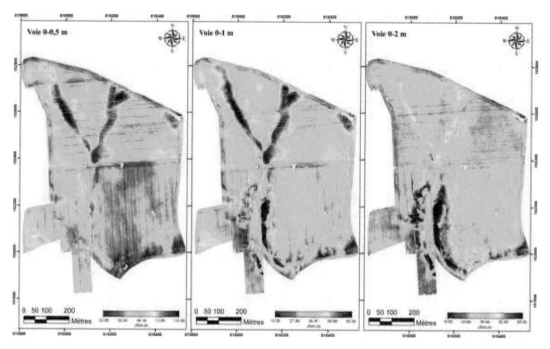

Plate 9.   Apparent Electrical resistivities ARP Channels 1, 2 and 3.

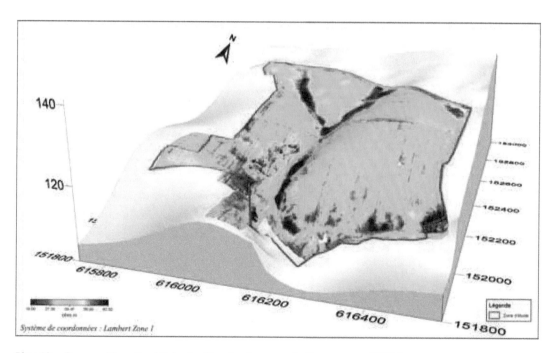

Plate 10.   Superposition of resistivimetry (1 m deep) over vertically exaggerated DEM.

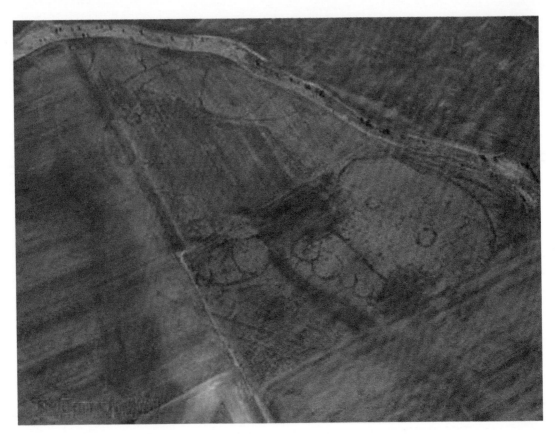

Plate 11.   Aerial photograph of Monte San Vicenzo at the Celone river. Cropmarks showing complex Neolithic ditched enclosures and Roman olive gardens O. Braasch 2002.

Plate 12a.   Masseria Anglisano 2005. Compilation of the oblique aerial photographs on the base of the orthophoto in GIS. Aerial photos by O. Braasch, GIS application and image processing by D. Gallo.

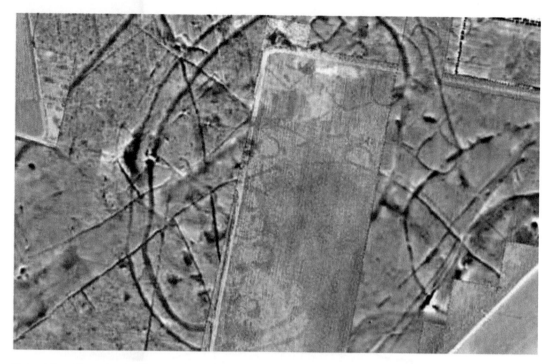

Plate 12b. Masseria Anglisano 2005. Same as Plate-12a, but with the addition of the magnetogram. Caesium-magnetometry by H. Becker with Scintrex Smartmag in quadro-sensor configuration.

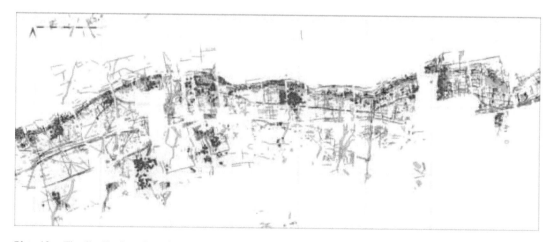

Plate 13. The distribution of Anglo-Saxon Grubenhäuser identified across all geophysical surveys.

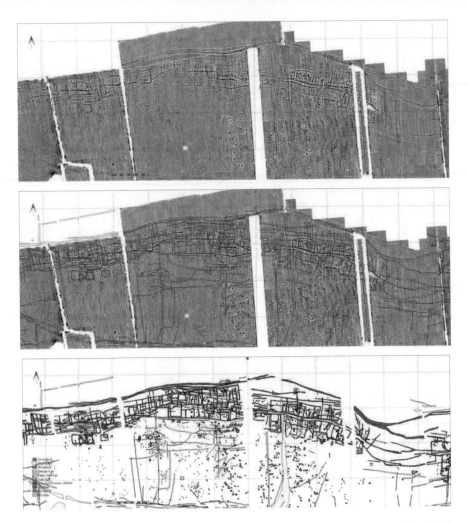

Plate 14. A section of the 'ladder settlement' shown through the geophysical results (top), overlain with the digitised polygon plot (centre), and classified according to basic phase (bottom).

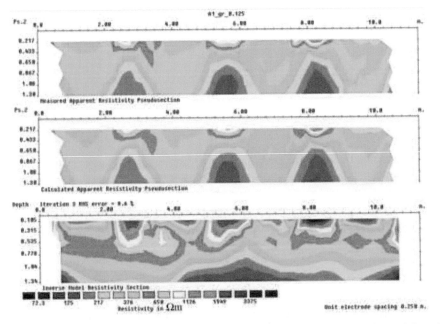

Plate 15. An electrical image over a Neolithic pit alignment in Staffordshire, UK. (A) is observed data plotted as a pseudosection, (B) is the pseudosection computed from a model and (C) is an image or model showing the true depth and the true feature resistivity.

Plate 16.   A three dimensional visualization of the Sunburst GPR survey results.

Plate 17.   Saï Island (Sudan), excavation of the detected features.

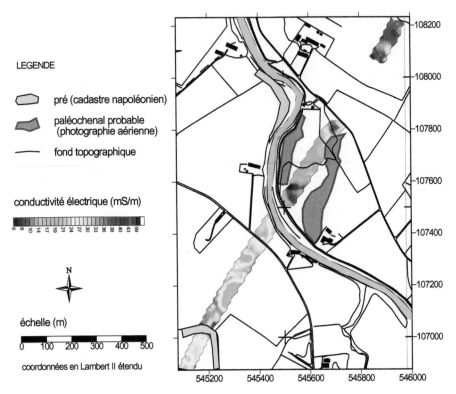

Plate 18.   A66 motorway route (Hautes Pyrénées, France) apparent magnetic susceptibility map, EM38 apparatus in VCP configuration.

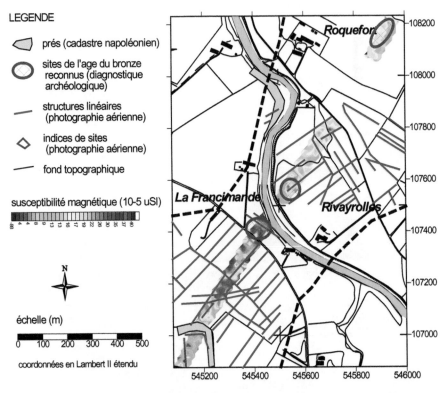

Plate 19.   A66 motorway route (Hautes Pyrénées, France) apparent conductivity map, EM38 apparatus in VCP configuration.

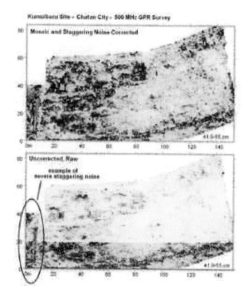

Plate 20.   Shown in bottom diagram is the staggering noise time slice which also has a sharp change in reflection above the line y = 20 m.

Plate 21.   Ikime Kofun Burial #7, Miyazaki City, Japan. Radargrams and 3D topo warped time slices along with excavation photos of the Ikime Burial #7 are shown.

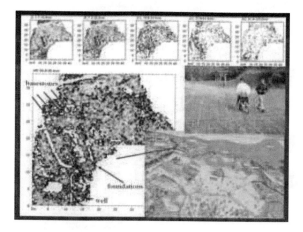

Plate 22.   Nanao City Castle, Ishikawa Ken, Japan. Example of overlay analysis. 2D time slices from the complete 3D dataset are first normalized and the relative-strongest-reflectors from each map down to the desired depth range are chosen and then overlaid onto a single map.

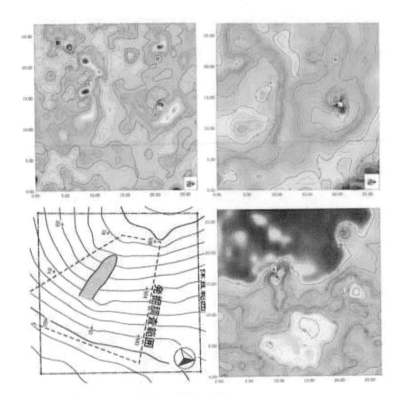

Plate 23. Geophysical survey results and the are: Goshogawara Kiln Site.

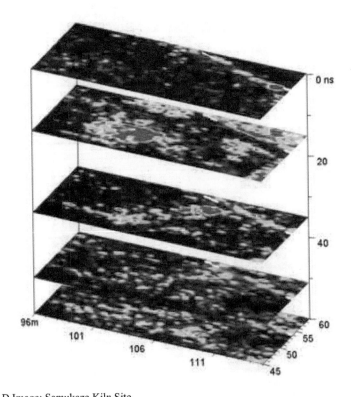

Plate 24. GPR 3-D Image: Samukaze Kiln Site.

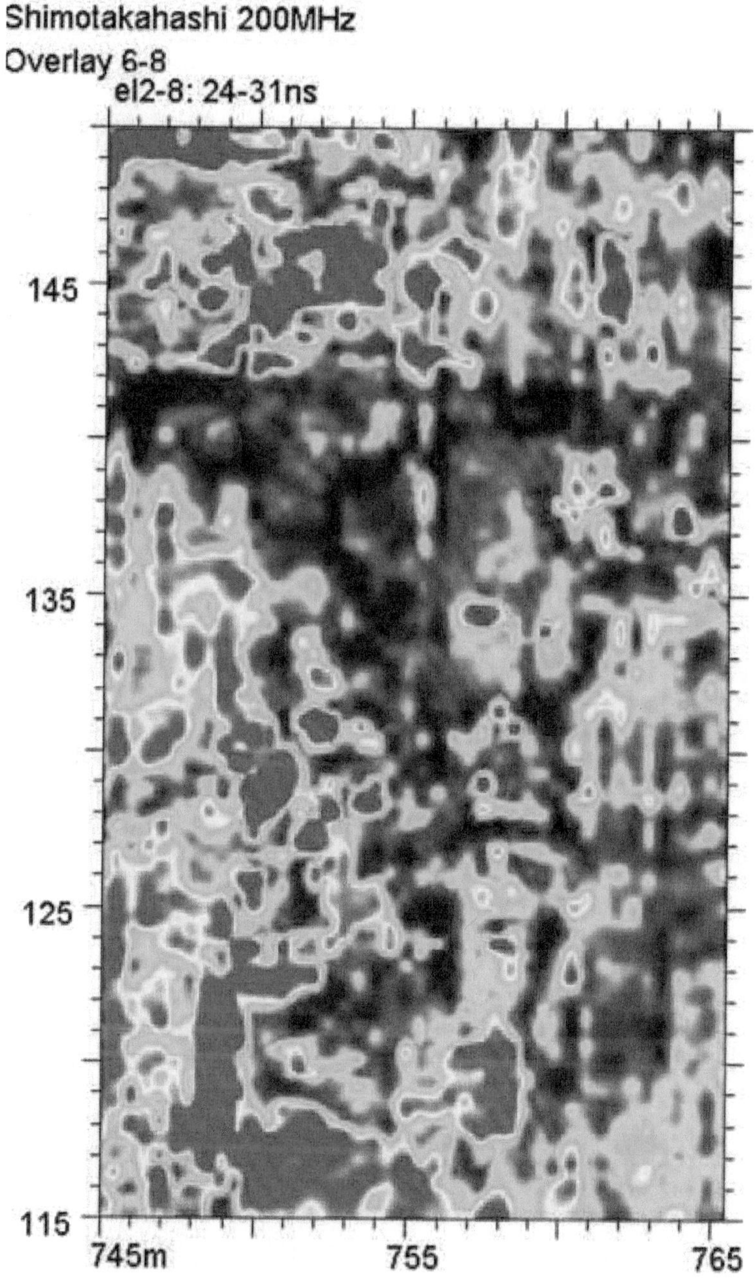

Shimotakahashi 200MHz
Overlay 6-8
el2-8: 24-31ns

Plate 25.   Details of building (200 MHz antenna): Shimotakahashi Local Government Office Site.

Plate 26.   Ground view of the test site.

Plate 27.　Geophysical survey at Aiali-test site. From top left: GPS SIR 3000; GSSI Terravision (Geostudi Astier); Helmut Becker using the magnetometer GEM Overhouser; Foerster DLG Kartograph multi-probe gradiometer; field lecture; survey with the electomagnetometer EM-31; setting up the electomagnetometer EM-38 before survey; EM-38 survey; Gianfranco Morelli processing the ERT data; Michel Dabas setting up the geoelectrical instrument (syscal-pro); Michel Dabas explaining field data acquisition to a group of students; differential GPS and total station survey; Helmut Becker Scintrex Smartmag SM4G-special system in various sensor configurations (portable and trolley); Automatic Resistivity Profiler (ARP© Terranova).

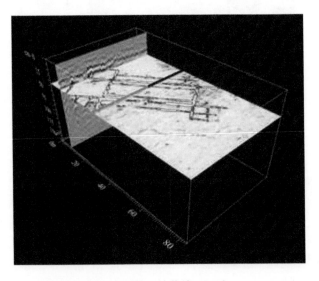

Plate 28.　3D example fence plot generated for site AB at Aiali (Grosseto).

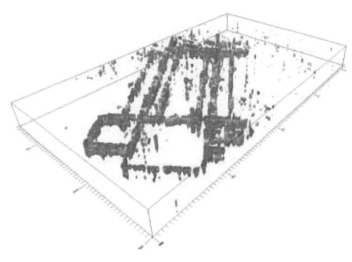

Plate 29.　3D isosurface render for the villa located below site A and B.

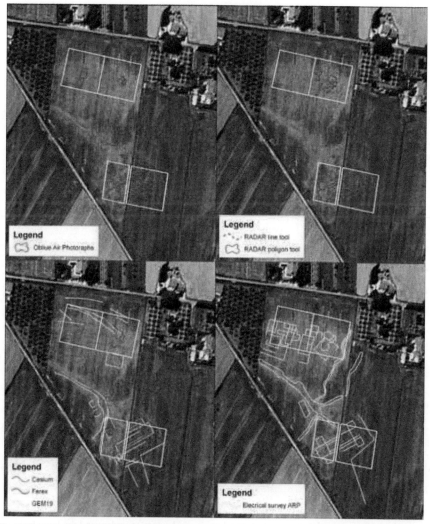

Plate 30.　Top left: archaeological interpretation of the features visible from the air. Top right: archaeological interpretation of three different sets of magnetic data—fluxgate, overhouser and caesium. Bottom left: archaeological interpretation of the GPR survey. Bottom right: interpretation of the archaeologyical features visible in the fast electrical imaging data set.

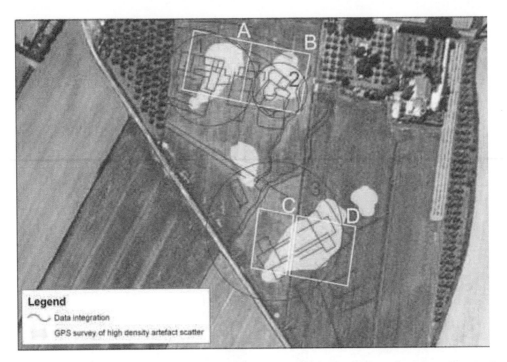

Plate 31. Integrated interpretation. Notice the striking difference between the GPS survey of the higher density of artifact scatters and the integrated remote sensing datasets.

Plate 32. Landscape between Grosseto and Roselle and related areas where we applied very high intensity and integrated survey methods. Form a landscape-based approach it should be easy to understand the huge lost of information due to the absence of information between "sites". Missing the connective tissue of our landscape we definitively lost evidences of relationships.